The Urban Book Series

The Urban Book Series is a resource for urban studies and geography research worldwide. It provides a unique and innovative resource for the latest developments in the field, nurturing a comprehensive and encompassing publication venue for urban studies, urban geography, planning and regional development.

The series publishes peer-reviewed volumes related to urbanization, sustainability, urban environments, sustainable urbanism, governance, globalization, urban and sustainable development, spatial and area studies, urban management, transport systems, urban infrastructure, urban dynamics, green cities and urban landscapes. It also invites research which documents urbanization processes and urban dynamics on a national, regional and local level, welcoming case studies, as well as comparative and applied research.

The series will appeal to urbanists, geographers, planners, engineers, architects, policy makers, and to all of those interested in a wide-ranging overview of contemporary urban studies and innovations in the field. It accepts monographs, edited volumes and textbooks.

Indexed by Scopus.

Liang Zhou • Bo Yuan
David López-Carr • Fengning Hu

Urban Surface Thermal Environments

A Metropolitan Area Perspective

Springer

Liang Zhou
Lanzhou Jiaotong University
Lanzhou, Gansu, China

Bo Yuan
Nanjing University
Nanjing, Jiangsu, China

David López-Carr
University of California, Santa Barbara
Santa Barbara, CA, USA

Fengning Hu
City University of Macau
Macau, China

ISSN 2365-757X ISSN 2365-7588 (electronic)
The Urban Book Series
ISBN 978-3-031-94498-7 ISBN 978-3-031-94499-4 (eBook)
https://doi.org/10.1007/978-3-031-94499-4

This Springer imprint is published by the registered company Springer Nature Switzerland AG
The registered company address is: Gewerbestrasse 11, 6330 Cham, Switzerland

Preface

Climate change has emerged as one of the most significant and complex environmental challenges facing the planet and humanity. It is arguably the most crucial issue the world must address in the coming decades. Beginning in 2016, with the multinational Paris Agreement, significant interventions have begun, on both local and global levels, to combat climate change. However, according to the Sixth Assessment Report of the Intergovernmental Panel on Climate Change, despite ever-increasing awareness and multifaceted countermeasures to slow global warming and its devastating ramifications, the anthropogenic increase in global surface temperature continues at an alarming rate. Consider the following facts: Beginning in 1850, it took 50 years for the global surface temperature to increase 1.1 °C; beginning in 2011, that same 1.1 °C temperature increase occurred over a mere 9 years. The pervasive increase in global surface temperature, coupled with the rapidly accelerating rate of temperature increase, is already affecting both macro- and microclimate patterns, leading to unprecedented extremes of drought and wildfires; mega-storms and flooding; melting polar icecaps; and rising sea levels—all of which threaten human health, economic growth, and crop yields. If left unchecked, continued increases in global surface temperature will cause an even broader range of climate events in the decades ahead, including potential devastation of coastal and small island communities, and irrevocable destruction of delicate terrestrial and marine ecosystems. In short, the frequency, intensity, and costs of extreme climate events are all increasing exponentially, and the consequences are relevant for each and every one of us. Moving forward, we must learn to plan our cities far more thoughtfully, so as to "beat the heat" and support equitable human thermal comfort, while simultaneously preserving the very foundations of our long-term, climate-dependent survival.

This book focuses on the characteristics and drivers of the urban surface thermal environment, the heat risk faced by urban residents, and strategies for mitigating the urban heat island effect in the Xi'an Metropolitan Area. It proposes a series of analytical frameworks, models, and action strategies aimed at understanding the complex dynamics, spatial heterogeneity, and interaction among multiple contributing factors shaping the urban thermal environment. To achieve these objectives, this book is structured into ten chapters. Chapter 1 introduces two challenges affecting the urban thermal environment: Climate change and urbanization. We provide a comprehensive review of the drivers and impacts of two challenges, with an emphasis on how they influence the urban thermal environment. Chapter 2 is an overview of the progress of the urban thermal environment research and systematically synthesizes key knowledge and methodological advances in the field. In the following chapters, we propose four innovative research analytical approaches to explore the spatiotemporal characteristics (Chap. 3) and drivers (Chaps. 4, 5, 6, and 7) of the urban thermal environment, the heat risk for urban residents (Chap. 8), and mitigation strategies of the urban heat island effect (Chap. 9).

- First, in Chap. 3, the authors describe the urban landscape change and surface temperature evolution of the Guanzhong Plain urban agglomeration from 2000 to 2018, providing a detailed analysis of spatial and temporal patterns of the urban thermal environment (UTE).
- Second, in Chaps. 4–7, they thoroughly investigate and clearly outline the drivers of surface urban heat island (SUHI) and propose a comprehensive analytical framework to capture its hierarchical characteristics and the interactions among key influencing factors, providing valuable guidelines for "beating the heat" in high-density urban domains.
- Third, in Chap. 8, the authors provide new insights into identifying populations potentially vulnerable to high temperatures, emphasizing the pivotal role of local climate zones in future strategic planning aimed at mitigating recurring extreme heat events.
- Fourth, in Chap. 9, and perhaps most importantly, the authors discuss how a combination of aggressive urban construction strategies, the deployment of multifaceted active cooling measures, and the expansion of urban green infrastructure at multiple levels (i.e., from single-family home, to residential community, to street block, to entire city), can collectively moderate urban heat and help slow the rise in global surface temperatures over the coming critical decades. The chapter concludes with recommendations for future research and related policy development.

This book aims to provide informed analyses to policymakers, public officials, research administrators, scientists, and students to support evidence-based decision-making. In addition, this book will also contributes to advancing the United Nations Sustainable Development Goals (e.g., SDG11), and promotes efforts in climate change mitigation, mega-city governance, and the pursuit of social equity and environmental justice. Furthermore, it highlights innovative policy approaches in China, including the new urbanization strategy, improvements in the human settlement environment, and the application of carbon-peaking and carbon-neutrality techniques. This book is likely to provide a typical regional case study in plain areas that will be of great help to other cities of similar climate types throughout the world. The authors dedicate their work to the ongoing search for, and dissemination of, relevant climate-change knowledge to serve the public interest.

Lanzhou, Gansu, China — Liang Zhou
Nanjing, Jiangsu, China — Bo Yuan
Santa Barbara, CA, USA — David López-Carr
Macau, China — Fengning Hu

Acknowledgments

This book was supported by the National Natural Science Foundation of China (No.42271214), the Open Research Program of the International Research Center of Big Data for Sustainable Development Goals (No. CBAS2023ORP04), the Top Leading Talents Project in Gansu Province, and the Intergovernmental international science and technology innovation cooperation program under National key research and development plan (2024YFE0198600). The authors are responsible for all errors and interpretations. We would like to thank all the authors and reviewers for their contribution to this book. Thanks also go to Ron Doering for his generous support of this book.

We would like to express our appreciation to a lot of people: Mengqiu Cui, Liping Zhang, Jeffary Magnes, Xin Zhao, Kai Jia, Leiqi Tan, Chunzhu Wei, Qian Zhang, and Hong Chen for their careful review and recommendations on some chapters. We would also like to thank Haowen Yan, Chenghu Zhou, Tao Che, Chunlin Huang, Michael Peterson, and Shaohua Wang. Additionally, we would like to thank Lanzhou Jiaotong University and the University of California, Santa Barbara, for providing great research environment for us.

Contents

About the Authors

Liang Zhou is a Professor at Lanzhou Jiaotong University. His research focuses on urban thermal environments, urban geography, simulation, disaster risk, mountain city governance, and sustainable urban development. He has conducted postdoctoral and visiting research at the Institute of Geographic Sciences and Natural Resources Research, Chinese Academy of Sciences, as well as at the University of California, Santa Barbara (UCSB), and Indiana University Bloomington (IUB). He has published over 160 peer-reviewed papers and received multiple academic awards and scientific honors. As principal investigator, he has led five research projects funded by the National Natural Science Foundation of China (NSFC). He also serves as Editorial Director of the international journal *Journal of Geovisualization and Spatial Analysis*. To date, he has supervised more than 40 master's and doctoral students. His research supports urban planning, climate adaptation, and the achievement of Sustainable Development Goals through the application of geospatial technologies.

Bo Yuan is currently pursuing his PhD at Nanjing University. His research focuses on the impacts of compound climate extremes on vegetation vulnerability and ecosystem recovery, with additional interests in urban climate and remote sensing of the environment.

David López-Carr is Professor of Geography at the University of California, Santa Barbara, where he directs the Human-Environment Dynamics Lab (HED). López-Carr received a PhD in Geography from the University of North Carolina in 2002, where he subsequently held a NIH postdoctoral fellowship in Biostatistics in the School of Public Health and Carolina Population Center. He has lived, worked, and traveled in over 70 countries worldwide and speaks Spanish, Portuguese, Italian, French, and rudimentary Q'eqchí Maya.

Professor López-Carr uses mixed methods to examine potential win-win outcomes for people and the environment. His research specifically focuses on links among population, health, rural livelihoods, agriculture, and marine and forest resource use and conservation through ongoing projects in California, Latin

America, Africa, and Asia. He has authored over 230 scientific publications. With co-authors, he was awarded the 2024 Ecological Society of America's (ESA) Sustainability Science Award and the Frontiers Planet Prize from the US National Academy of Sciences. His scholarship has been conducted with colleagues and students thanks to funding from over 50 fellowships, grants, and awards from NASA, NOAA, NSF, NIH, the Mellon, two Fulbright Fellowships, and numerous other sources. López-Carr was awarded the Research Excellence Award from the Population Specialty Group and the Human Dimensions of Global Change Specialty Group of the Association of American Geographers (AAG). He was also presented the AAG Excellence in Mentoring Award and the Conference of Latin American Geography Teaching Award. Named a National Academy of Sciences Kavli Frontiers of Science Fellow in 2013, in 2014 he was elected an American Association for the Advancement of Science (AAAS) Fellow "for advancing our scientific understanding of the coupled process of human population dynamics and environmental change."

Website: www.geog.ucsb.edu/~carr/.

Fengning Hu is currently pursuing his PhD at the City University of Macau. His research focuses on remote sensing-based mapping and analysis of urban logistics infrastructure and urban climate.

Abbreviations

AHSD	Average architecture height standard deviation
AI	Aggregation index
AR	Aspect ratio
AR6	Sixth Assessment Report
AREA_MN	Mean patch area
AT	Air temperature
ATC	Annual temperature cycle
ATCH	Hybrid ATC
BCR	Building coverage ratio
BLEST	BLEnd spatiotemporal temperatures
BRT	Boosted regression tree
CFD	Computational fluid dynamic
CKFM	Combination of kernel-driven and fusion-based method
CLDAS	China Land Data Assimilation System
CNN	Convolutional neural networks
CO_2	Carbon dioxide
CPCs	Circulation pattern classifications
DEM	Digital elevation model
DTC	Diurnal temperature cycle
EB	Energy balance
ECOSTRESS	Ecosystem Spaceborne Thermal Radiometer Experiment on Space Station
ED	Edge density
EEA	European Environment Agency
EU ETS	European Emission Trading System
FAR	Floor area ratio
FGC	Fractional grass coverage
FTC	Fractional tree coverage
GCR	Green coverage ratio
GHGs	Greenhouse gases
GLASS	Global Land Surface Satellite

GSL	Growing season length
Gt	Gigatons
HANTS	Harmonic analysis of time series
HI	Heat index
HRE	Height of roughness elements
HW	Heat wave
IDW	Inverse distance weighting
IPCC	Intergovernmental Panel on Climate Change
IS	Impervious surface
LCT	Local contour tree
LCZ	Local climate zone
LPI	Largest patch index
LRAR	Low-rise buildings area ratio
LST	Land surface temperature
LSTM	Long short-term memory
LULC	Land use/land cover
MBA	Mean building area
MBH	Mean building height
ML	Machine learning
MPS	Mean patch size
MRT	Mean radiant temperature
NDVI	Normalized difference vegetation index
NP	Neighboring-pixel
NPP	Net primary productivity
NTL	Nighttime light
OECD	Organisation for Economic Co-operation and Development
PAISUHII	Percentage of area with increasing SUHII
PD	Patch density
PET	Physiological equivalent temperature
PF	Planning families
PLAND	Percent of landscape
PMW	Passive microwave
RF	Random forest
SADFAT	Spatiotemporal adaptive data fusion algorithm for temperature mapping
SDGs	Sustainable Development Goals
SEB	Surface energy budget
SHAP	SHapley Additive exPlanations
SHAPE_AM	Area-weighted mean shape index
SNAPP	Science for Nature and People Partnership
SRES	Special Report on Emissions Scenarios
ST	Surface temperature
STARFM	Spatial and temporal adaptive reflectance fusion model
STDFA	Spatial and temporal data fusion approach
STITFM	Spatiotemporal integrated temperature fusion model

SUHI	Surface urban heat island
SUHII	Surface urban heat island intensity
SURFRAD	Surface Radiation Budget Observing Network
SVF	Sky view factor
SVM	Support vector machine
TES	Temperature emissivity separation
UGS	Urban green space
UHI	Urban heat island
UHII	Urban heat island intensity
UHS	Urban heat sink
UN	United Nations
UNFCCC	United Nations Framework Convention on Climate Change
US	United States
UTE	Urban thermal environment
WRF	Weather Research and Forecasting
WTs	Weather typing classifications
WUDAPT	World Urban Database and Access Portal Tools
XGBoost	Extreme gradient boosting
XXMA	Xi'an–Xianyang Metropolitan Area

Chapter 1
Background

1.1 Climate Change

Human-induced climate change is the largest and most pervasive threat to the natural environment and societies that the world has ever experienced. The poorest countries are bearing the heaviest burden of its consequences. No part of the world is immune to the devastating impacts of climate change. Rising temperatures are leading to environmental degradation, natural disasters, extreme weather events, food and water shortages, economic disruptions, conflicts, and even terrorism. Sea levels are on the rise, the Arctic is melting, coral reefs are dying, oceans are becoming more acidic, and forests are engulfed in flames. These observable changes provide undeniable evidence of global warming and its severe consequences (IPCC, 2007).

One of the most evident changes is global warming caused by increased atmospheric concentrations of CO_2, CH_4, and N_2O. Greenhouse gas (GHG) emissions have already made a certain level of global temperature rise and other changes to the climate inevitable. In order to avert catastrophic health impacts and prevent millions of climate change-related deaths (Table 1.1), the world must limit temperature rise to 1.5 °C according to *the Paris Agreement*. However, global heating of even 1.5 °C/2.0 °C is not considered safe, and every additional tenth of a degree of warming will seriously impact people's lives and health. The people whose health is being harmed first and worst by the climate crisis are those who contribute least to its causes. With the poorest people largely uninsured, health shocks and stresses already push around 100 million people into poverty every year (Hallegatte et al., 2016).

Climate change directly and indirectly affects poverty. *No Poverty*, i.e., eradicating poverty in all its forms everywhere, although it is the first goal of the 17 sustainable development goals (SDGs) (Fig. 1.1) (UN, 2015), climate change makes achieving the goal more challenging. The effects of climate change are hitting many

L. Zhou et al., *Urban Surface Thermal Environments*, The Urban Book Series,
https://doi.org/10.1007/978-3-031-94499-4_1

Table 1.1 Global disasters have caused losses in the past decade. (Source: Global disaster data platform)

Year	Number of people affected (10^4 people)	Number of dead and missing (people)	Direct economic loss (10^6 USD)
2012	11,627.6	11,668	1,849,410.4
2013	9938.6	22,159	1,403,162.5
2014	14,202.4	20,922	1,124,018.3
2015	43,105.1	24,308	1,001,926.2
2016	20,544.5	10,714	1,670,675.7
2017	10,421.5	11,906	3,604,710.7
2018	8398.4	15,004	1,441,552.8
2019	10,919.9	22,515	1,069,861.4
2020	9853.3	15,353	1,812,564.9
2021	10,786.9	14,442	2,537,975.6
2022	18,216.6	12,471	2,198,591.0

Fig. 1.1 The 17 sustainable development goals (Source: United Nations)

of the world's poorest people first and hardest. Extreme weather patterns, natural hazards, and the resulting food and water shortages due to climate change pose significant threats to the lives of those living in poverty. Moreover, the more impoverished individuals are, the more challenging it becomes to recover from adverse events, such as failed harvests, damaged homes, and health crises. It has been estimated that, by 2030, climate change could push more than 120 million people into poverty (Rozenberg & Hallegatte, 2015). Climate change, together with other factors, also increases the risk of infectious disease outbreaks like the COVID-19 pandemic (IPCC, 2021), which reversed the steady progress of poverty reduction over the past 25 years. The global poverty rate witnessed a steep rise from 8.3% in 2019 to 9.2% in 2020, representing the first increase in extreme poverty since 1998 and the largest increase since 1990, setting back poverty reduction by about 3 years (Jensen, 2022).

Climate change (Table 1.2) refers to changes in the long-term pattern of behavior of the atmosphere over millennia or, more recently, natural processes or human activity. The Intergovernmental Panel on Climate Change (IPCC) defined it as any change in climate over time, whether due to natural variability (e.g., variations in the solar cycle) or as a result of human activity. In recent years, climate change more often refers to changes in the earth–ocean–atmosphere system resulting from the GHG emissions from anthropogenic sources, and it has led to a series of irreversible impacts on natural and human systems.

1.1.1 Drivers

Global warming is one of the most evident phenomena of climate change. Other changes in our planet caused by climate change, such as rising sea levels, shrinking mountain glaciers, and accelerating ice melting in Greenland, are all the consequences of warming. The Earth's surface is warming up, and one of the most related factors is human activities, such as changes in land cover and of GHG. GHG emissions are the major human-influenced drivers of climate change. Since the Industrial Revolution, the world has been heavily reliant on the over-exploration of natural resources in forms of extensive use of fossil fuels (i.e., coal, oil, and gas), deforestation, and other forms of land use change. Human activities have been the main

Table 1.2 The various definitions of climate change

Sources	Contents
Simple definition	Climate change is the global phenomenon of climate transformation characterized by the changes in the usual climate of the planet (regarding temperature, precipitation, and wind) that are especially caused by human activities
Official definition from NASA	A broad range of global phenomena created predominantly by burning fossil fuels, which add heat-trapping gases to the earth's atmosphere. These phenomena include the increased temperature trends described by global warming, but also encompass changes such as sea-level rise; ice mass loss in Greenland, Antarctica, the Arctic, and mountain glaciers worldwide; shifts in flower/plant blooming; and extreme weather events
Climate change definition from US geological survey	Global warming is just one aspect of climate change. In fact, global warming refers to specifically the rise in global temperatures primarily due to the increasing concentrations of GHGs in the atmosphere. Also, climate change refers to long-term changes in the measures of climate, including precipitation, temperature, and wind patterns
Global warming definition from climate.Gov	Global warming refers only to the Earth's rising surface temperature, while climate change includes warming and the "side effects" of warming—like melting glaciers, heavier rainstorms, or more frequent drought. Ultimately, this means that global warming is one side of the much larger problem of human-caused climate change

Note: Some definitions are referred at the link: https://youmatter.world/en/definition/climate-change-meaning-definition-causes-and-consequences/

driver of climate change since the 1800s, contributing to the ever-increasing concentration of GHGs (e.g., CO_2 and CH_4) into Earth's atmosphere (IPCC, 2007). These gases mainly come from natural systems and human activities (Fig. 1.2).

Natural systems encompass a wide range of events and components, including forest fires, earthquakes, oceans, permafrost, wetlands, mud volcanoes, and volcanoes. In contrast, human activities are primarily linked to energy production, industrial processes, transportation, food systems, alterations in land use, and water disposal. For instance, people utilize gasoline for powering vehicles, coal for heating, and landfills as major sources of methane emissions (Miller et al., 2013). These gases trap heat from longwave radiation emitted from the Earth's surface, leading to an increase in the Earth surface's average temperature, commonly referred to as global warming. This warming trend has persisted despite some global cooling observed in the early 1970s. Some scientists have hypothesized that pollutants emitted into the atmosphere may have a cooling effect by increasing the atmospheric absorption of solar radiation (Peterson et al., 2008). However, as the temporary cooling phase concludes and temperatures resume their upward trajectory, a minority of scientists express concerns about planet cooling.

Significantly, while smog remains suspended in the air for weeks, CO_2 can persist in the atmosphere for centuries. Notably, after 1988, climate change began to exhibit striking and unusual patterns. The summer of 1988 set temperature records, although many even hotter summers have occurred since. In that same year, the United States experienced widespread droughts and wildfires. Subsequently, daily heat records have been broken more than 2.3 million times at weather stations across the nation, surpassing cold records by half a million times. To date, human activities have been responsible for approximately 1.0 °C (±0.2 °C) of global

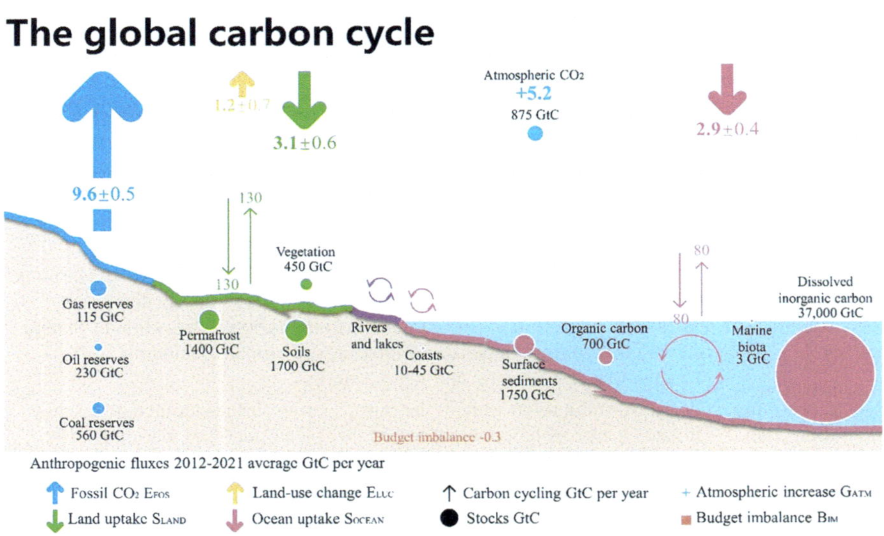

Fig. 1.2 Global carbon budget 2022 (Friedlingstein et al., 2022)

warming above pre-industrial levels. If current emission rates continue, it is expected that global temperatures could reach 1.5 °C between 2030 and 2052 (IPCC, 2018). The warming effect from anthropogenic emissions, spanning from the pre-industrial era to the present, will persist for centuries to millennia and will continue to instigate long-term alterations in the climate system, including sea-level rise and associated impacts. However, these emissions alone are unlikely to cause global warming of 1.5 °C. Other emissions from seven distinct sources have also contributed to accelerated global warming (Fig. 1.3).

Generating Electricity The generation of electricity and heat through the burning of fossil fuels contributes to huge global emissions. Most electricity is still generated by burning coal, oil, or natural gas, producing CO_2 and N_2O in the processes. Globally, just over a quarter of electricity comes from wind, solar, and other renewable sources, which emit little or no GHGs or pollutants into the air compared to fossil fuels. Thermal power accounts for 60% of the global power generation struc-

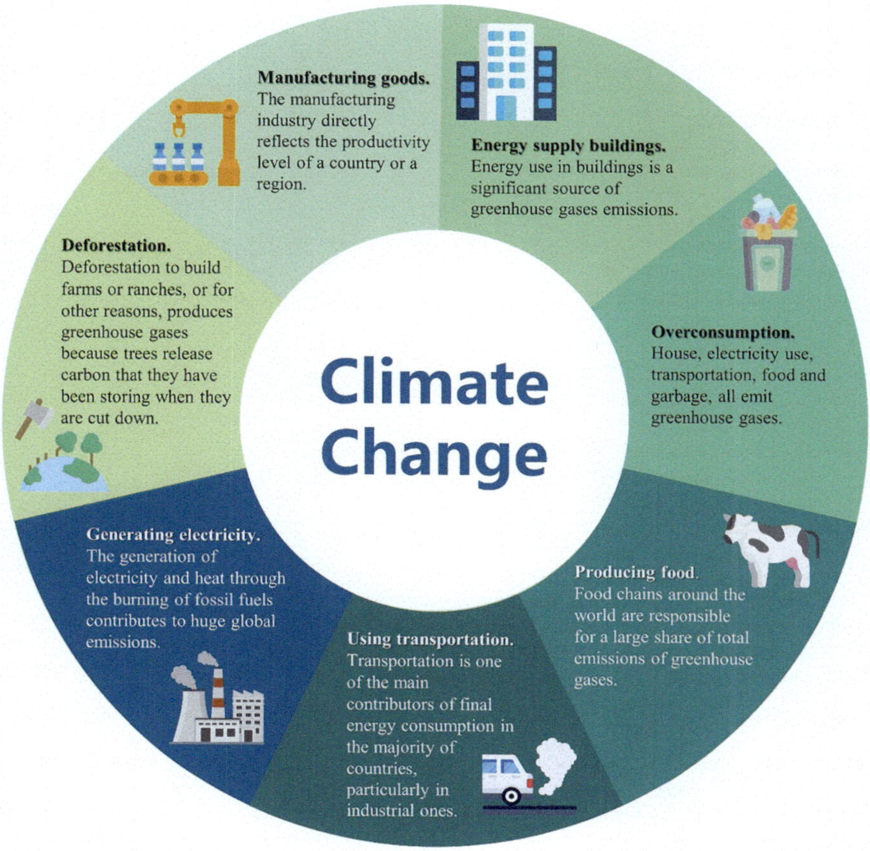

Fig. 1.3 Drivers of climate change

ture and is the dominant form of power generation due to the stable electricity generation compared to the power from renewable sources. With greater use of and access to electricity, especially in developing countries, emissions from electricity generation have been taking up a larger share of total CO_2 emissions globally (Goh et al., 2018). In 2015, electricity generation contributed to 37.6% of total global energy-related CO_2 emissions, up from 30.5% in 1990 (IEA, 2018). In China, the CO_2 emissions from electricity generation accounted for almost 40% of the national total in 2017, with a rapid increase from 935 Mt in 1997 to 3511 Mt in 2017 despite a slight reduction during the period 2013–2015 (Wang et al., 2021). In 2021, the EU electricity sector was 45% less GHG intensive than in 1990, yet 20% more than in 2020 (Fig. 1.4).

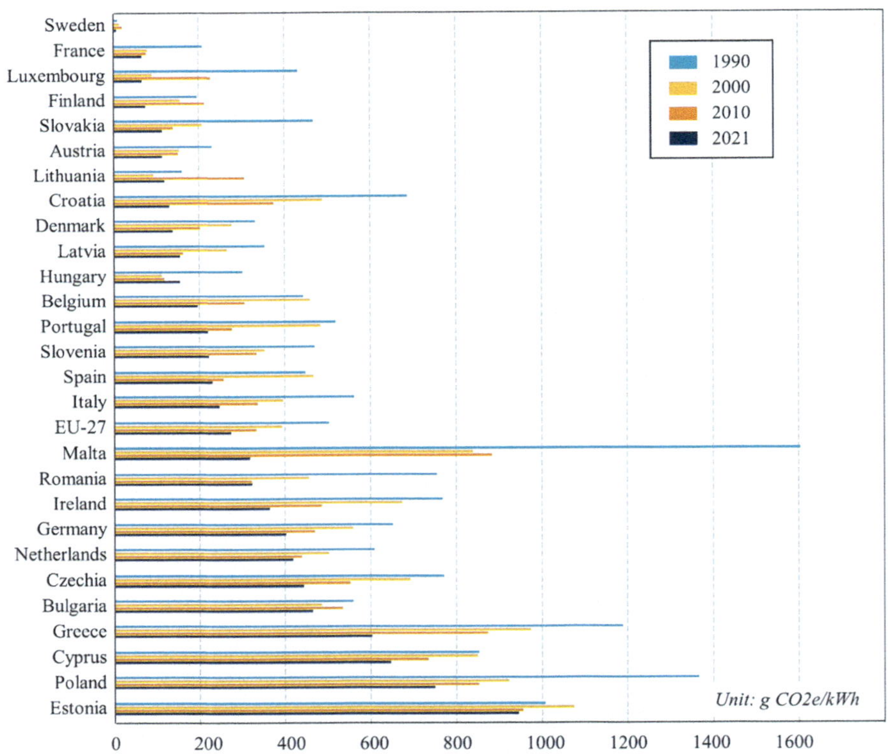

Fig. 1.4 GHG emission intensity of electricity generation in Europe at country level. GHG emission intensity (g CO_2 e/kWh) is calculated as the ratio of CO_2 equivalent emissions from public electricity production (as a share of CO_2 equivalent emissions from public electricity and heat production related to electricity production) and gross electricity production. (Source: National emissions reported to the UNFCCC and to the EU Greenhouse Gas Monitoring Mechanism provided by the European Environment Agency (EEA))

Manufacturing Goods The manufacturing industry directly reflects the productivity level of a country or a region. However, it produces a substantial amount of CO_2 emissions in the process of creating economic benefits (Yang et al., 2020). GHG emissions from manufacturing and industry come primarily from burning fossil fuels to provide energy to make cement, steel, electronics, plastic products, clothing, and other goods (Fig. 1.5). Mining and other industrial activities also release GHGs, as does the construction industry. Machines used in manufacturing processes often run on coal, oil, or natural gas for energy; some materials, such as plastics, are made from chemicals found in fossil fuels.

Manufacturing is one of the largest sources of global GHG emissions. The manufacturing industry is responsible for about 29% of the total direct CO_2 emissions from the industrial sector (IEA, 2016; IPCC, 2014). In Japan, emissions from the manufacturing industry accounted for approximately one-quarter (24.4%) of the total in 2015. If emissions derived from the energy conversion sector were allocated to energy (i.e., electricity and heat) consumers, emissions from the manufacturing industry would have accounted for approximately 40% of the total (Matsumoto et al., 2019). In 2015, industrial sector installations included in the European Emission Trading System (EU ETS) emitted 574 Mt. CO_2-equivalent GHG emissions (Pfaff et al., 2018), including the production of clinker, lime, ammonia, blast furnace operations, refineries, and others.

Deforestation Deforestation to build farms or ranches, or for other reasons (Fig. 1.6), produces GHGs because trees release carbon that they have been storing when they are cut down. About 12 million hectares of forest are destroyed each year.

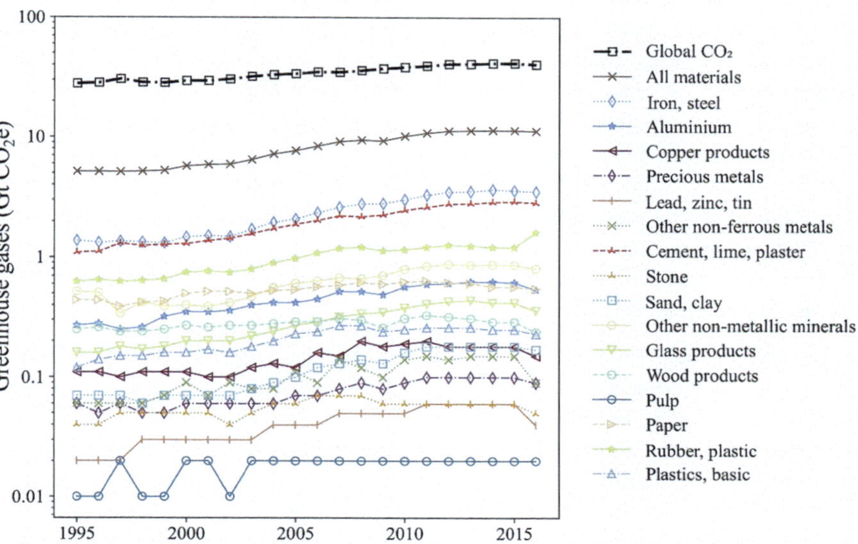

Fig. 1.5 GHG emissions in the production of materials over the period 1995–2015 (Hertwich, 2021)

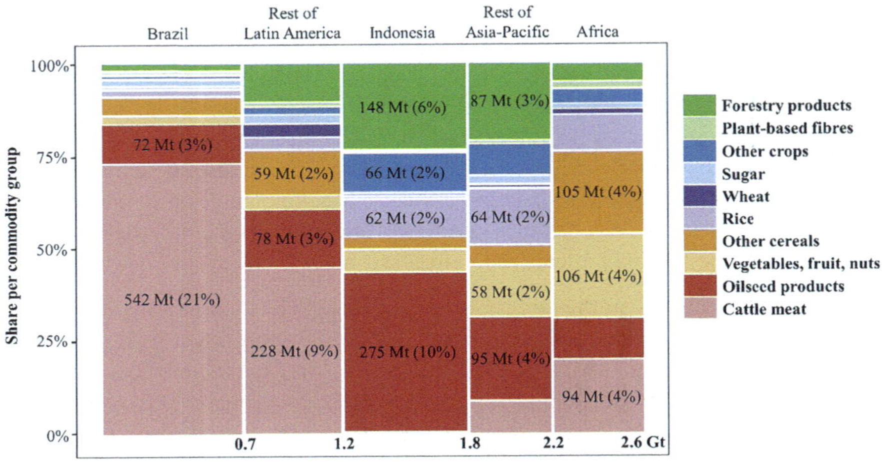

Fig. 1.6 Emission sources for deforestation-related CO_2 emissions are diverse and vary by region. Emissions embodied in production are quantified for each commodity group within each country (here summarized by region). A region's width on the *x*-axis corresponds to the embodied emissions produced in that region, while the y-axis shows the share of emission attributed to each commodity group within each region, implying that the rectangles within the plot are scaled according to the emissions embodied in each region-commodity combination. The percentages within the rectangles indicate the share of the total (2.6 Gt CO_2 year^{-1}) embodied emissions. For forestry products, the results show emissions associated with tree-plantation expansion, but not emissions due to clearing purely for timber without subsequent land use expansion (Pendrill et al., 2019)

Because forests absorb CO_2, deforestation also limits nature's ability to stop CO_2 from being released into the atmosphere. Deforestation, combined with agriculture and other land use changes, generates about a quarter of all global GHG emissions from these activities. Individually, deforestation contributes 6–17% of global anthropogenic CO_2 emissions to the atmosphere (Van Der Werf et al., 2009). The net carbon emissions from deforestation and degradation in the tropics, including the draining and burning of peat swamps in SE Asia, averaged ~1.4 (±0.5) PgC year^{-1} over the period 1990–2010. Most (60–90%) of the emissions were from deforestation (Houghton, 2012). Tropical forests play a crucial role in the global carbon cycle, with carbon stocks varying between 193 and 229 Pg (Baccini et al., 2012; Saatchi et al., 2011), representing about 54% of the global aboveground carbon stock (Liu et al., 2015b).

The area of these forests, however, declined by 10%, from 19.65 million km^2 in 1990 to 17.70 million km^2 in 2015, because of land use and land cover changes (Keenan et al., 2015). For about 30 developing countries, including Brazil, Bolivia, Indonesia, Myanmar, and Zambia, deforestation and forest degradation are the largest sources of CO_2. During the 2003–2015 period in the Brazilian Amazon, CO_2 committed emissions from forest fires (5904 Tg) (Aragão et al., 2018) and edge effects (2068 Tg) (Silva et al., 2020a) reached 88% of the gross deforestation emissions (9108 Tg) (Aragão et al., 2018). The CO_2 emissions resulting from degradation are not all immediate. Degraded forests continue to emit more CO_2 than they absorb for many years and are becoming significant carbon sources (Silva et al., 2020b).

Transportation Transportation is one of the main contributors to final energy consumption in most countries, particularly in industrial ones. This sector, as the second top CO_2 emitter sector among other sectors, plays an important role in the global carbon emission level (Solaymani, 2019). Most cars, trucks, ships, and airplanes run on fossil fuels for energy (Fig. 1.7). This makes transportation a major source of GHG emissions, especially CO_2. On-road vehicles emit the most because their internal combustion engines burn petroleum-based products, such as gasoline. However, emissions from ships and aircraft continue to grow. Transportation accounts for approximately a quarter of global energy-related carbon emissions. Trends indicate that energy consumption by transportation will increase significantly in the coming years. It is believed that CO_2 emissions from the transport sector are growing faster than total CO_2 emissions. Total CO_2 emissions from fossil fuel increased approximately 38% from 20.9 gigatons (Gt) in 1990 to 28.8 Gt in 2007, of which the emissions from the transport sector rose from 4.58 Gt in 1990 to 6.63 Gt in 2007, an increase of 45% (Wright, 1986). The total transport sector in 2007 accounted for 23% (worldwide) and 27% (OECD countries) of the total CO_2 emissions, most of which can be attributed to the road transport sector.

World transport energy use is projected to grow at 2% per year, with the highest rates of growth in emerging economies, while total transport energy use and carbon emissions are projected to be about 80% higher than the current levels by 2030 (IPCC, 2012). The transport sector is 95% dependent on oil and accounts for 60% of all oil consumption. However, road transport particularly automobiles dominates the share of CO_2 emissions in comparison with air and rail transport sectors (Saboori et al., 2014). During 2000–2015, transport carbon emissions increased in the United

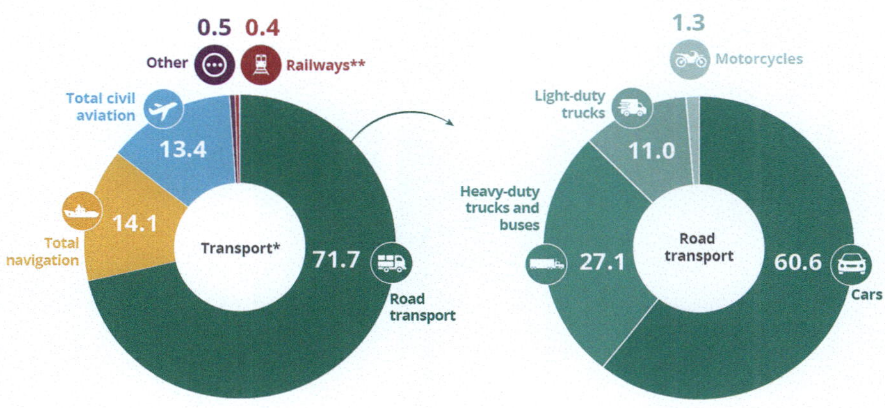

Fig. 1.7 GHG emissions from transport in the EU in 2022. Unit: Gt. (Source: European Environment Agency)

States, China, India, Canada, Russia, and Brazil but decreased in Japan (Solaymani, 2019). In most other Asian countries, the transport sector already accounts for a substantial share of total national CO_2 emissions.

Food Production Food chains around the world are responsible for a large share of total emissions of GHGs, and it is estimated that about one-fourth of all GHG emissions may be caused by the global food system (Mrówczyńska-Kamińska et al., 2021). Food production systems as a group are very heterogeneous, the range of products is huge, and production systems vary within product groups as well. However, there are some common traits (Havas et al., 2015). The process of producing food emits CO_2, CH_4, and other GHGs in a variety of ways. These include deforestation and clearing of land for farming and grazing, digestion of food by cattle and sheep, production and use of fertilizers and manure to grow crops, and often the use of energy sources such as fossil fuels to power agricultural equipment or fishing boats (Fig. 1.8). For vegetable products, N_2O is often the most important emission. The same holds true for monogastric animals (pork and poultry), while for ruminants, methane is often the dominating gas emitted. Methane and nitrous oxide are very potent GHGs, with methane having a weighting factor of 25 times CO_2 and 298 times N_2O (IPCC, 2007). For seafood products, the climate impact of products from capture fisheries is dominated by fossil CO_2 emissions from fuel use

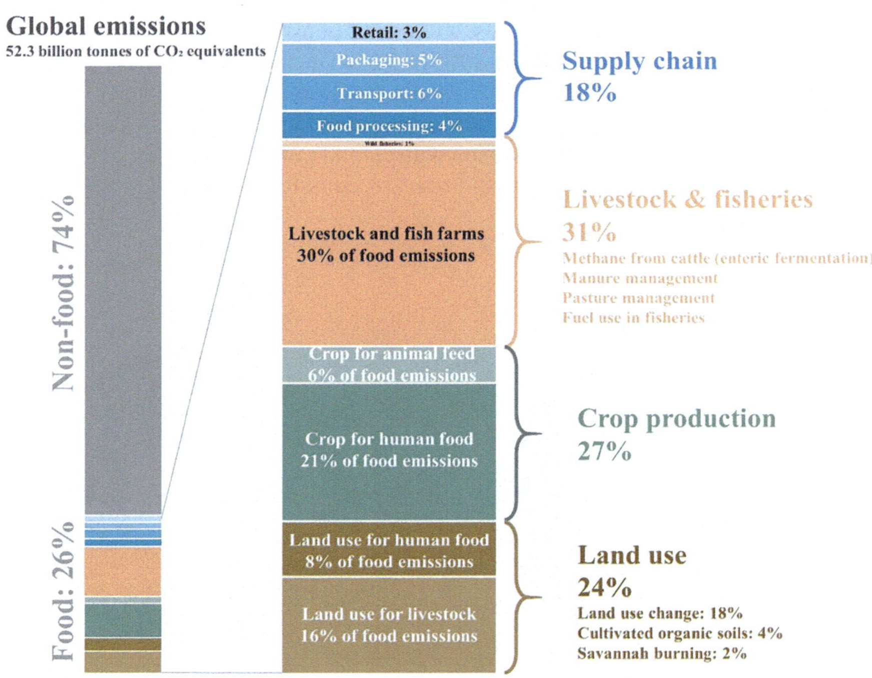

Fig. 1.8 Global GHG emissions from food production. (Source: Our World in Data)

on fishing boats. All these activities make food production a major contributor to climate change. In addition, packaging and distributing food also emits GHGs.

Emissions in food production systems vary between continents and countries (Vermeulen et al., 2012), which are largely due to the directions of production, i.e., the differentiation between the unit emissions from animal production and plant production. A study on the economic aspects in terms of emissions per production value concluded that the environmental performance of food production shows significantly higher emission intensity from animal production compared to plant production (Vetter et al., 2017) due to high emissions at the breeding stage, as well as a higher degree of processing at a later stage of production (O'Donoghue et al., 2019). Based on the energy intensity in a food production system, Pelletier et al. (Pelletier et al., 2011) reported that a significant source of GHG emissions in food production in the United States includes emissions from energy consumption in the processing industry, including cooking, chilling, and freezing, which accounts for 15–20% of the total energy consumption in the US food system, also including distribution.

Energy Supply Buildings Energy use in buildings is a significant source of GHG emissions. Residential and commercial buildings consume more than half of the world's electricity. Because these buildings still use coal, oil, and natural gas for heating and cooling, they emit significant amounts of GHGs (Fig. 1.9) (Lu & Lai,

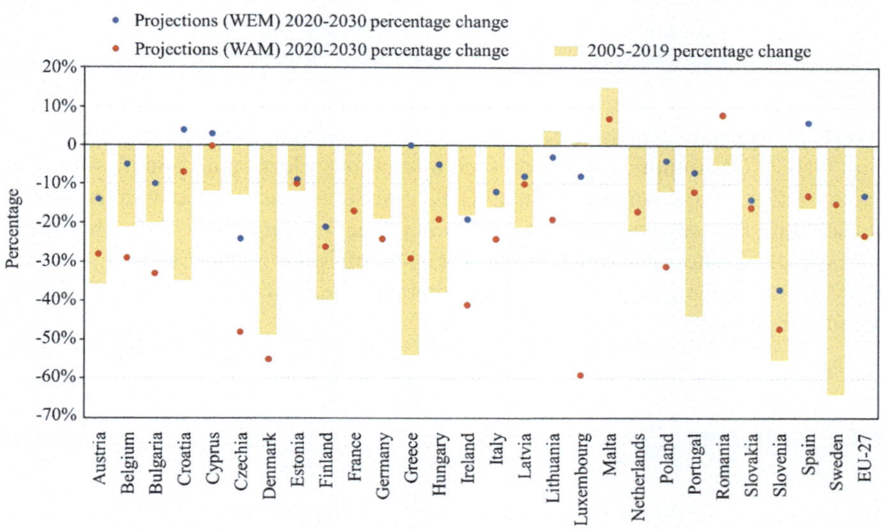

Fig. 1.9 GHG emissions from energy use in buildings by country. WEM and WAM are two types of projections: Projections considering the (current) existing domestic policies and measures (WEM, with existing measures) and projections taking also into account additional (planned) domestic policies and measures (WAM, with additional measures). (Data sources: National emissions reported to the UNFCCC and to the EU Greenhouse Gas Monitoring Mechanism provided by the European Environment Agency (EEA))

2020). In recent years, growing energy demand for heating and cooling as air conditioning ownership has increased, along with increased electricity use for lighting, appliances, and connected devices, which has led to a rise in energy-related CO_2 emissions from buildings. Energy use in the buildings sector was responsible for 7.85 Gt CO_2 emissions in 2002, 33% of the global total of energy-related emissions. Buildings were also responsible for approximately 1.5 Gt CO_2 equivalent emissions from fluorinated gases. The commonly used IPCC Special Report on Emissions Scenarios (SRES) project growth of these emissions to 11 Gt (B2 scenario) and 15.6 Gt CO_2 (A1 scenario) by 2030 (Nakicenovic et al., 2000), with the share of the sector remaining at approximately 34% of the total.

In Organization for Economic Co-operation and Development (OECD) countries, buildings account for about 35–40% of national CO_2 emissions from the use of fossil fuels. This includes direct on-site emissions from the use of natural gas and oil (primarily) for heating and cooling, as well as emissions from the generation of that portion of the electricity used in buildings (buildings account for about 50% of total electricity consumption in OECD countries) (Ürge-Vorsatz et al., 2007). Buildings consume 30–40% of the yearly primary energy in developed countries and approximately 15–25% in developing countries (Li et al., 2015). In the United States, buildings account for around 40% of primary energy consumption, and therefore 40% of the total US CO_2 emissions and 7.4% of the total global CO_2 emissions. More narrowly, residential buildings comprise 21% of the US energy consumption and are therefore responsible for about 20% of the total US CO_2 emissions (EIA, 2018). In 2018, the US buildings sector was responsible for 36% of national energy-related CO_2 emissions, making it a critical component of the Mid-Century Strategy.

Overconsumption House, electricity use, transportation, food, and garbage, all emit GHGs. Consumption of goods such as clothing, electronics, and plastics also emits GHGs. A large portion of global GHG emissions is associated with individual households. The way we live has a profound impact on our planet. The richest people bear the greatest responsibility: The GHG emissions of the richest 1% of the world's population are greater than the total emissions of the poorest 50% of the world's population.

1.1.2 Impacts

The Sixth Assessment Report (AR6) of the United Nations IPCC found that climate impacts are at the high end of previous estimates, with all parts of the world being affected (IPCC, 2021). The report identified 127 different negative impacts of climate change, some of which are irreversible. At least 3.3 billion people, about 40% of the world population, now fall into the most serious category of "highly vulnerable," with the worst effects in the developing world (IPCC, 2021).

The 1980s was once reported as the warmest decade in Earth's history; however, the temperature of the 1990s broke the record. The mean global temperature increased by 0.6 °C (±0.2 °C, 99% CL) during the twentieth century (IPCC, 2021). Nowadays, the Earth is about 1.1 °C warmer than it was in the late 1800s and the record is constantly being broken with the increase in temperature. Warming has been most intense in the summer and in the Northern Hemisphere. Human-induced climate change, including more frequent and intensive extreme events, has caused adverse impacts and related losses and damages to nature and people. The rise in weather and climate extremes has led to some irreversible impacts on natural and human systems (Fig. 1.10).

1.1.2.1 Impacts on the Natural System

Snow, Ice, and Frozen Ground (Including Permafrost) Due to climate change, snow cover has decreased in most regions, especially in spring and summer. Northern Hemisphere snow cover observed by satellite over the 1966–2005 period decreased in every month except November and December, with a stepwise drop of 5% in the annual mean in the late 1980s (Fig. 1.11). Additionally, the maximum extent of seasonally frozen ground in the Northern Hemisphere decreased by about 7% from 1901 to 2002, with up to a 15% decrease in spring (Dqg et al., 2005). Decreases in snow, ice, and frozen ground (including permafrost) will contribute to changes in wildfire. The fire weather season is projected to lengthen and fire regimes are expected to encroach into tundra regions (IPCC, 2021). These changes have threatened certain Arctic and Antarctic ecosystems, including those in sea ice biomes, as well as top predators in the food chain. Loss of sea ice negatively affects organisms that rely on ice, e.g., emperor penguins and polar bears (Chown et al., 2022; Wege et al., 2021).

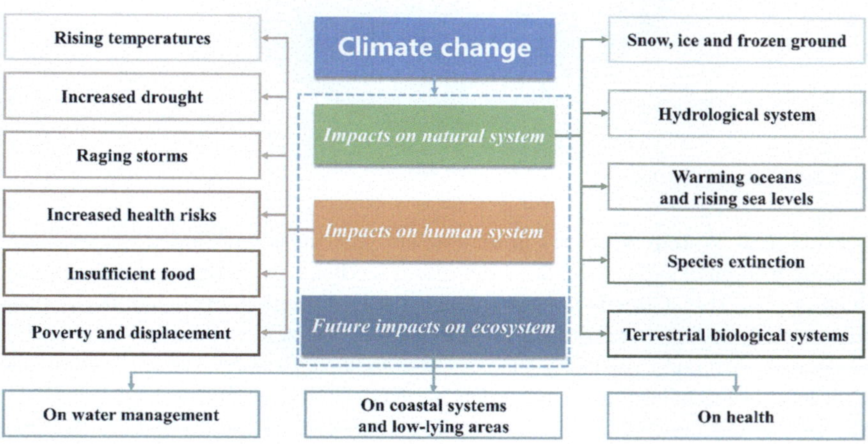

Fig. 1.10 The various impacts of climate change

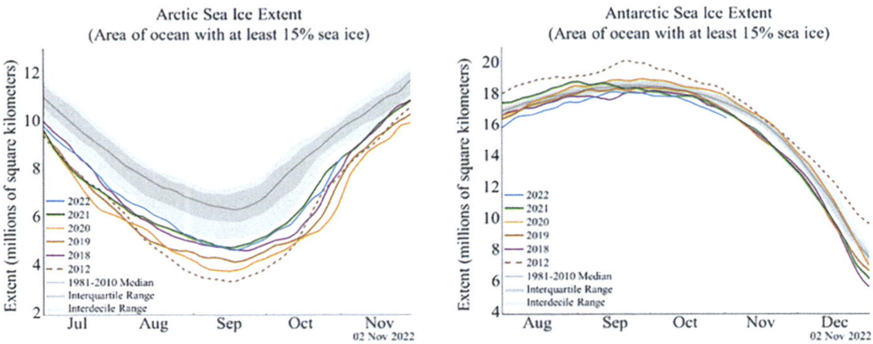

Fig. 1.11 The impacts of climate change. (Source: National Snow and Ice Data Center)

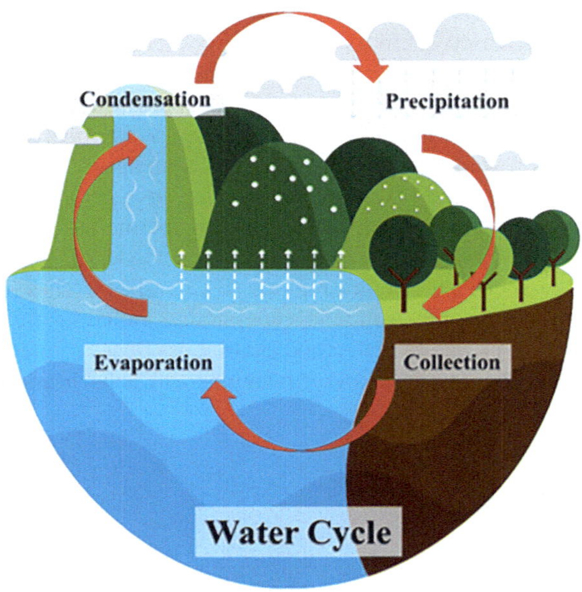

Fig. 1.12 The water cycles

Hydrological System Changes in climate are intricately interlinked with changes to the hydrological cycle—the most important feedback cycle in the climate system (Stagl et al., 2014). Climate change impacts the hydrological cycle and the stationarity of hydrologic systems (Fig. 1.12), leading to increased runoff and earlier spring peak discharge in many glacier- and snow-fed rivers. These changes can result in extreme natural disasters, such as heavy rainfall and floods. In July 2021, extreme rainfall became floods that washed away centuries-old homes in Europe, triggered landslides in Asia, and inundated subways in China. Conversely, climate warming will lead to an increase in the evaporative demand of the air, or "potential evaporation" (Stagl et al., 2014). Increased evapotranspiration due to rising temperature without enhancement in precipitation has heightened the intensity and frequency of hydrological droughts (Sheffield et al., 2012), which further have seriously negative impacts on various sectors, such as agriculture, urban life, industry, and hydropower generation, which depend mainly upon surface water resources (Vasiliades et al., 2011).

Warming Oceans and Rising Sea Levels The oceans have absorbed more than 90% of the heat from GHGs caused by human activities, but it is taking a toll on our oceans: 2021 set a new record for ocean heating (Nunez, 2019). Over the past 20 years, the rate of warming has increased throughout the ocean, regardless of depth. As a consequence of the thermal expansion of warming oceans and the addition of freshwater from melting continental ice sheets, the global sea level has risen (Fig. 1.13) (Han et al., 2010). Like mountain glaciers, increased heat is causing the massive ice sheets that cover Greenland and Antarctica to melt more quickly. The contribution made by these two ice sheets has greatly increased since the 1990s due to increased outflow caused by the warming of immediate adjacent oceans (Carvalho & Wang, 2019), which has threatened the densely populated and low-lying coastal areas and island communities. In addition, the oceans continue to absorb CO_2, making it warmer and more acidic. The oceans are currently 26% more acidic than they were before the world became industrialized, with the rate of acidification increasing faster than we have ever witnessed before, endangering marine life and coral reefs.

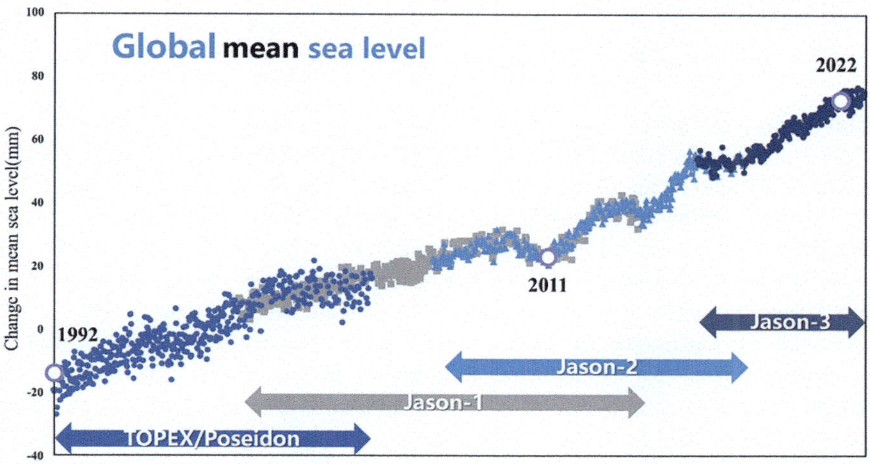

Fig. 1.13 Changes in mean sea level (mm) from 1992 to 2022. (Source: National Oceanic and Atmospheric Administration)

Species Extinction Climate change has become an important driving factor for species extinction (Thomas et al., 2004). This poses a substantial threat to the survival of both terrestrial and marine species, with the risks intensifying as temperatures continue to rise. Currently, climate change impacts 19% of the species listed on the IUCN Red List of Threatened Species (Foden et al., 2009). The influence of climate change on species extinction is accelerating, with species disappearing globally at a rate 1000 times faster than any other time in human history. This alarming trend puts approximately one million species at risk of extinction in the coming decades. Threats like forest fires, extreme weather events, pest invasions, and disease outbreaks are all intricately linked to climate change. Moreover, climate change is driving changes within species themselves. Rising temperatures have brought about ecological shifts, including changes in migration patterns. Behavioral changes, such as earlier breeding times, have also been observed. Physiological shifts are noticeable, including imbalanced female-to-male sex ratios and genetic changes such as hybridization (interbreeding as species' habitats change), collectively posing a threat to their survival (Foden et al., 2009).

Terrestrial Biological Systems Over the last three to five decades, climate change has caused notable impacts on terrestrial biological systems, such as the earlier spring and summer phenology and longer growing seasons in mid- and higher latitudes, bird migration and advance egg-laying dates, and poleward and upward shifts in ranges in plant and animal species (IPCC, 2007). For example, some studies show that leaf unfolding and flowering in spring and summer have, on average, advanced by 1–3 days per decade in Europe, North America, and Japan over the last 30–50 years (Chmielewski & Rotzer, 2001; Menzel, 2003). Earlier flowering implies an earlier start of the pollen season, which means pollinators and birds may arrive too late to feed.

1.1.2.2 Impacts on Human Systems

Rising Temperatures Temperature serves as one of the most direct indicators of climate change. As concentrations of GHGs continue to rise, global surface temperatures are also on an upward trajectory. Notably, the past decade, spanning from 2011 to 2020, holds the record for being the warmest (Fig. 1.14). Since the 1980s, each decade has surpassed the previous one in terms of warmth. This upward trend is accompanied by a surge in hot weather and heat waves, which are becoming increasingly common as global temperatures escalate. In particular, the year 2022 witnessed an onslaught of extreme heat across nearly the entire Northern Hemisphere. Heat waves persisted throughout the year, affecting regions like Europe, Americas, China, Australia, and the Indian subcontinent. Many of these heat waves have led to continent-wide wildfires, resulting in the tragic loss of thousands of lives due to heat-related causes. It is crucial to note that wildfires are more likely to ignite and spread rapidly in hotter weather, which is exacerbated by the combination of dry fuels and strong winds (Cardil et al., 2021). Additionally, heat

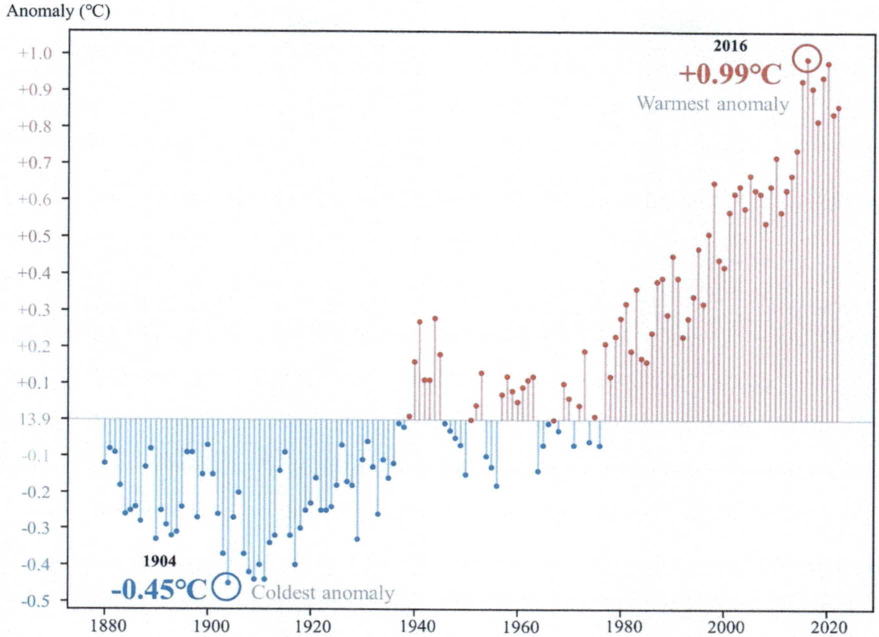

Fig. 1.14 Global land and ocean temperature anomalies, compared to the 1910–2000 average (13.9 °C). (Data source: NOAA National Centers for Environmental Information)

waves can lead to a higher incidence of heat-related illnesses and make outdoor work significantly more challenging (Shayegh et al., 2021). In 2022, an extreme heat wave in the UK was responsible for an estimated 1458 excess deaths among individuals over the age of 65, excluding those related to COVID-19. This figure represents the highest excess mortality observed during heat waves since the introduction of the heat wave plan for England in 2004 (Antonelli et al., 2022).

Increasing Drought Events Drought affects both surface and groundwater resources and can result in reduced water supplies, deterioration of water quality, crop failures, reduced rangeland productivity, reduced power generation, disturbance of riparian habitats, suspension of recreational activities, and impacts on a range of economic and social activities (Mishra & Singh, 2010; Riebsame et al., 2019). Climate change is changing the availability of water resources, making them scarce in more areas (Fig. 1.15). Global warming exacerbates water scarcity in areas that are short of water and increases the risk of agricultural and ecological droughts. Agricultural droughts are generally defined as a deficit in soil moisture and can affect plant growth and crop harvests. Severe agricultural drought events will cause large damage to agriculture and the social economy, posing threats to food security and human society (Mishra & Singh, 2010).

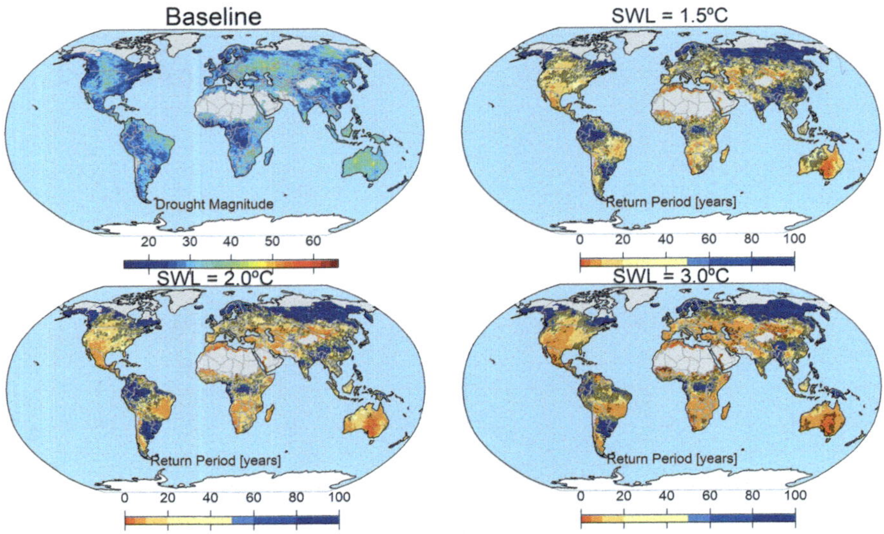

Fig. 1.15 Drought magnitude for 50-year return period droughts (baseline) and the differences between return periods of 50-year droughts for different warming levels (1.5, 2.0, and 3.0 °C). Changes that are not statistically significant at the 10% level are shaded out in black. Territories excluded from the analysis are masked in gray (Naumann et al., 2018)

In 1978–2003, drought-driven crop losses in China cost the nation 0.5–3.3% of its agricultural GDP (Pandey et al., 2004). Non-agricultural impacts of drought are often dispersed across larger regions and populations (Van Dijk et al., 2013). In 2016, the Science for Nature and People Partnership (SNAPP) (Jiang et al., 2021) defined ecological droughts as episodic deficits in available water induced by climate and human factors during the vegetation growth period. This will increase the vulnerability of ecosystems. According to one estimate, Australia's Millennium Drought cost nearly AU$810 million to mitigate ecological losses and replace vital ecosystem services (Banerjee et al., 2013). Droughts can also trigger devastating dust storms, which can carry billions of tons of sand to all continents. Accompanied by desert expansion, an amount of land is no longer available for growing food. Besides, extreme and long-time droughts have the potential for particularly severe damage to human economies, livelihoods, and health—even beyond their impact on natural resources and ecosystems (IPCC, 2014).

Raging Storms In many regions, devastating storms are becoming more destructive and occurring more frequently. As temperatures rise, more water evaporates, exacerbating extreme rainfall and flooding and triggering more destructive storms (Chisholm et al., 2021). The frequency and extent of tropical storms, i.e., cyclones, hurricanes, and typhoons, usually form on the surface of warm waters in the ocean. Such storms, as they impinge violently on the shoreline, may result in extensive damage to the local population and assets as a consequence of extreme high-water levels, flooding, and erosion processes (Mel et al., 2022; Rodríguez-Ramírez et al.,

2015). For instance, sea storms, in the short term, can devastate crops, and infrastructures, and take the lives of humans and livestock. In the long term, they may trigger significant erosion processes and losses of wide areas of land (Van Gelder et al., 2000).

Increased Health Risks Climate change is the greatest health threat facing humanity. Through air pollution, disease, extreme weather events, forced displacement, mental health stress, and increased hunger and malnutrition in places where people cannot grow or find enough food, climate change is harming human health (Fig. 1.16). In 2021, there were 432 disastrous events related to natural hazards worldwide, resulting in 10,492 deaths, 101.8 million people affected, and approximately 252.1 billion US$ of economic losses (CRED, 2022). Besides, changing weather conditions amplify the spread of disease and extreme weather events increase the number of deaths. Infectious diseases, pollution, and climate change seem to be connected in different ways, affecting each other, as there exist indications that variations of some meteorological parameters associated with climate change might favor the epidemics (Marazziti et al., 2021; Ruiz et al., 2010). For instance, a growing bulk of studies revealed an association between diffusion, incidence, morbidity, and mortality of the current COVID-19 pandemic and air pollutants, especially air PM levels (Conticini et al., 2020; Petrosillo et al., 2020). Another meta-analysis found that approximately a third of all workers exposed to occupational heat stress had negative effects, including an increased risk of hyperthermia and cardiovascular failure or collapse and the increased risk of acute kidney disease (Flouris et al., 2018).

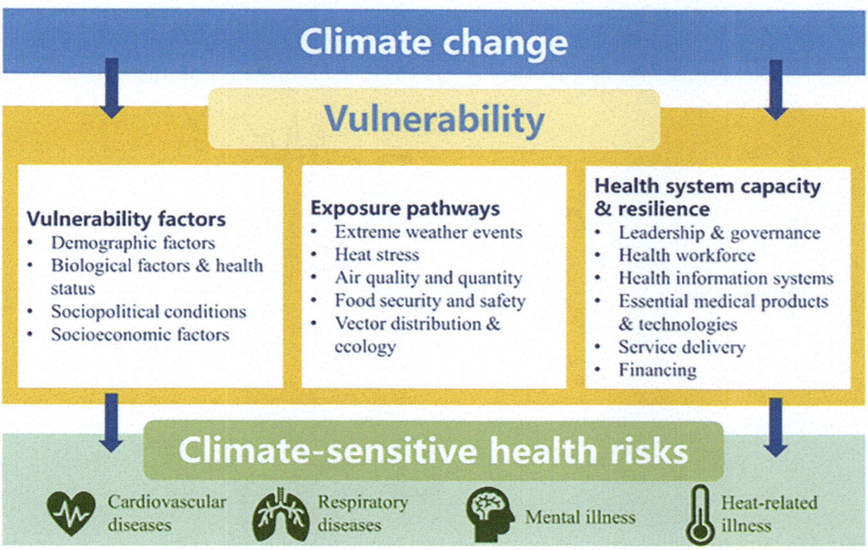

Fig. 1.16 An overview of climate-sensitive health risks. (Source: World Health Organization)

Insufficient Food Global hunger is currently rising. The proportion of undernourished people worldwide increased to 10.6% in 2015 and then to 11% in 2016 and it reached an estimated 821 million in 2017, which is around one in nine people (FAO, 2017). Climate change and the frequency of extreme weather events are both contributing to the increase in global hunger and malnutrition. Fisheries, crops, and livestock may be damaged or less productive. Ocean acidification is becoming more severe and the ocean resources that feed billions of people are in jeopardy. Changes in the snow and ice cover in many Arctic regions are disrupting the food supply from livestock, hunting, and fishing. Heat stress can reduce freshwater and grassland needed for grazing, leading to lower crop yields and affecting livestock. In addition, most developing countries face severe climate-related threats that affect crop production due to their geographical location (IPCC, 2014). These poor countries mainly located in tropical and semi-desert areas are exposed to rising temperatures and water scarcity and face insufficient food risk due to frequent crop failures, few technology inputs, and limited access to infrastructure (Fig. 1.17) (IPCC, 2007). It is therefore a great challenge for regions such as sub-Saharan Africa which often rely heavily on food aid due to the lack of purchasing power needed to improve their food security (Molotoks et al., 2021).

Poverty and Displacement People around the world are experiencing both subtle and stark effects of climate change. For example, floods can wash away urban slums and destroy homes and livelihoods, droughts affect crop harvests, etc. This further pushes more people into poverty. Particularly, these people already burdened by poverty and oppression, e.g., the people living in sub-Saharan Africa and South Asia

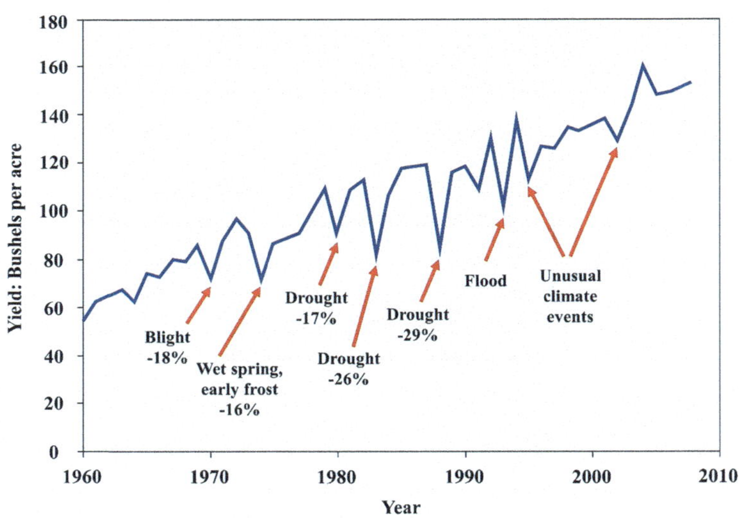

Fig. 1.17 Despite technological improvements that increase corn yields, extreme weather events have caused significant yield reductions in some years. (Source: the U.S. Global Change Research Program (2009))

often suffer the harshest consequences while having the least ability to cope. Over the past decade (2010–2019), an average of 23.1 million people were displaced each year by weather-related events, with many more likely to fall into poverty as a result (World Meteorological Organization, 2021).

Climate change will push up to 130 million people into poverty over the next 10 years if we take no actions, unraveling hard-won development gains, and could cause over 216 million people to migrate within their own countries by 2050 across the six World Bank regions (i.e., sub-Saharan Africa, East Asia and the Pacific, North Africa, South Asia, Latin America, Eastern Europe, and Central Asia) (Clement et al., 2021). Most refugees come from countries that are most vulnerable and unprepared to adapt to the effects of climate change. Some 9.8 million displacements, largely due to hydrometeorological hazards and disasters, were recorded during the first half of 2020, mainly concentrated in South and Southeast Asia and the Horn of Africa (Adger et al., 2020). More seriously, these people who experienced hydrometeorological events may also be subject to repeated and frequent displacement, leaving little time for recovery between one shock and the next.

Future Impacts on Ecosystems Climate change impacts biodiversity and is likely to intensify over the next few decades unless substantive mitigation efforts are implemented (IPCC, 2018). Biodiversity loss and degradation to, damage to, and transformation of ecosystems are key risks for every region due to past global warming and will continue to escalate with every increment of global warming. In terrestrial ecosystems, 3–14% of species assessed will likely face very high risk of extinction at global warming levels of 1.5 °C, increasing up to 3–18% at 2 °C, 3–29% at 3 °C, 3–39% at 4 °C, and 3–48% at 5 °C. In ocean and coastal ecosystems, the risk of biodiversity loss ranges between moderate and very high by 1.5 °C global warming level and is moderate to very high by 2 °C but with more ecosystems at high and very high risk and increases too high to very high across most ocean and coastal ecosystems by 3 °C. Very high extinction risk for endemic species in biodiversity hotspots is projected to at least double from 2% between 1.5 and 2 °C global warming levels and to increase at least tenfold if warming rises from 1.5 to 3 °C (IPCC, 2021).

Water Management The impacts of climate change on the water cycle are manifested by its influence on when, where, and how much precipitation falls. A larger amount of water evaporation caused by global warming will lead to higher levels of atmospheric water vapor and more frequent, heavy, and intense rains in the coming years (Society, 2022). On the one hand, amounts of water vapor will lead to floods since more water will fall than vegetation and soil can absorb. Excessive runoff eventually travels to larger bodies of water like lakes, estuaries, and the ocean, polluting the water supply and limiting water access for humans and ecosystems. Projected increases in direct flood damages are higher by 1.4–2 times at 2 °C and 2.5–3.9 times at 3 °C compared to 1.5 °C global warming without adaptation. At global warming of 4 °C, approximately 10% of the global land area is projected to face increases in both extremely high and low river flows in the same location (IPCC, 2021).

On the other hand, at approximately 2 °C global warming, snowmelt water availability for irrigation is projected to decline in some snowmelt-dependent river basins by up to 20%, and global glacier mass loss of 18 ± 13% is projected to diminish water availability for agriculture, hydropower, and human settlements in the mid and long term, with these changes projected to double with 4 °C global warming. Challenges for water management will be exacerbated in the near, mid, and long term, depending on the magnitude and rate and regional details of future climate change and will be particularly challenging for regions with constrained resources for water management.

Coastal Systems and Low-Lying Areas These areas are experiencing the adverse consequences of hazards related to sea level, ocean temperature, and ocean acidity. Due to sea-level rise, coastal systems and low-lying areas will increasingly experience adverse impacts, such as submergence, coastal flooding, and coastal erosion. The population potentially exposed to a 100-year coastal flood is projected to increase by about 20% if global mean sea level rises by 0.15 m relative to 2020 levels; this exposed population doubles at a 0.75 m rise in mean sea level and triples at 1.4 m without population change and additional adaptation. Warming of coastal waters will continue with significant negative consequences for coastal ecosystems (Bennett et al., 2015). There is evidence of southern expansion of habitat ranges for tropical species into southern waters and southward migration of cool-water species. Besides, although there are no impacts that have been attributed as yet to acidification, continued CO_2 emissions increases may adversely affect carbonate-dependent organisms, including cold-water corals, calcareous algae, crustaceans, and mollusks (He & Silliman, 2019).

Health Climate change and related extreme events will significantly increase ill health and premature deaths in the near and long term. Between 2030 and 2050, climate change is expected to cause approximately 250,000 additional deaths per year due to malnutrition, malaria, diarrhea, and heat stress (Fadda, 2020). Globally, population exposure to heat waves will continue to increase with additional warming, with strong geographical differences in heat-related mortality without additional adaptation. Climate-sensitive food-borne, water-borne, and vector-borne disease risks are projected to increase under all levels of warming without additional adaptation. Dengue risk will increase with longer seasons and a wider geographic distribution in Asia, Europe, Central and South America, and sub-Saharan Africa, potentially putting additional billions of people at risk by the end of the century. Mental health challenges, including anxiety and stress, are expected to increase under further global warming in all assessed regions, particularly for children, adolescents, the elderly, and those with underlying health conditions.

Biodiversity Biodiversity, the unique variety of life on our planet, underpins our cultural, economic, and social well-being. However, human-induced changes to ecosystems and the extinction of species have been more rapid in the past 50 years than at any time in human history (Ceballos et al., 2015). By the end of the century, climate change is likely to become one of the most significant drivers of biodiversity loss (Max Finlayson, 2018). Climate change is forcing biodiversity to adapt either through shifting habitats, changing life cycles, or the development of new physical traits. Besides, the loss of biodiversity across the world is also having a severe impact on people. It was reported that the loss of coastal habitats that provide a natural buffer against extreme weather events has put 100–300 million people at an increased risk of floods and hurricanes (Larigauderie & Mooney, 2010). In addition to climate change driving biodiversity loss, biodiversity loss, in turn, hampers the world's ability to tackle and adapt to climate change. Around half of the CO_2 released by human activities is absorbed by the land and ocean; the ecosystems and the biodiversity they contain are responsible for this carbon removal service. However, biodiversity loss reduces ecosystem functioning, which can, in turn, affect the land and ocean's ability to absorb CO_2 (Teixidó et al., 2018).

1.1.3 Climate Targets and Actions

Reducing carbon emissions is no longer enough to halt the impacts of climate change. Many countries are realizing it is time to start adapting to a warming world. Effective adaptation relies not only on governments but also on the continuous and active participation of stakeholders, including national, regional, multilateral, and international organizations, as well as the public and private sectors, civil society, and other pertinent parties. Additionally, the efficient management of knowledge is crucial. Adaptation to the impacts of climate change can be implemented across various regions, and sectors, and at various levels.

Parties to the United Nations Framework Convention on Climate Change (UNFCCC) and its Paris Agreement recognize that adaptation is a global challenge faced by all with local, subnational, national, regional, and international dimensions. Adaptation is a vital part of the long-term global response to climate change, aimed at protecting people, livelihoods, and ecosystems. Parties acknowledge that adaptation action should follow a country-driven, gender-responsive, participatory, and fully transparent approach, considering vulnerable groups, communities, and ecosystems, and should be based on and guided by the best available science. Where appropriate, these efforts should also incorporate traditional knowledge, indigenous knowledge, and local knowledge systems, with a view to integrating adaptation into relevant socioeconomic and environmental policies and actions (Fig. 1.18).

The inaugural World Climate Conference held in Geneva in 1979 marked a pivotal moment as one of the earliest conferences addressing the detrimental effects of climate change. During this event, technical and scientific experts examined the most up-to-date knowledge on climate change and variability arising from both natural and human factors. They also delved into the projected future impacts and associated risks. As a result, a set of recommendations was formulated to guide

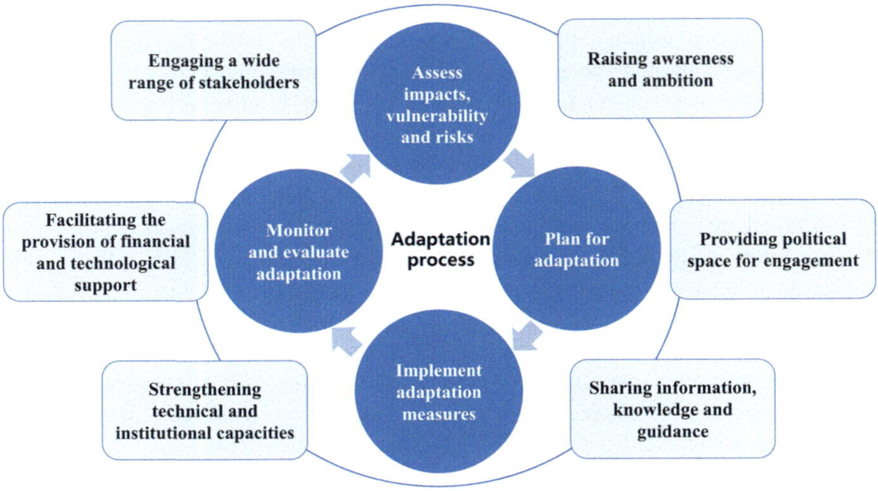

Fig. 1.18 Graphically, the adaptation cycle under the UN climate change regime, including four general components, which are individually featured under the SPOTLIGHT section. (Source: UNFCCC)

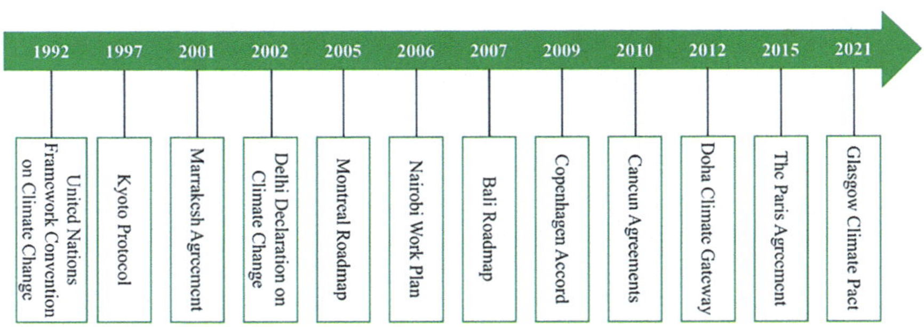

Fig. 1.19 Climate actions

future actions (WMO, 1979). One of the most significant steps taken to address climate change came in 1992, with the adoption of UNFCCC (Fig. 1.19). This convention came into effect in 1994 and set the target of reducing emissions by the year 2000 to levels achieved in the previous decade. It outlined commitments for all participating parties, placing substantial responsibilities on developed countries to implement national policies aimed at curbing anthropogenic emissions and enhancing GHGs sinks.

Additionally, in 1997, the Kyoto Protocol was adopted during the third UNFCCC conference of the parties (COP-3) and started in 2005. This protocol introduced two project-based mechanisms, known as the Clean Development Mechanism and Joint Implementation Mechanism, designed to mitigate emissions. The Kyoto Protocol laid the groundwork for carbon emissions trading, along with the concept of carbon pricing. One of the most notable emissions trading schemes is the European Emissions Trading Scheme, which, upon implementation, is projected to reduce GHG emissions by approximately 1.58% annually (Villoria-Sáez et al., 2016). After a decade of operation, it is estimated that roughly 23.43% of emissions reduction could be achieved compared to a scenario without implementation.

During the twenty-first UNFCCC conference of the parties (COP-21) held in Paris in 2015, the Paris Agreement was adopted and came into force in 2016. The primary goal of the Paris Agreement is to limit the global temperature increase to **2 °C** by 2100 and make efforts to restrict it to **1.5 °C**. The agreement seeks to achieve the global peaking of GHG emissions as soon as possible to strike a balance between human-induced emissions and GHG sinks and reservoirs between 2050 and 2100. Furthermore, in 2015, all United Nations Member States adopted the 17 SDGs as part of the 2030 agenda for sustainable development. This agenda outlined a 15-year plan to accomplish these goals. Among these, *SDG 13: Climate Action* aims to take urgent measures to combat climate change and its impacts. This includes enhancing resilience and adaptive capacity to climate-related hazards and natural disasters in all countries (SDG 13.1), as well as promoting mechanisms for enhancing the capacity for effective climate change-related planning and management, particularly in the least developed countries and small island developing states. This effort also emphasizes focusing on women, youth, and local marginalized communities (SDG 13.B).

The main objective of these actions is to reduce carbon emissions, with a focus on the balance between carbon emissions and carbon sequestration. Addressing the escalating climate change crisis and achieving carbon neutrality in the near future have become paramount global tasks. The Chinese government has set clear targets, aiming to reach its *carbon peak* by 2030 and achieve *carbon neutrality* by 2060. This endeavor seeks to gradually achieve net-zero CO_2 emissions. This initiative is designed to enhance the Paris Agreement and potentially accelerate global progress toward the temperature targets of 1.5 and 2 °C.

1.1.4 Barriers to Climate Change Adaptation

It is unequivocal that human activities have heated our climate. Recent changes are rapid, intensifying, and unprecedented over centuries to thousands of years. With each additional increment of warming, these changes will become larger, resulting in long-lasting, irreversible implications, in particular for sea-level rise. Here we outline the most common barriers to adapting to climate change experienced by individuals, organizations, and governments.

Barriers Experienced by Individuals People who have a low personal understanding of climate change will find it difficult to plan for its impacts or even to accept that others should plan to adapt. This may be because they feel that too little is known about climate science, or that the science is not "settled," or that there needs to be more certainty about the science before actions can take place, or that the risks are small and action can be delayed (CoastAdapt, 2017). Personal factors such as the role of traditional knowledge, political affiliation, educational background, and trust placed in different information sources affect risk perception and subsequent decision-making (IPCC, 2014; Klein et al., 2015). This category of barriers is crucial for the first three phases of the climate risk assessment of adaptation processes and is likely to manifest itself in other forms, such as the lack of political will, in the later implementation phases if not overcome in the beginning (Weyrich, 2016).

Barriers Experienced by Organizations Resources are essential in each planning and management stage (Weyrich, 2016) and long-term allocation, especially of financial resources, is a signal of recognition by leadership that these issues are prioritized. However, organizations often have limited financial, technical, and human resources to plan for a complex and contested issue (e.g., adapting to climate change). Lack of leadership within organizations and agencies charged with adapting to climate change is also a frequently cited barrier, while a lack of political will among government officials has also been noted (Hamin et al., 2014). The barrier of competing priorities may reflect resource deficits but may also indicate conflicting values, within an organization or at the community or governmental level (Aylett, 2015). A lack of locally relevant and practical information about potential climate impacts may be compounded by a lack of technical expertise to interpret climate change projections for the local area. Organizations that lack access to information will not act appropriately. In this case, climate information is a public good; private actors tend to underprovide it and so there is a role for governments and public authorities to support its production and dissemination (FCFA, 2018).

Barriers Experienced by Government Climate adaptation finance is pivotal to implementing adaptation measures. Estimated adaptation costs in developing countries are five to ten times greater than current public adaptation finance flows, and the adaptation finance gap is widening (UNEP, 2021). Political or policy barriers, such as a reluctance to take on debt to invest in adaptation, could prevent spending on adaptation. However, limited fiscal space could exacerbate these barriers, making it harder for governments to commit funds to adaptation. Besides, regulatory requirements, such as those for insurance or permitting (McGuire, 2018) or the lack of legal basis for actions or official remit by specific actors, can impede movement on adaptation issues (Measham et al., 2011). The most important strategies to overcome governance and institutional constraints involve small, incremental policy, planning and management changes, as well as efforts to change existing governance structures to create space for continuing the adaptation process (Ekstrom & Moser, 2014).

1.2 Urbanization

1.2.1 Urbanization Process

Urbanization refers to the mass movement of populations from rural to urban settings (Fig. 1.20) and the consequent physical changes to urban settings. The process of urban formation is the permanent concentration of a large population in relatively small areas. The most urbanized regions are Northern America with 82% of its population living in urban areas (as of 2018), Latin America and the Caribbean (81%), Europe (74%), and Oceania (68%). Asia has about 50% level of urbanization and Africa remains mostly rural, with only 43% of its population living in urban areas.

1.2.2 Drivers

A steady stream of human migration out of the countryside, and into swelling metropolitan centers, has shaken up the world's power dynamic in just decades. Since the mid-twentieth century, the world population has more than tripled in size, rising from around 2.5 billion in 1950 to almost 7.9 billion in 2021 (UNDESA, 2021). It is expected that over a third of the projected urban growth between now and 2050 will occur in just three countries: India, China, and Nigeria. By 2050, it is estimated to result in more than 6 billion inhabitants (United Nations, 2015) and is projected

Fig. 1.20 Population growth in urban areas. (Data source: the United Nations)

that India could add 416 million urban dwellers, China 255 million, and Nigeria 189 million. This rapid urbanization poses significant challenges to sustainable cities, including issues related to climate resilience, poverty, and inequality, as well as the impacts of the pandemic era (Fig. 1.21).

Population Boom The major reasons for the increasing urban population are rural to urban migration, including international migration to a lesser extent, and the re-classification or expansion of existing city boundaries to include populations that were hitherto classified as being resident outside the city limits. These are estimated to contribute about 60% of the region's urban growth, while natural increase accounts for 40%. On Tuesday, 15 November 2022, the world's population hit 8 billion people. It is estimated that the global population is projected to increase to around 9.8 billion by 2050, and more than 68% world's population will live in cities (6.7 billion) (Fig. 1.22) and 32% in rural areas (3.1 billion) (Ritchie & Roser, 2018).

Rural–Urban Migration Primary driving forces of rural–urban migration include the opportunities and services offered in urban areas—especially jobs and education, while in some cases, conflict, land degradation, and exhaustion of natural resources in rural areas are also important. The patterns of rural–urban migration may be city-specific, reflecting, among other things, changes in the city's economic base, labor market, and age structure. They also reflect social, economic, and political changes within the region and nation and are influenced by economic factors in the surrounding and distant rural areas, such as landowning structure, agricultural practices and crop prices, and overall rural productivity. Most evidence suggests that increasing the income and level of education of rural populations accelerates migration and this phenomenon, coupled with the greater access to urban areas, has

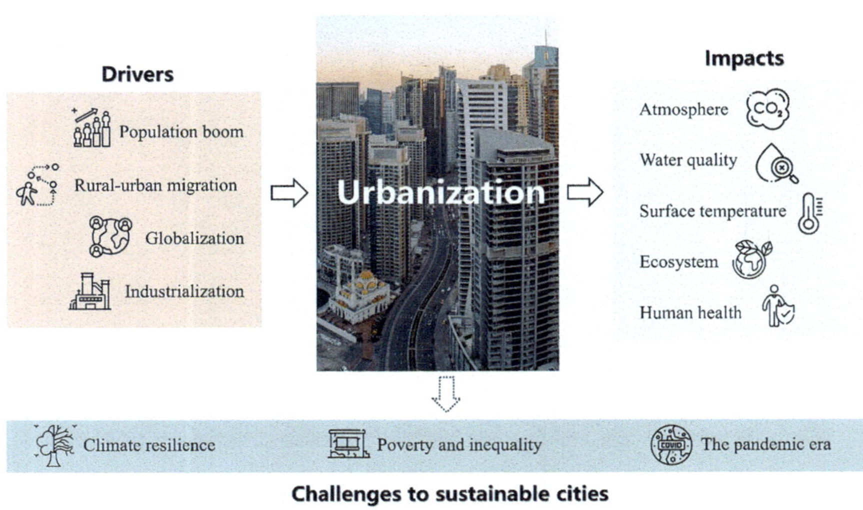

Fig. 1.21 Drivers, impacts, and challenges of urbanization

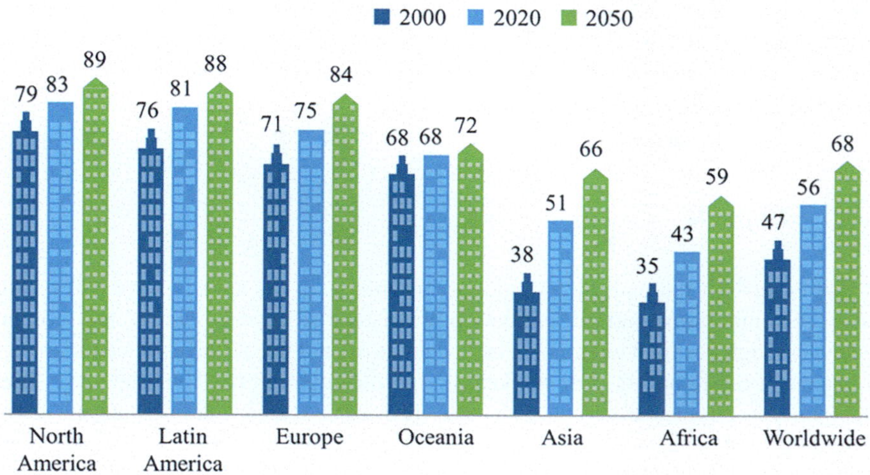

Fig. 1.22 World's population living in cities across different regions. (Source: United Nations)

led to an inevitable increase in rural populations seeking employment opportunities in urban areas.

Globalization The steady increase in the level of urbanization reflects the fact that the size of the world's economy has grown many times and has also changed from one dominated by relatively closed national economies or trading blocs to one where most countries have more open economies and where production and the services they need, including financial services, are increasingly integrated internationally. These trends appear to be strengthening, reinforced especially by the freer and faster flows of information and knowledge under the impact of new information technologies. Technology has increased the already dominant economic role and importance of urban areas worldwide, indicating the growing importance of cities in the global economy. A new process of urbanization is driven by international trade, global investment, and technological flows, all facilitated by advances in information technology (Hollands, 2008). The most striking forms of agglomeration today are super-urban agglomerations or global city regions, which are the sites of dense masses of interrelated economic activities and are the engines of national and global economies.

Industrialization Industrialization is one of the main catalysts of urbanization. The technological explosion of the Industrial Revolution led to a significant acceleration in the process of urbanization. Industrialization is a trend representing a shift from the old agricultural economics to a novel non-agricultural economy, which creates a modernized society. Through the industrial revolution, more people have been attracted to move from rural to urban areas on account of improved employment opportunities and to work in modern sectors in job categories that stir economic developments (Wang & Su, 2019). As industrialization creates economic growth, the demand for improved education and public works agencies in urban areas increases, which unquestionably concentrates large numbers of skilled workers and their families in cities.

1.2.3 Impacts

Probably most of the major environmental problems of the next century will result from the continuation and intensification of existing problems that currently do not receive enough political attention. The problems are not necessarily noticed in many countries or then nothing is done even though the situation has been detected. The most emerging issues are climate change, freshwater scarcity, deforestation, freshwater pollution, and population growth. These problems are very complex and their interactions are hard to define. It is very important to examine problems through the social–economic–cultural system. While interconnections between environmental problems are now better known, we still lack exact information on how the issues are linked, to what degree they interact and what are the most effective measures (Uttara et al., 2012).

Urbanization does not have only local environmental impacts but also large so-called ecological footprints beyond their immediate vicinity (Allen, 2003). Intensive and extensive exploitation of natural resources to support urban economy includes excessive extraction of energy resources (including fuelwood), quarrying and excavation of sand, gravel, and building materials at large scales, and over extraction of water. These all contribute to the degradation of the natural support systems and irreversible loss of critical ecosystem functions, such as the hydrological cycle, carbon cycle, and biological diversity, in addition to conflicts with rural uses of such limited resources. Other effects can be felt further afield such as pollution of waterways, and long-range air pollution that impacts human health as well as vegetation and soils at a considerable distance (Fig. 1.23).

Atmosphere Among the attributable sources of fine particulate emissions, urbanization is an important contributor. The research on the relationship between socioeconomic factors and $PM_{2.5}$ in the urbanization process centralized more on economic growth, population growth, and industrial transformation. Socioeconomic factors are closely related to $PM_{2.5}$ concentrations in urban agglomerations, but two different views exist. One is that urbanization level rise accompanied by excessive population concentration and increased exhaust emissions from motor vehicles results in more critical urban $PM_{2.5}$ pollution. The other is that those metropolises with high economic levels and high population density would pay more attention to pollution control and prevention, high-tech companies would be willing to innovate in technology, and government fiscal expenditures on pollution control could effectively alleviate serious air pollution problems.

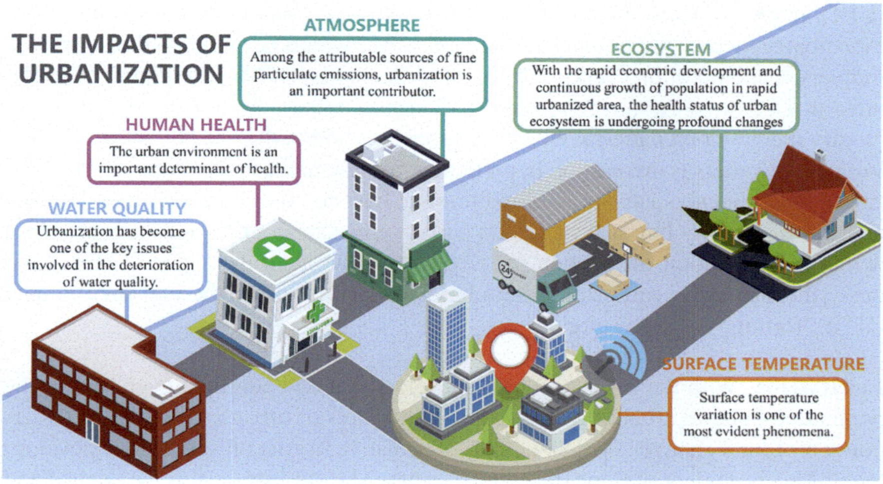

Fig. 1.23 The impacts of urbanization

Water Quality As an important part of watershed management, urbanization has become one of the key issues involved in the deterioration of water quality. Urbanization has changed the predominant land use/land cover type from natural vegetation to impervious surface (Knouft & Chu, 2015), resulting in increased surface runoff and modified stream channel morphology. The chemical conditions of streams in urban areas can be influenced by runoff from a variety of sources, including rooftops, parking lots, golf courses, construction sites, and residential areas, which all contribute to the impervious surface area (Müller et al., 2020). Impervious surfaces prevent rainfall from infiltrating into soil and ground water, leading to increased runoff to streams, which subsequently leads to changes in water quality. Moreover, the rising population and socioeconomic activities have increased the discharge of untreated effluent during urban expansion and eventually degraded the quality of stream water. All these changes brought about by urbanization have major impacts on the hydrologic regime and habitat quality, ultimately causing ecological degradation and biodiversity loss.

Surface Temperature Surface temperature variation is one of the most evident phenomena. Land surface temperature modifies the air temperature of the atmospheric boundary layer and is a key component in the surface energy balance. So land surface temperature is very important for the study of urban climates (Voogt & Oke, 2003). Changes in land surface temperature have significant effects on local climates. By covering the landscape with buildings, roads, parking lots, and other impervious surfaces, urban areas usually have higher solar radiation absorption and a greater thermal conductivity and capacity for releasing heat storage during the day at night. This procedure generally leads to the phenomenon that urban areas are warmer than the surrounding rural areas, which is referred to as an urban heat island

(UHI). In recent years, the global UHI effect at middle and high latitudes has been exacerbated, accompanied by extreme heat wave, which has increased in intensity, frequency, and duration (Perkins-Kirkpatrick & Lewis, 2020). The higher temperatures in UHI increase air conditioning demands, raise pollution levels, and may modify precipitation patterns, etc. As summers get hotter, the need for space cooling from city dwellers is increasing. In particular, the rise in the utilization rate of refrigeration equipment such as air conditioning has not only increased the energy demand of the city but also increased carbon emissions, which has seriously hindered the realization of "carbon peak" and "carbon neutrality" goals. More seriously, the increase in carbon emissions has worsened the urban thermal environment (UTE), forming a vicious circle.

Ecosystems With the rapid economic development and continuous growth of the population in rapid urbanized areas, the health status of urban ecosystems is undergoing profound changes, which will have a negative impact on sustainable development. This is evident in Bangalore, India, where shifting administrative control and geographical expansion of city boundaries happen against a backdrop of rapid population growth and social inequalities. The city's struggle to accommodate the needs for infrastructure and settlements has had serious consequences for the city's trees, parks, lakes, and other urban ecosystems (D'Souza & Nagendra, 2011). Future challenges such as water scarcity and temperature increases related to climate change are likely to be exacerbated further if more trees and lakes are lost. The ongoing urbanization further severely reduces the ecosystem's capacity to provide benefits for human well-being but also by limiting people's participation in the management and protection of these green spaces.

Urbanization in coastal areas often leads to the destruction of sensitive ecosystems and can also alter the hydrology of coasts and their natural features such as mangrove swamps, reefs, and beaches that serve as barriers to erosion and form important habitats for species (Adhikari & Hansen, 2018). According to the Millennium Ecosystem Assessment report, more than 60% of global ecosystem services are significantly degraded, with human activity being the main driver (MEA, 2005). To safeguard coastal ecosystems and human well-being, the United Nations clearly stated "Conserve and sustainably use the oceans, seas, and marine resources for sustainable development" as a sustainable development goal (SDG 14) (UN, 2015), in which there were 10 sub-goals. Among them, SDGs 14.1 and 14.2 proposed to reduce marine pollution caused by human activities on land and to sustainably manage and protect marine and coastal ecosystems from the perspective of land–sea coordination development, respectively.

Human Health The urban environment is an important determinant of health. Urbanization, in developed countries, generally affects health positively (Welander et al., 2015). Lifestyle changes in cities, the aging of the population, and the increase in the early diagnosis of diseases with the developments in health services have caused the diseases to shift toward chronic diseases. Urbanization, however, as beneficial to human health is not always confirmed. Factors such as rapid, unplanned, and irregular urbanization, as well as air and water pollution, poor quality housing, inadequate hygiene, traffic accidents, and noise adversely affect health. When the urban population continues to increase over a short period of time, the growth rate of medical resources often lags far behind. This creates a "crowded effect" on the residents of the city and adversely affects health (Qin et al., 2013). In particular, uncontrolled urbanization has been associated in some contexts with pollution, social isolation, overcrowding, changes in dietary and physical activity patterns, inadequate service capacity to provide drinking water, sanitation, and waste disposal. All of these increase the risk of harming population health.

Pollutions, such as smoke, fine particle matters, and exhaust fumes, have a proven relationship to respiratory diseases (e.g., chronic bronchitis, asthma, and emphysema). *Heat islands*, during heat waves in summer, high urban temperatures are deadly, especially for heat-vulnerable populations such as children and the elderly. Extreme heat leads to heat cramps, heatstroke, exhaustion, and even death. Heat exposure is also associated with several non-fatal health outcomes, including heatstrokes, dehydration, loss of labor productivity, and decreased learning. Not only that, but urban high-temperature zones often result in the accumulation of exhaust fumes and toxic gases, posing a serious health risk to residents. *Noise* from social life, building construction, industry, and traffic disturbs people's sleep. Long-term exposure to noise will cause insomnia, tinnitus, dreaminess, memory loss, that is, neurasthenia. *Soil pollution* is mainly due to rapid population growth, rapid industrial development, solid waste constantly piled and dumped on the soil surface, harmful wastewater constantly infiltrating into the soil, and harmful atmospheric gases and drifting dust constantly landing in the soil with rainwater. In particular, soil pollutants such as heavy metal pollution and pesticides eventually affect human health through food, vegetables, fruits, etc.

1.2.4 Challenges to Sustainable Cities

The New Urban Agenda adopted at the UN Habitat III conference aims to guide urban development in the next two decades toward sustainable, livable, and resilient futures (Evans et al., 2016). Despite progress since then, the COVID-19 pandemic and other crises have posed huge challenges. Urban areas were particularly hard hit by the pandemic—underscoring the importance of stepping up efforts to build a more sustainable and equitable urban future. An increased involvement of science in these international urban policy processes and practices is called for (Bai et al., 2016; McPhearson et al., 2016). Here, we outline the three major challenges toward sustainable cities in the near future.

Climate Resilience Climate change is affecting every country on every continent. It is disrupting national economies and affecting lives, costing individuals, communities, and nations significantly today and will do so even more in the future. Climate action (*SDG 13*) affirms the urgency of acting to combat climate change and its impacts by calling for actions to strengthen resilience and adaptive capacity with respect to climate hazards; integrate climate change measures into national policies; and improve education, raise awareness and human and institutional capacity on climate change mitigation, adaptation, impact reduction, and early warning. Climate change is closely linked to other SDGs, such as *No Poverty* (SDG 1), *Zero Hungry* (SDG 2), and *Good Health and Well-Being* (SDG 3). If the frequency and intensity of climate hazards increase, it will be harder for countries to end poverty and hunger, achieve food security, improve nutrition, promote sustainable agriculture, and ensure healthy lives.

Poverty and Inequality Cities generate wealth but also concentrate poverty and inequality. From the overcrowded slums in the developing world to homelessness and pockets of destitution in the developed world, urban poverty and inequality take many forms (Habitat, 2022). The global development agenda gives prime of place to *No Poverty* (SDG 1), which calls for a world in which we "end poverty in all its forms everywhere." If urban poverty is not addressed, then this goal will remain elusive. The COVID-19 pandemic has resulted in the emergence of newly poor people: Those who would have exited poverty in the absence of the pandemic but remain poor; and those who have fallen into poverty on account of the pandemic (Habitat, 2022). Urban poverty and inequality remain two of the most intractable challenges confronting cities. Particularly, climate-related issues, such as drought, flooding, and heat waves, are increasing the risk to populations already vulnerable due to poverty and inequity.

The Pandemic Era Cities are profoundly impacted by the COVID-19 pandemic, which raises fundamental questions about sustainable and symbiotic urban development. The multidimensional nature of the pandemic has left an indelible mark on the outlook of the cities and has led to rethinking cities' development in different dimensions—social, cultural, economic, and environmental (Elavarasan et al., 2021). The COVID-19 pandemic has provoked a need to better integrate public health priorities into urban resilience efforts. Not only would this strengthen the overall public health system, it would also create an enabling environmental condition for all to lead healthy lifestyles (Megahed & Ghoneim, 2020). Also, the COVID-19 lockdown has highlighted and further deepened the inequalities present in urban environments around the world (Kumaravel et al., 2020). Not all people have had the same opportunities because not all homes are equally large and comfortable, not all workers are able to work remotely, and not all children have the digital means to keep up with their classes. Societies' housing stock deficits and deep digital divides have been evident in cities on all continents (Fernández de Losada, 2020). Guaranteeing the right to fair housing and economic recovery using all means to mitigate the social disruption caused by technology will be major challenges for cities in the context of the pandemic.

1.3 Interconnected Impacts of Urbanization and Climate Change

1.3.1 Impacts of Climate Change on Urban Sustainable Development

Historically, cities have been perceived as places of refuge from disasters and as buffers against environmental change. However, under the context of climate change, they are better described as hotspots of disasters and risk (UNDP, 2004). The consequences of climate change are the most significant challenges to achieving sustainable development in cities, and those consequences threaten to drag millions of people into inequality and grinding poverty. For example, many regions have experienced changes in the pattern and intensity of rainfall, resulting in more frequent and intense floods and droughts, as well as severe high temperatures (Ray & Rajib, 2019), which are the three main challenges facing urban sustainable development due to climate change. These natural hazards have developed into extreme events because of climate change and other factors and have been reported in many global cities recently. More seriously, these three hazards are among the most frequent, dangerous, and costly of all natural disasters, causing significant damage to infrastructure and threatening people's lives every year.

Drought Urban drought is more in relation to water scarcity, which can be both physical water scarcity and socioeconomic scarcity of water (Ray & Rajib, 2019). According to the World Bank, roughly 1.6 billion people live in countries with water scarcity, and that number is likely to increase in the future. In particular, Asia is already identified as the world's driest continent in per capita availability of freshwater. The impacts of climate change are expected to lead an additional one billion people in Asia to water-stressed conditions (Lolla, 2016). Asian cities would be the worst hit. Expanding urban population and changing lifestyles along with climate variability are making water resource management increasingly difficult particularly in water-stressed urban areas.

From mid-June to late August in 2022, an extreme heat wave spread across central, eastern, and southern China—the longest-running and most intense since records began in 1961. After this extreme heat wave, the Yangtze River basin in southern China experienced the worst drought on record between July and August (Fig. 1.24). The basin received up to 80% less rain than the 30-year average for that period, and temperatures were 2–4 °C higher than the average. The drought was so extreme that water levels in China's largest freshwater lake, Poyang Lake, dropped from 19 m in June to 9 m by the end of August (Mallapaty, 2022). The extreme drought event also led to electricity shortage in Sichuan Province. Some 70% or 80% of electricity in Sichuan comes from hydropower. However, water availability during the extreme drought is less than half of previous years and reservoirs of several large power stations approach their dead levels, which results in many factories shutting down to ensure that residents have access to enough electricity. Besides,

Fig. 1.24 The dry riverbed of the Jialing River in Chongqing on August 21, 2022. (Source: Photo by Xinhua News Agency reporter Wang)

drought also creates instability problems in infrastructure due to excess pressure exerted on building foundations, transportation systems, pipes, and joints.

Flood Urban floods have increased in recent years across the world, not only in the severity and frequency of short-duration but also in damage. The situation may be exacerbated in the future due to the global climate change impacts. Cities, or areas of cities that are located in valleys or on the lower parts of hillslopes, are often affected by the fast-flowing water generated by these extreme rainfall events. The impacts of extreme rainfall tend to be more severe in paved areas compared to areas with unpaved surfaces (Pour et al., 2020), especially in urban areas with highly dense impervious surfaces. Although casualties from urban floods are usually not great, structural and property losses and economic losses can be significant due to the population density in urban areas, which usually also contain properties of high value and a high concentration of economic activity (Pour et al., 2020). Therefore, urban floods are always more disastrous (Salimi & Al-Ghamdi, 2020), for example, devastating urban floods killed hundreds in both Zhengzhou and New York in 2021 (Figs. 1.25 and 1.26). More seriously in 2022, one-fifth of Pakistan's territory was flooded and 20 million people, one-ninth of Pakistan's total population, were affected by the floods to varying degrees. Floods have walloped people, infrastructure, houses, livestock, and farmland. This extreme flood event was "unprecedented" and caused a crisis of "unimaginable proportions".

High Temperature IPCC reported that the rise in surface temperature will tend to continue in the future with the GHG emissions (IPCC, 2007). Climate change

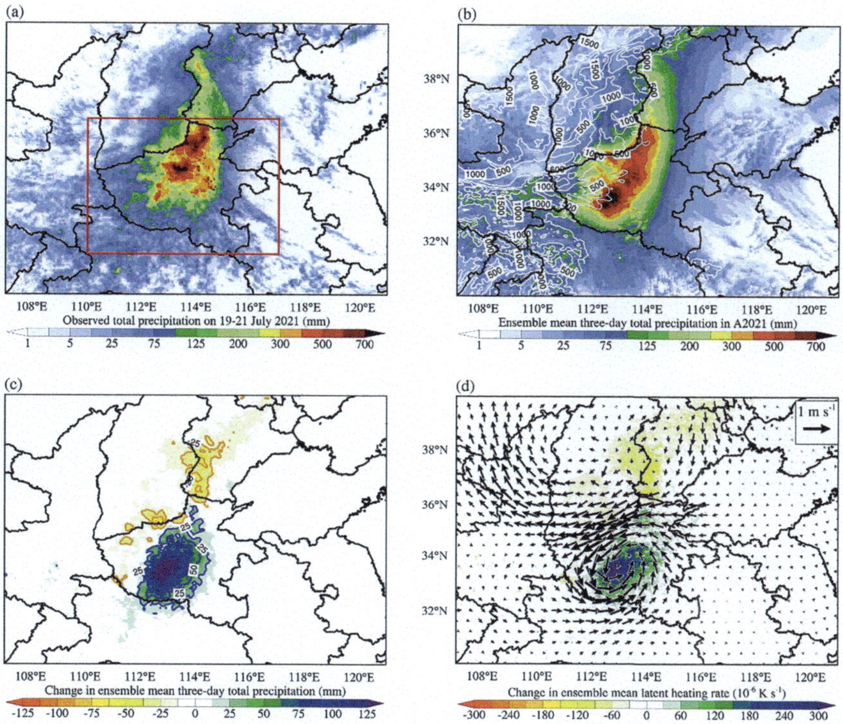

Fig. 1.25 The spatial pattern for the three-day rainfall amount in Zhengzhou, China (LST 0800 19 July–0800 22 July 2021) based on multi-source integrated precipitation product (**a**) and ensemble mean of A2021 (**b**). The red rectangle in (**a**) indicates the target region for event attribution. The white contours in (**b**) show the terrain elevation in the model. (**c**) and (**d**) show the ensemble mean differences between A2021 and N2021 in three-day rainfall totals and in three-day mean 500-hPa latent heat release and 700-hPa horizontal winds (Wang et al., 2022)

assessments consistently project increases in extremely hot days and decreases in extremely cold days (IPCC, 2011). Urban areas with highly dense populations will be at great risk under the compound effect of climate change and UHI. High temperature is the deadliest of all climate hazards. Extreme heat events cause more deaths in the United States than all other climate hazards combined. Increased body temperatures can lead to pyrexia, which if left untreated can lead to organ failure and death. While 0.35% of the overall U.S. mortality rate can be attributed to high temperatures (which cause about 8750 deaths per year), an updated study indicates that heat causes nearly 12,000 deaths and predicts that heat deaths could be as high as 100,000 in 2100 under an increased temperature scenario (Lay et al., 2021). In addition to health impacts, extreme heat threatens urban infrastructure, ecosystems, and economic development. Extreme heat affects the entire urban system, including public health, the economy, urban ecology, water and energy use, and infrastructure.

Fig. 1.26 Floods between 17 and 31 July 2021 in Henan Province, China. (Source: Aly Song/Alamy)

In the future, urban areas should become resilient to encounter climate change-related shocks (Leichenko, 2011).

Poverty It is evident that poverty remains a persistent issue in urban areas, both in the past and the foreseeable future. Globally, approximately 700 million people live below the poverty line of US$1.90 per day, with many more precariously hovering just above this threshold (Rozenberg & Hallegatte, 2015). Individuals living in poverty are particularly susceptible to various shocks, including those triggered by natural disasters like floods, droughts, and rising temperatures. The cumulative impact of these hazards places impoverished populations in increasingly dire straits. The vulnerability of impoverished individuals to the risks posed by natural disasters is of great concern, especially within the context of climate change. Climate change has the potential to alter the frequency, intensity, and spatial distribution of droughts, floods, and high temperatures (IPCC, 2012). The consequences of climate change could pose a significant obstacle to poverty eradication in the future (Hallegatte et al., 2016).

Empirical evidence suggests that floods, for instance, can disrupt household livelihoods and prospects, ultimately leading to an increase in local poverty levels due to income and asset losses (Rodriguez-Oreggia et al., 2013). Furthermore, the impact of natural disaster risk on poverty operates through both observable ex-post channels (losses incurred when a disaster strikes) and less obvious ex-ante channels. Households exposed to weather-related risks tend to reduce their investments in productive assets and opt for low-risk, low-return activities (Cole et al., 2013). This connection between exposure to natural hazards and poverty may establish a feedback loop in which impoverished households are left with no choice but to settle in high-risk zones, thereby facing heightened challenges in escaping poverty (Winsemius et al., 2018).

1.3.2 Impacts of Urbanization on Climate Change

Urbanization significantly contributes to human-induced climate change at a regional scale through altered land–atmosphere radiation balance (i.e., the balance between solar shortwave radiation and longwave radiation balance of land-atmosphere) and land–atmosphere interactions (i.e., including the exchange of energy, matter, and momentum at the land–atmosphere system.) over urban areas. These impacts are reflected by changes in climate elements, including temperature, precipitation, humidity, etc. Generally, urbanization affects climate change through the following aspects: (1) change of underlying surface, (2) GHG and aerosols emission, and (3) anthropogenic heat emission.

Changes in the Underlying Surface Urbanization has significantly changed the underlying surface, leading to a decrease in land surface albedo and an increase in land roughness. These changes result in reduced shortwave radiation reflected from the ground and increased absorption of shortwave radiation, contributing to higher urban temperatures and the urban heat island effect, promoting urban temperature increase and urban heat island effect. Due to the small vegetation proportion, building materials, and complex land surface geometry, urban areas have relatively lower albedo than other areas. Conversely, high land surface roughness increases the repeated reflection of shortwave radiation, thereby resulting in an increase in radiation absorption of facades.

Urbanization can also influence precipitation patterns with a complex mechanism, i.e., the factors that promote and restrain precipitation are present at the same time. On the one hand, the urbanization process increases sensible heat fluxes, convergence, convection, and atmospheric instability all contributing to increased precipitation in urban areas (Mitra et al., 2012; Niyogi et al., 2011). On the other hand, the impervious surface of urban areas can largely reduce the moisture available for evaporation, which hinders the formation of convective clouds and acts as a suppressant to rainfall. In addition, external urban elements such as seasonal factors, regional urban patterns, weather conditions, and urban artificial vegetation also have an impact on the local climate of cities.

GHG and Aerosol Emissions In the radiation balance of the land–atmosphere system, the GHGs provide an insulation effect for the earth by absorbing longwave radiation emitted from the Earth's surface. However, GHG emissions from anthropogenic activities (fossil fuel, land use change, urban landfills, etc.) exceed the threshold, break the radiation balance, and produce a positive radiative forcing, which promotes the increase in surface and air temperature. Besides, the conversion of vegetation to urban land impedes the uptake of CO_2 by vegetation. The warming effect of GHG emissions affects urban meteorological conditions and thereby affects the diffusion capacity of the atmosphere, which is not conducive to pollutant dispersion (Liu et al., 2015a, b). The increasing temperature also promotes the transformation of secondary organic aerosols (Xue et al., 2019), which aggravates air pollution.

Anthropogenic Heat Emission Anthropogenic heat refers to the sensible and latent heat fluxes released to the atmosphere caused by human activities, which are mainly from four sources: Buildings, transportation, metabolism, and industry (Chen et al., 2020). Rapid global urbanization has made energy consumption and urban population problems more prominent, and the resulting anthropogenic heat emissions exert a great impact on the local urban climate (Huber & Knutti, 2012). Cities as human gathering places with frequent and high-intensity human activities bring a large amount of anthropogenic heat emissions, which intensify the local urban climate change. According to the statistics, the global anthropogenic heat emissions generated from energy consumption in 2016 are about 6.04 × 1020 J, and the anthropogenic heat emissions in the past 20 years have increased by 52.16% (Energy Information Administration, 2007). A recent study shows that future emissions and urbanization would make extreme compound events two to five times more frequent (2090s versus 2010s). There will be no slowdown of the large-scale warming due to the future ongoing anthropogenic emissions and climate change will be faster and more furious than anticipated in the next 20 years (Xu et al., 2018).

1.4 Urban Thermal Environments under Climate Change and Urbanization

1.4.1 Urban Thermal Environments

The term UTE encompasses the heat-related physical conditions that influence human perceptions of warmth and cold, health, survival, and overall development. Essentially, it revolves around key factors such as land surface temperature (LST) and air temperature (AT) within urban areas, atmospheric parameters like air humidity, wind speed, and atmospheric turbidity, as well as subsurface attributes including land use cover types, heat capacity, emissivity, and albedo. Additionally, it considers surface solar radiation as an integral component of the physical environment system that directly impacts human well-being and activities. One of the most recognized phenomena within the UTE is the urban heat island (UHI) effect, where urban AT and LST are notably higher compared to the temperatures in surrounding suburban or rural areas. This localized temperature difference serves as a clear indicator and manifestation of the UTE. Consequently, the quality of the UTE is primarily assessed by evaluating the thermal comfort experienced by urban residents. Any decline in the UTE implies a reduction in the thermal comfort of city inhabitants.

The deterioration of the UTE can be attributed to multiple factors, but it is predominantly linked to changes in land cover and land use associated with urbanization. These changes alter subsurface characteristics such as surface roughness, albedo, and the level of greenery within cities (Estoque et al., 2017). Notably, the reduction in latent heat flux resulting from the proliferation of impermeable surfaces in green spaces impacts the surface energy balance and consequently influences outdoor temperatures. Furthermore, the process of urbanization driven by population growth has exacerbated anthropogenic heat emissions and the greenhouse effect. These factors contribute significantly to the worsening of the UTE (Li et al., 2020).

1.4.2 Impacts of Climate Change on UTE

The most evident phenomenon of the impacts of climate change on UTE is heat waves or extreme heat events. Extreme heat can have pronounced and potentially dramatic public health implications, including heatstroke, dehydration, and exacerbation of existing medical conditions, like cardiovascular and cerebrovascular disease, diabetes, chronic obstructive pulmonary disease, pneumonia, and asthma, and increased mortality (Heaviside et al., 2017; Tan et al., 2010). Heat waves also lead to physical and mental stress, as well as increased susceptibility to food and vector-borne infectious diseases (Shahmohamadi et al., 2011). These adverse health effects are likely to further increase in the future, due to climate change (Hajat et al., 2014). Climate change will likely lead to more frequent, more severe, and prolonged heat waves during the summer months.

The evidence that climate change will increase ambient temperatures in the UK has focused attention on how to protect against the health risks of summer heat. In England and Wales, heat waves in 2003 and 2006 were associated with 2000 and 680 excess deaths, respectively (Johnson et al., 2005). These will not be exceptional events by mid-century, and potential vulnerability to similar heat waves is expected to increase as the population ages (Hajat et al., 2014).

In North America, the extreme heat event at the end of June 2021 caused hundreds of casualties and mountain fires that severely disrupted ecosystems. Moreover, the heat event formed a local "heat dome," which hindered heat dissipation, resulting in persistent high temperatures in many areas. During the heat wave, over 1000 deaths occurred due to the direct consequences of the heat wave in Canada (Potestio, 2021). In the United States, the death toll was lower but still in the hundreds; most of the deaths occurred among the elderly (Isaksen et al., 2015). The heat wave also strained and damaged the region's infrastructure. It caused the sidewalks to buckle due to the thermal expansion of concrete (Smoyer Tomic, 2003). More severe, the heat event sparked significant fires across the West Coast and Western Canada, which eventually burned around 600 km^2. Attribution analysis found this was a

1000-year weather event, making 150 times more likely by climate change (Philip et al., 2021). Climate change has the capacity to modify the climatic potential for UHIs and duration of extreme heat events and exacerbated further the amplitude of urban overheating due to the synergetic effects between urban heat islands and heat waves (Pyrgou et al., 2020).

1.4.3 *Impacts of Urbanization on UTE*

Over the last few decades, urban expansion has been the dominant form of urban development at local, regional, and global scales (Li et al., 2018). Urban expansion has significantly changed characteristics of the urban environment and transformed natural ecosystems into semi-natural or even non-natural ecosystems (Peng et al., 2016). The extensive expansion of impervious surface is one of the most representative symbols of urbanization (Hu et al., 2022). This process leads to land use and land cover (LULC) changes on limited horizontal land resources and a rise of building heights in the vertical direction (Mahtta et al., 2019), which directly alters the thermal properties of the land cover, changes the energy budget at the ground surface, changes the surrounding atmospheric circulation characteristics, generates a great amount of anthropogenic waste heat, and leads to a series of changes in the urban environmental system (Huang et al., 2009).

UHI UHI (Fig. 1.27) occurs when cities replace natural land cover with dense concentrations of pavement, buildings, and other surfaces that absorb and retain heat. It is generally defined as the difference in temperature between urban areas and rural areas (or surrounding areas) and generally has four types according to differences in

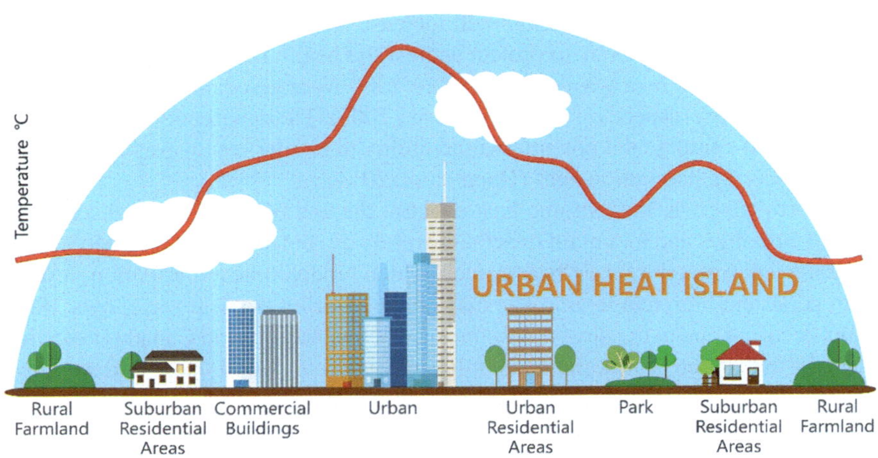

Fig. 1.27 Schematic picture of the urban heat island effect. (Source: World Meteorological Organization)

urban and rural cooling and warming rates at the surface, in the substrate, and in the air (Oke et al., 2017).

Canopy UHI Traditionally, the UHI was defined as the difference in near-surface air temperature between the urban core and its rural surroundings. This is now commonly known as the canopy UHI (CUHI), as air temperature measurements are typically taken at a height of 1.5–2 m above the ground, within the urban canopy layer (Oke, 1976).

Surface UHI The satellite products allow researchers to remotely measure the radiometric LST, or skin temperature, over continuous space (Gallo et al., 1995). This creates a new proxy of the UHI effect called the surface UHI (SUHI), which represents the difference in LST between the urban area and its surrounding rural area. Thanks to the worldwide coverage of polar-orbiting satellites, estimating both intra- and inter-urban variability in SUHI is much easier compared to CUHI (Qian et al., 2022). In addition to CUHI and SUHI, UHIs have also been defined for the boundary layer, known as the boundary layer UHI, and for the subsurface. The positive SUHI influences both the boundary layer, leading to higher and warmer boundary layers over urban areas (Oke, 1995), and urban subsurfaces (Menberg et al., 2013). Globally, over the past 15 years, 87% of urban clusters exhibited a positive surface UHI during the daytime and 93% during the nighttime (Chakraborty & Lee, 2019). Recently, the increase in UHI intensity (UHII) and extremely high-temperature events worldwide have affected urban dwellers' physical and mental health and raised heat-related morbidity and mortality (Burke et al., 2018). More seriously, UHI causes high electricity consumption by increasing the demand for indoor cooling, which causes a feedback effect resulting in an increased LST in urban regions (Santamouris et al., 2015).

Heat Exposure Heat exposure refers to a heat-induced impact on systems (e.g., natural ecosystems, urban ecosystems, and socioeconomic systems) and their components (e.g., people, livelihoods, and infrastructures), which are associated with heat properties (Voelkel et al., 2018). Heat exposure changes with temperatures, UHIs, and heat waves, while, to a degree, exposure is also affected by the composition of the system and its components (Bernhard et al., 2015). Higher temperatures, stronger UHIs, or longer-lasting heat waves generally elevate heat exposure (Kjellstrom, 2014). Different characteristics of the system components (e.g., quantity, density, and quality) contribute to different degrees of heat exposure risk (Hass & Ellis, 2019).

UHI causes an increased daytime temperature and reduced nighttime cooling leading to the elevation of many different diseases such as physical distress, breathing problems, tiredness, heatstroke, and heat-related death (Tan et al., 2010). The compound effects of increasing urbanization, climate change, and aging populations will lead to escalating health risks for urban populations in the future (Patz et al., 2005). Older populations, those suffering from various health problems, outdoor workers, and floating populations are at higher risk of these problems (Hsu et al., 2021). Exposure to extreme heat also reduces labor productivity, as people feel heat stress and it damages their well-being, causing substantial economic losses (Zander et al., 2018).

1.4.4 Beating the Heat for a Comfortable UTE

The effects of global warming and rapid urbanization are felt most acutely in cities (Waltz, 2014). The world's cities are warming at twice the global average rate due to the UHI effect. Even a small shift in average annual temperatures can mean a large increase in the number of extremely hot days. An increasingly warming world poses a major threat to human systems (e.g., heat exposure, heatstroke, morbidity, and mortality) and the natural system (e.g., drought, wildfire, species extinction, and ocean acidification). Human thermal comfort is one of the most important issues for planners to address as communities across the globe continue to get hotter (Keith & Meerow, 2022). Particularly those with urban poor communities face long-term challenges in ensuring the thermal comfort of their inhabitants. These challenges are partly a result of the direct and indirect impacts of climate change and urbanization and are often compounded by preexisting vulnerability (ADB, 2014).

Resilient cities have already been on the agenda for years, following the UN's *SDG 3* (i.e., good health and well-being) and *SDG 11* (i.e., sustainable cities and communities). However, climate change is exacerbating the UHI effect, which makes it more difficult to achieve sustainability and resilience planning goals. Planning strategies to address urban heat must keep the complex interactions of heat with urban systems and its long-term impacts in mind to avoid maladaptation—adaptation actions that have unintended negative consequences (Keith & Meerow, 2022). It is noted that improving the comfort of the UTE or enhancing urban heat resilience requires holistic strategies that recognize the connections between different urban systems and across scales. Hence, planning strategies need efforts to mitigate heat in urban areas (e.g., through vegetation or design of the built environment) and manage heat risks (e.g., through emergency response or social services) (Keith & Meerow, 2022). As the effects of climate change are being felt, we must urgently plan our cities through a deep understanding of the *characteristics* and *drivers* of the UTE, *heat risk* of urban residents, and *mitigation* of UHI effects across scales to resist the heat and better support equitable human thermal comfort in communities today and in the future.

References

ADB. (2014). *Urban climate change resilience: A synopsis*. ADB: Asian Development Bank.

Adger, W. N., Crépin, A.-S., Folke, C., Ospina, D., Chapin, F. S., Segerson, K., Seto, K. C., Anderies, J. M., Barrett, S., Bennett, E. M., Daily, G., Elmqvist, T., Fischer, J., Kautsky, N., Levin, S. A., Shogren, J. F., van den Bergh, J., Walker, B., & Wilen, J. (2020). Urbanization, migration, and adaptation to climate change. *One Earth, 3*, 396–399. https://doi.org/10.1016/j.oneear.2020.09.016

Adhikari, A., & Hansen, A. J. (2018). Land use change and habitat fragmentation of wildland ecosystems of the north Central United States. *Landscape and Urban Planning, 177*, 196–216. https://doi.org/10.1016/j.landurbplan.2018.04.014

Allen, A. (2003). Environmental planning and management of the peri-urban interface: Perspectives on an emerging field. *Environment and Urbanization, 15*, 135. https://doi.org/10.1630/095624703101286402

Antonelli, M., Pujol, J. C., Spector, T. D., Ourselin, S., & Steves, C. J. (2022). Risk of long COVID associated with delta versus omicron variants of SARS-CoV-2. *Lancet, 399*, 2263–2264.

Aragão, L. E. O. C., Anderson, L. O., Fonseca, M. G., Rosan, T. M., Vedovato, L. B., Wagner, F. H., Silva, C. V. J., Silva Junior, C. H. L., Arai, E., Aguiar, A. P., Barlow, J., Berenguer, E., Deeter, M. N., Domingues, L. G., Gatti, L., Gloor, M., Malhi, Y., Marengo, J. A., Miller, J. B. … Saatchi, S. (2018). 21st century drought-related fires counteract the decline of Amazon deforestation carbon emissions. *Nature Communications*, 9, 1–12. https://doi.org/10.1038/s41467-017-02771-y

Aylett, A. (2015). Institutionalizing the urban governance of climate change adaptation: Results of an international survey. *Urban Climate, 14*, 4–16. https://doi.org/10.1016/j.uclim.2015.06.005

Baccini, A., Goetz, S. J., Walker, W. S., Laporte, N. T., Sun, M., Sulla-Menashe, D., Hackler, J., Beck, P. S. A., Dubayah, R., Friedl, M. A., Samanta, S., & Houghton, R. A. (2012). Estimated carbon dioxide emissions from tropical deforestation improved by carbon-density maps. *Nature Climate Change, 2*, 182–185. https://doi.org/10.1038/nclimate1354

Bai, X., Surveyer, A., Elmqvist, T., Gatzweiler, F. W., Güneralp, B., Parnell, S., Prieur-Richard, A. H., Shrivastava, P., Siri, J. G., Stafford-Smith, M., Toussaint, J. P., & Webb, R. (2016). Defining and advancing a systems approach for sustainable cities. *Current Opinion in Environment Sustainability, 23*, 69. https://doi.org/10.1016/j.cosust.2016.11.010

Banerjee, O., Bark, R., Connor, J., & Crossman, N. D. (2013). An ecosystem services approach to estimating economic losses associated with drought. *Ecological Economics, 91*, 19–27. https://doi.org/10.1016/J.ECOLECON.2013.03.022

Bennett, S., Wernberg, T., Harvey, E. S., Santana-Garcon, J., & Saunders, B. J. (2015). Tropical herbivores provide resilience to a climate-mediated phase shift on temperate reefs. *Ecology Letters, 18*, 714–723. https://doi.org/10.1111/ele.12450

Bernhard, M. C., Kent, S. T., Sloan, M. E., Evans, M. B., McClure, L. A., & Gohlke, J. M. (2015). Measuring personal heat exposure in an urban and rural environment. *Environmental Research, 137*, 410–418. https://doi.org/10.1016/j.envres.2014.11.002

Burke, M., González, F., Baylis, P., Heft-Neal, S., Baysan, C., Basu, S., & Hsiang, S. (2018). Higher temperatures increase suicide rates in the United States and Mexico. *Nature Climate Change, 8*, 723–729. https://doi.org/10.1038/s41558-018-0222-x

Cardil, A., Rodrigues, M., Ramirez, J., De-Miguel, S., Silva, C. A., Mariani, M., & Ascoli, D. (2021). Coupled effects of climate teleconnections on drought, Santa Ana winds and wildfires in southern California. *Science of The Total Environment, 765*, 142788. https://doi.org/10.1016/j.scitotenv.2020.142788

Carvalho, K. S., & Wang, S. (2019). Characterizing the Indian Ocean Sea level changes and potential coastal flooding impacts under global warming. *Journal of Hydrology, 569*, 373–386. https://doi.org/10.1016/j.jhydrol.2018.11.072

Ceballos, G., Ehrlich, P. R., Barnosky, A. D., García, A., Pringle, R. M., & Palmer, T. M. (2015). Accelerated modern human-induced species losses: Entering the sixth mass extinction. *Science Advances, 1*, e1400253. https://doi.org/10.1126/sciadv.1400253

Chakraborty, T., & Lee, X. (2019). A simplified urban-extent algorithm to characterize surface urban heat islands on a global scale and examine vegetation control on their spatiotemporal variability. *International Journal of Applied Earth Observation and Geoinformation, 74*, 269–280. https://doi.org/10.1016/j.jag.2018.09.015

Chen, Q., Yang, X., Ouyang, Z., Zhao, N., Jiang, Q., Ye, T., Qi, J., & Yue, W. (2020). Estimation of anthropogenic heat emissions in China using cubist with points-of-interest and multi-

source remote sensing data. *Environmental Pollution, 266*, 115183. https://doi.org/10.1016/j.envpol.2020.115183
Chisholm, L., Talbot, T., Appleby, W., Tam, B., & Rong, R. (2021). Projected changes to air temperature, sea-level rise, and storms for the Gulf of Maine region in 2050. *Elementa, 9*. https://doi.org/10.1525/elementa.2021.00059
Chmielewski, F. M., & Rotzer, T. (2001). Response of tree phenology to climate change across Europe. *Agricultural and Forest Meteorology, 108*, 101–112. https://doi.org/10.1016/S0168-1923(01)00233-7
Chown, S. L., Leihy, R. I., Naish, T. R., Brooks, C. M., Convey, P., Henley, B. J., Mackintosh, A., Phillips, L. M., Kennicutt, M. C. I., & Grant, S. M. (2022). Antarctic climate change and the environment: A decadal synopsis and recommendations for action. Sci. Comm. Antarct. Res. Cambridge, United Kingdom. www.scar.org.
Clement, V., Rigaud, K. K., De Sherbinin, A., Jones, B., Adamo, S., Schewe, J., Sadiq, N., & Shabahat, E. (2021). *Groundswell part 2: Acting on internal climate migration*. World Bank Group.
CoastAdapt. (2017). *Barriers to adapting to climate change [WWW Document]*. CoastAdapt.
Cole, S., Giné, X., Tobacman, J., Topalova, P., Townsend, R., & Vickery, J. (2013). Barriers to household risk management: Evidence from India. *American Economic Journal: Applied Economics, 5*, 104–135. https://doi.org/10.1257/app.5.1.104
Conticini, E., Frediani, B., & Caro, D. (2020). Can atmospheric pollution be considered a co-factor in extremely high level of SARS-CoV-2 lethality in northern Italy? *Environmental Pollution, 261*, 114465. https://doi.org/10.1016/j.envpol.2020.114465
CRED. (2022). *2021 Disasters in numbers*. CRED. https://doi.org/10.1787/eee82e6e-en
D'Souza, R., & Nagendra, H. (2011). Changes in public commons as a consequence of urbanization: The Agara Lake in Bangalore, India. *Environmental Management, 47*, 840–850. https://doi.org/10.1007/s00267-011-9658-8
Dqg, Z., Ohdg, P. D., & Vhyhuh, W. R. (2005). Observations: Changes in snow. *Ice and Frozen Ground Chapter, 4*(2), 2004.
EIA (2018). Energy information administration: Annual energy outlook 2018 with projections to 2050. J. Phys. A Math. Theor.
Ekstrom, J. A., & Moser, S. C. (2014). Identifying and overcoming barriers in urban climate adaptation: Case study findings from the San Francisco Bay Area, California, USA. *Urban Climate, 9*, 54–74. https://doi.org/10.1016/j.uclim.2014.06.002
Elavarasan, R. M., Pugazhendhi, R., Shafiullah, G. M., Irfan, M., & Anvari-Moghaddam, A. (2021). A hover view over effectual approaches on pandemic management for sustainable cities—The endowment of prospective technologies with revitalization strategies. *Sustainable Cities and Society, 68*, 102789. https://doi.org/10.1016/j.scs.2021.102789
Energy Information Administration (2007). Choice Rev. Online 44, 44-3624-44-3624. https://doi.org/10.5860/choice.44-3624
Estoque, R. C., Murayama, Y., & Myint, S. W. (2017). Effects of landscape composition and pattern on land surface temperature: An urban heat Island study in the megacities of Southeast Asia. *The Science of the Total Environment, 577*, 349–359. https://doi.org/10.1016/j.scitotenv.2016.10.195
Evans, B., Elisei, P., Rosenfeld, O., Roll, G., Figueiredo, A., & Keiner, M. (2016). HABITAT III—toward a new urban agenda. *DISP, 52*, 86. https://doi.org/10.1080/02513625.2016.1171053
Fadda, J. (2020). Climate change: An overview of potential health impacts associated with climate change environmental driving forces. https://doi.org/10.1007/978-3-030-18488-9_8.
FAO. (2017). The state of food security and nutrition in the world. *International Journal of Phytoremediation*.
FCFA (2018). Overcoming the barriers to climate change adaptation 1–8.
Fernández de Losada, A. (2020). Managing a global pandemic. Towards more sustainable and resilient urban future. Cidob Rep. "Cities Front. Manag. coronavirus Cris.

Flouris, A. D., Dinas, P. C., Ioannou, L. G., Nybo, L., Havenith, G., Kenny, G. P., & Kjellstrom, T. (2018). Workers' health and productivity under occupational heat strain: A systematic review and meta-analysis. *Lancet Planet Health, 2*, e521–e531. https://doi.org/10.1016/S2542-5196(18)30237-7

Foden, W., Mace, G. M., Vié, J.-C., Angulo, A., Butchart, S. H. M., DeVantier, L., Dublin, H. T., Gutsche, A., Stuart, S., Turak, E., et al. (2009). Species susceptibility to climate change impacts. Wildl. a Chang. world—an Anal. 2008 IUCN Red List Threat. species 77.

European Commission: Directorate-General for Climate Action, adelphi, Ecofys, Fraunhofer, GreenStream, Pfaff, M., Eichhammer, W., Pfluger, B., Fleiter, T.Herbst, A., Impact on the environment and the economy of technological innovations for the Innovation Fund (IF) – In the fields of – Energy-intensive industries, renewables, carbon capture and storage / use (CCS/CCU), energy storage, Publications Office, 2018, https://data.europa.eu/doi/10.2834/068719

Friedlingstein, P., O'Sullivan, M., Jones, M.W., Andrew, R.M., Gregor, L., Hauck, J., Le Quéré, C., Luijkx, I.T., Olsen, A., Peters, G.P., Peters, W., Pongratz, J., Schwingshackl, C., Sitch, S., Canadell, J.G., Ciais, P., Jackson, R.B., Alin, S.R., Alkama, R. … Zheng, B., 2022. Global carbon budget 2022. *Earth System Science Data* 14, 4811–4900. https://doi.org/10.5194/essd-14-4811-2022.

Gallo, K. P., Tarpley, J. D., McNab, A. L., & Karl, T. R. (1995). Assessment of urban heat islands: A satellite perspective. *Atmospheric Research, 37*, 37–43. https://doi.org/10.1016/0169-8095(94)00066-M

Goh, T., Ang, B. W., & Xu, X. Y. (2018). Quantifying drivers of CO2 emissions from electricity generation—Current practices and future extensions. *Applied Energy, 231*, 1191–1204. https://doi.org/10.1016/j.apenergy.2018.09.174

Habitat, U.N. (2022). World cities report 2022: Envisaging the future of cities.

Hajat, S., Vardoulakis, S., Heaviside, C., & Eggen, B. (2014). Climate change effects on human health: Projections of temperature-related mortality for the UK during the 2020s, 2050s and 2080s. *Journal of Epidemiology and Community Health, 68*, 641–648. https://doi.org/10.1136/jech-2013-202449

Hallegatte, S., Bangalore, M., Bonzanigo, L., Fay, M., Kane, T., Narloch, U., Rozenberg, J., Treguer, D., & Vogt-Schilb, A. (2016). *Shock waves: Managing the impacts of climate change on poverty, shock waves: Managing the impacts of climate change on poverty*. World Bank. https://doi.org/10.1596/978-1-4648-0673-5

Hamin, E. M., Gurran, N., & Emlinger, A. M. (2014). Barriers to municipal climate adaptation: Examples from coastal Massachusetts smaller cities and towns. *Journal of the American Planning Association, 80*, 110–122. https://doi.org/10.1080/01944363.2014.949590

Han, W., Meehl, G. A., Rajagopalan, B., Fasullo, J. T., Hu, A., Lin, J., Large, W. G., Wang, J. W., Quan, X. W., Trenary, L. L., Wallcraft, A., Shinoda, T., & Yeager, S. (2010). Patterns of Indian Ocean Sea-level change in a warming climate. *Nature Geoscience, 3*, 546–550. https://doi.org/10.1038/ngeo901

Hass, A. L., & Ellis, K. N. (2019). Using wearable sensors to assess how a heatwave affects individual heat exposure, perceptions, and adaption methods. *International Journal of Biometeorology, 63*, 1585–1595. https://doi.org/10.1007/s00484-019-01770-6

Havas, L., Ballweg, J., Penna, C., & Race, D. (2015). *Food production and emissions of greenhouse gases*. Energy Effic.

He, Q., & Silliman, B. R. (2019). Climate change, human impacts, and coastal ecosystems in the Anthropocene. *Current Biology, 29*, R1021–R1035. https://doi.org/10.1016/j.cub.2019.08.042

Heaviside, C., Macintyre, H., & Vardoulakis, S. (2017). The urban Heat Island: Implications for health in a changing environment. *Current Environmental Health Reports*. https://doi.org/10.1007/s40572-017-0150-3

Hertwich, E. G. (2021). Increased carbon footprint of materials production driven by rise in investments. *Nature Geoscience, 14*, 151–155. https://doi.org/10.1038/s41561-021-00690-8

Hollands, R. G. (2008). Will the real smart city please stand up? Intelligent, progressive or entrepreneurial? *City, 12*, 303. https://doi.org/10.1080/13604810802479126

Houghton, R. A. (2012). Carbon emissions and the drivers of deforestation and forest degradation in the tropics. *Current Opinion in Environment Sustainability, 4*, 597. https://doi.org/10.1016/j.cosust.2012.06.006
Hsu, A., Sheriff, G., Chakraborty, T., & Manya, D. (2021). Disproportionate exposure to urban heat Island intensity across major US cities. *Nature Communications, 12*, 2721. https://doi.org/10.1038/s41467-021-22799-5
Hu, D., Meng, Q., Schlink, U., Hertel, D., Liu, W., Zhao, M., & Guo, F. (2022). How do urban morphological blocks shape spatial patterns of land surface temperature over different seasons? A multifactorial driving analysis of Beijing, China. *International Journal of Applied Earth Observation and Geoinformation, 106*, 102648. https://doi.org/10.1016/j.jag.2021.102648
Huang, S., Taniguchi, M., Yamano, M., & Wang, C. H. (2009). Detecting urbanization effects on surface and subsurface thermal environment—a case study of Osaka. *The Science of the Total Environment, 407*, 3142–3152. https://doi.org/10.1016/j.scitotenv.2008.04.019
Huber, M., & Knutti, R. (2012). Anthropogenic and natural warming inferred from changes in earth's energy balance. *Nature Geoscience, 5*, 31–36. https://doi.org/10.1038/ngeo1327
IEA. (2016). Statistics: CO2 emissions from fuel combustion. *International Journal of Electrochemical Science.*
IEA (2018). CO2 emissions from fuel combustion: Overview. Int. Energy Agency.
IPCC. (2007). *Climate change 2007: Impacts, adaptation and vulnerability: Contribution of working group II to the fourth assessment report of the intergovernmental panel.* Cambridge University Press.
IPCC. (2011). *Summary for policymakers. Special Report on Renewable Energy Sources and Climate Change Mitigation.* Cambridge University Press. https://doi.org/10.5860/CHOICE.49-6309
IPCC, 2012: Managing the Risks of Extreme Events and Disasters to Advance Climate Change Adaptation. A Special Report of Working Groups I and II of the Intergovernmental Panel on Climate Change [Field, C. B., V. Barros, T. F. Stocker, D. Qin, D. J. Dokken, K. L. Ebi, M. D. Mastrandrea, K. J. Mach, G. -K. Plattner, S. K. Allen, M. Tignor, and P. M. Midgley (eds.)]. Cambridge University Press, Cambridge, UK, and New York, NY, USA, 582 pp.
IPCC (2014). Part A: Global and sectoral aspects. (Contribution of Working Group II to the Fifth Assessment Report of the Intergovernmental Panel on Climate Change). Clim. Chang. 2014 Impacts, Adapt. Vulnerability.
IPCC (2018). Summary for policymakers. In: Global warming of 1.5°C., An IPCC Special Report on the impacts of global warming of 1.5°C above pre-industrial levels and related global greenhouse gas emission pathways, in the context of strengthening the global response to the threat of climate change, sustainable development
IPCC. (2021). *Climate change 2021: Summary for policy makers.* Cambridge Univ. Press.
Isaksen, T. B., Yost, M. G., Hom, E. K., Ren, Y., Lyons, H., & Fenske, R. A. (2015). Increased hospital admissions associated with extreme-heat exposure in King County, Washington, 1990-2010. *Reviews on Environmental Health, 0*, 51. https://doi.org/10.1515/reveh-2014-0050
Jensen, L. (2022). *The sustainable development goals report 2022.* United Nations.
Jiang, T., Su, X., Singh, V. P., & Zhang, G. (2021). A novel index for ecological drought monitoring based on ecological water deficit. *Ecological Indicators, 129*, 107804. https://doi.org/10.1016/j.ecolind.2021.107804
Johnson, H., Kovats, S., McGregor, G., Stedman, J., Gibbs, M., & Walton, H. (2005). The impact of the 2003 heat wave on daily mortality in England and Wales and the use of rapid weekly mortality estimates. *Eurosurveillance, 10*, 15–16. https://doi.org/10.2807/esm.10.07.00558-en
Keenan, R. J., Reams, G. A., Achard, F., de Freitas, J. V., Grainger, A., & Lindquist, E. (2015). Dynamics of global forest area: Results from the FAO global Forest resources assessment 2015. *Forest Ecology and Management, 352*, 9. https://doi.org/10.1016/j.foreco.2015.06.014
Keith, L., & Meerow, S. (2022). Planning for urban heat resilience. American Planning Association.

Kjellstrom, T. (2014). Impact of climate conditions on occupational health and related economic losses: A new feature of global and urban health in the context of climate change. *Asia-Pacific Journal of Public Health, 28*, 28S–37S. https://doi.org/10.1177/1010539514568711
Klein, R. J. T., Midgley, G. F., Preston, B. L., Alam, M., Berkhout, F. G. H., Dow, K., Shaw, M. R., Gitay, H., Thurlow, J., Buob, S., & Thomas, A. (2015). Adaptation opportunities, constraints, and limits, in: Climate change 2014 impacts, adaptation and vulnerability: Part A: Global and sectoral aspects https://doi.org/10.1017/CBO9781107415379.021.
Knouft, J. H., & Chu, M. L. (2015). Using watershed-scale hydrological models to predict the impacts of increasing urbanization on freshwater fish assemblages. *Ecohydrology, 8*, 273. https://doi.org/10.1002/eco.1506
Kumaravel, S. K., Subramani, R. K., Jayaraj Sivakumar, T. K., Madurai Elavarasan, R., Manavalanagar Vetrichelvan, A., Annam, A., & Subramaniam, U. (2020). Investigation on the impacts of COVID-19 quarantine on society and environment: Preventive measures and supportive technologies. *3 Biotech, 10*, 393. https://doi.org/10.1007/s13205-020-02382-3
Larigauderie, A., & Mooney, H. A. (2010). The intergovernmental science-policy platform on biodiversity and ecosystem services: Moving a step closer to an IPCC-like mechanism for biodiversity. *Current Opinion in Environment Sustainability, 2*, 9. https://doi.org/10.1016/j.cosust.2010.02.006
Lay, C. R., Sarofim, M. C., Vodonos Zilberg, A., Mills, D. M., Jones, R. W., Schwartz, J., & Kinney, P. L. (2021). City-level vulnerability to temperature-related mortality in the USA and future projections: A geographically clustered meta-regression. *Lancet Planetary Health, 5*, e338–e346. https://doi.org/10.1016/S2542-5196(21)00058-9
Leichenko, R. (2011). Climate change and urban resilience. *Current Opinion in Environment Sustainability, 3*, 164. https://doi.org/10.1016/j.cosust.2010.12.014
Li, X., Wu, W., & Yu, C. W. F. (2015). Energy demand for hot water supply for indoor environments: Problems and perspectives. *Indoor and Built Environment, 24*, 5. https://doi.org/10.1177/1420326X14564285
Li, G., Sun, S., & Fang, C. (2018). The varying driving forces of urban expansion in China: Insights from a spatial-temporal analysis. *Landscape and Urban Planning, 174*, 63–77. https://doi.org/10.1016/j.landurbplan.2018.03.004
Li, W., Han, C., Li, W., Zhou, W., & Han, L. (2020). Multi-scale effects of urban agglomeration on thermal environment: A case of the Yangtze River Delta megaregion, China. *The Science of the Total Environment, 713*, 136556. https://doi.org/10.1016/j.scitotenv.2020.136556
Liu, H., Ma, W., Qian, J., Cai, J., Ye, X., Li, J., & Wang, X. (2015a). Effect of urbanization on the urban meteorology and air pollution in Hangzhou. *Journal of Meteorological Research, 29*, 950–965. https://doi.org/10.1007/s13351-015-5013-y
Liu, Y. Y., Van Dijk, A. I. J. M., De Jeu, R. A. M., Canadell, J. G., McCabe, M. F., Evans, J. P., & Wang, G. (2015b). Recent reversal in loss of global terrestrial biomass. *Nature Climate Change, 5*, 470–474. https://doi.org/10.1038/nclimate2581
Lolla, S. V. T. (2016). Path planning and adaptive sampling in the Coastal Ocean. Massachusetts Inst. Technol.
Lu, M., & Lai, J. (2020). Review on carbon emissions of commercial buildings. *Renewable and Sustainable Energy Reviews*. https://doi.org/10.1016/j.rser.2019.109545.
Mahtta, R., Mahendra, A., & Seto, K. C. (2019). Building up or spreading out? Typologies of urban growth across 478 cities of 1 million+. *Environmental Research Letters, 14*, 124077. https://doi.org/10.1088/1748-9326/ab59bf
Mallapaty, S. (2022). China's extreme weather challenges scientists trying to study it. *Nature, 609*, 888. https://doi.org/10.1038/d41586-022-02954-8
Marazziti, D., Cianconi, P., Mucci, F., Foresi, L., Chiarantini, I., & Della Vecchia, A. (2021). Climate change, environment pollution, COVID-19 pandemic and mental health. *The Science of the Total Environment, 773*, 145182. https://doi.org/10.1016/j.scitotenv.2021.145182

Matsumoto, K., Shigetomi, Y., Shiraki, H., Ochi, Y., Ogawa, Y., & Ehara, T. (2019). Addressing key drivers of regional CO2 emissions of the manufacturing industry in Japan. *The Energy Journal, 40*, 233. https://doi.org/10.5547/01956574.40.si1.kmat

Max Finlayson, C. (2018). Millennium ecosystem assessment. In *The wetland book: I: Structure and function, management, and methods* (pp. 355–359). WRI: World Resources Institute. https://doi.org/10.1007/978-90-481-9659-3_81

McGuire, C. J. (2018). Examining legal and regulatory barriers to climate change adaptation in the coastal zone of the United States. *Cogent Environmental Science, 4*. https://doi.org/10.1080/23311843.2018.1491096

McPhearson, T., Parnell, S., Simon, D., Gaffney, O., Elmqvist, T., Bai, X., Roberts, D., & Revi, A. (2016). Scientists must have a say in the future of cities. *Nature, 538*, 165. https://doi.org/10.1038/538165a

MEA (2005). Ecosystems and human Well-being: Wetlands and water synthesis (World Resources Institute Washington, DC, USA), Millennium Ecosystem Assessment.

Measham, T. G., Preston, B. L., Smith, T. F., Brooke, C., Gorddard, R., Withycombe, G., & Morrison, C. (2011). Adapting to climate change through local municipal planning: Barriers and challenges. *Mitigation and Adaptation Strategies for Global Change, 16*, 889–909. https://doi.org/10.1007/s11027-011-9301-2

Megahed, N. A., & Ghoneim, E. M. (2020). Antivirus-built environment: Lessons learned from Covid-19 pandemic. *Sustainable Cities and Society, 61*. https://doi.org/10.1016/j.scs.2020.102350

Mel, R. A., Feudo, T. L., Miceli, M., Sinopoli, S., & Maiolo, M. (2022). A coupled modelling system to assess the effect of Mediterranean storms under climate change. *Natural Hazards and Earth System Sciences*, 1–31.

Menberg, K., Bayer, P., Zosseder, K., Rumohr, S., & Blum, P. (2013). Subsurface urban heat islands in German cities. *The Science of the Total Environment, 442*, 123–133. https://doi.org/10.1016/j.scitotenv.2012.10.043

Menzel, A. (2003). Plant phenological anomalies in Germany and their relation to air temperature and NAO. *Climatic Change, 57*, 243–263. https://doi.org/10.1023/A:1022880418362

Miller, S. M., Wofsy, S. C., Michalak, A. M., Kort, E. A., Andrews, A. E., Biraud, S. C., Dlugokencky, E. J., Eluszkiewicz, J., Fischer, M. L., Janssens-Maenhout, G., Miller, B. R., Miller, J. B., Montzka, S. A., Nehrkorn, T., & Sweeney, C. (2013). Anthropogenic emissions of methane in the United States. *Proceedings of the National Academy of Sciences of the United States of America, 110*, 20018–20022. https://doi.org/10.1073/pnas.1314392110

Mishra, A. K., & Singh, V. P. (2010). A review of drought concepts. *Journal of Hydrology, 391*, 202. https://doi.org/10.1016/j.jhydrol.2010.07.012

Mitra, C., Shepherd, J. M., & Jordan, T. (2012). On the relationship between the premonsoonal rainfall climatology and urban land cover dynamics in Kolkata city, India. *International Journal of Climatology, 32*, 1443–1454. https://doi.org/10.1002/joc.2366

Molotoks, A., Smith, P., & Dawson, T. P. (2021). Impacts of land use, population, and climate change on global food security. *Food and Energy Security, 10*, e261. https://doi.org/10.1002/fes3.261

Mrówczyńska-Kamińska, A., Bajan, B., Pawłowski, K. P., Genstwa, N., & Zmyślona, J. (2021). Greenhouse gas emissions intensity of food production systems and its determinants. *PLoS One, 16*, e0250995. https://doi.org/10.1371/journal.pone.0250995

Müller, A., Österlund, H., Marsalek, J., & Viklander, M. (2020). The pollution conveyed by urban runoff: A review of sources. *The Science of the Total Environment, 709*, 136125. https://doi.org/10.1016/j.scitotenv.2019.136125

Nakicenovic, N., Alcamo, J., Davis, G., Vries, B. D., Fenhann, J., Gaffin, S., ... & Zhou, D. (2000). Special report on emissions scenarios.

Naumann, G., Alfieri, L., Wyser, K., Mentaschi, L., Betts, R. A., Carrao, H., Spinoni, J., Vogt, J., & Feyen, L. (2018). Global changes in drought conditions under different levels of warming. *Geophysical Research Letters, 45*, 3285–3296. https://doi.org/10.1002/2017GL076521

Niyogi, D., Pyle, P., Lei, M., Arya, S. P., Kishtawal, C. M., Shepherd, M., Chen, F., & Wolfe, B. (2011). Urban modification of thunderstorms: An observational storm climatology and model case study for the Indianapolis urban region. *Journal of Applied Meteorology and Climatology, 50*, 1129–1144. https://doi.org/10.1175/2010JAMC1836.1

Nunez, C. (2019). Sea level rise, explained. Natl. Geogr. Mag.

O'Donoghue, C., Chyzheuskaya, A., Grealis, E., Kilcline, K., Finnegan, W., Goggins, J., Hynes, S., & Ryan, M. (2019). Measuring GHG emissions across the Agri-food sector value chain: The development of a bioeconomy input-output model. *International Journal of Food and System Dynamics, 10*, 55–85. https://doi.org/10.18461/ijfsd.v10i1.04

Oke, T. R. (1976). The distinction between canopy and boundary-layer urban heat Islands. *Atmosphere (Basel), 14*, 268–277. https://doi.org/10.1080/00046973.1976.9648422

Oke, T. R. (1995). The heat Island of the urban boundary layer: Characteristics, causes and effects. In *Wind climate in cities* (pp. 81–107). Springer. https://doi.org/10.1007/978-94-017-3686-2_5

Oke, T. R., Mills, G., Christen, A., & Voogt, J. A. (2017). Urban climates, Urban climates. https://doi.org/10.1017/9781139016476.

Pandey, S., Bhandari, H., Sharan, R., Taunk, S. K., Naik, D., Prapertchob, P., Ding, S., Pandey, S., Bhandari, H., Sharan, R., Taunk, S. K., Naik, D., Prapertchob, P., & Ding, S. (2004). Economic costs of drought and Rice farmers risk coping mechanisms: A cross country comparative analysis. https://doi.org/10.22004/AG.ECON.19966.

Patz, J. A., Campbell-Lendrum, D., Holloway, T., & Foley, J. A. (2005). Impact of regional climate change on human health. *Nature, 438*, 310–317. https://doi.org/10.1038/nature04188

Pelletier, N., Brodt, S., Garnett, T., Audsley, E., Henriksson, P., Kendall, A., Kramer, K. J., Murphy, D., Nemecek, T., & Troell, M. (2011). Energy intensity of agriculture and food. *Artic. Annual Review of Environmental Resources*. https://doi.org/10.1146/annurev-environ-081710-161014

Pendrill, F., Persson, U. M., Godar, J., Kastner, T., Moran, D., Schmidt, S., & Wood, R. (2019). Agricultural and forestry trade drives large share of tropical deforestation emissions. *Global Environmental Change, 56*, 1–10. https://doi.org/10.1016/j.gloenvcha.2019.03.002

Peng, J., Xie, P., Liu, Y., & Ma, J. (2016). Urban thermal environment dynamics and associated landscape pattern factors: A case study in the Beijing metropolitan region. *Remote Sensing of Environment, 173*, 145–155. https://doi.org/10.1016/j.rse.2015.11.027

Perkins-Kirkpatrick, S. E., & Lewis, S. C. (2020). Increasing trends in regional heatwaves. *Nature Communications, 11*, 3357. https://doi.org/10.1038/s41467-020-16970-7

Peterson, T. C., Connolley, W. M., & Fleck, J. (2008). The myth of the 1970s global cooling scientific consensus. *Bulletin of the American Meteorological Society, 89*, 1325–1338. https://doi.org/10.1175/2008BAMS2370.1

Petrosillo, N., Viceconte, G., Ergonul, O., Ippolito, G., & Petersen, E. (2020). COVID-19, SARS and MERS: Are they closely related? *Clinical Microbiology and Infection, 26*, 729. https://doi.org/10.1016/j.cmi.2020.03.026

Philip, S. Y., Kew, S. F., Oldenborgh, G. J. Van, Yang, W., Vecchi, G. A., Anslow, F. S., Li, S., Seneviratne, S. I., Luu, L. N., Arrighi, J., Singh, R., Aalst, V., Hauser, M., Schumacher, D. L., Marghidan, C. P., Ebi, K. L., Vautard, R., Tradowsky, J., Coumou, D., … Otto, F. E. L. (2021). Rapid attribution analysis of the extraordinary heatwave on the Pacific Coast of the US and Canada June 2021. World Weather Attrib 119–123.

Potestio, M. (2021). *Estimate on number of suspected heat-related deaths rises to 808*. Toronto Star. Retrieved July 22, 2021, from https://www.thestar.com/news/Canada/2021/07/21/estimate-on-number-of-suspected-heat-related-deaths-rises-to-808.html

Pour, S. H., Wahab, A. K. A., Shahid, S., Asaduzzaman, M., & Dewan, A. (2020). Low impact development techniques to mitigate the impacts of climate-change-induced urban floods: Current trends, issues and challenges. *Sustainable Cities and Society*. https://doi.org/10.1016/j.scs.2020.102373

Pyrgou, A., Hadjinicolaou, P., & Santamouris, M. (2020). Urban-rural moisture contrast: Regulator of the urban heat Island and heatwaves' synergy over a Mediterranean city. *Environmental Research, 182*, 109102. https://doi.org/10.1016/j.envres.2019.109102

Qian, Y., Chakraborty, T. C., Li, J., Li, D., He, C., Sarangi, C., Chen, F., Yang, X., & Leung, L. R. (2022). Urbanization impact on regional climate and extreme weather: Current understanding, uncertainties, and future research directions. *Advances in Atmospheric Sciences, 39*, 819. https://doi.org/10.1007/s00376-021-1371-9

Qin, X., Li, L., & Hsieh, C. R. (2013). Too few doctors or too low wages? Labor supply of health care professionals in China. China. *Econometric Reviews, 24*, 150. https://doi.org/10.1016/j.chieco.2012.12.002

Ray, B., & Rajib, S. (2019). *Urban drought*. Springer.

Riebsame, W. E., Changnon, S. A., & Karl, T. R. (2019). Drought and natural resources management in the United States: Impacts and implications of the 1987–89 drought, drought and natural resources Management in the United States: Impacts and implications of the 1987–89 drought. https://doi.org/10.4324/9780429045332.

Ritchie, H., & Roser, M. (2018). Urbanization. Our world data.

Rodriguez-Oreggia, E., de la Fuente, A., de la Torre, R., & Moreno, H. A. (2013). Natural disasters, human development and poverty at the municipal level in Mexico. *Journal of Development Studies, 49*, 442–455. https://doi.org/10.1080/00220388.2012.700398

Rodríguez-Ramírez, A., Pérez-Asensio, J. N., Santos, A., Jiménez-Moreno, G., Villarías-Robles, J. J. R., Mayoral, E., Celestino-Pérez, S., Cerrillo-Cuenca, E., López-Sáez, J. A., León, Á., & Contreras, C. (2015). Atlantic extreme wave events during the last four millennia in the Guadalquivir estuary, SW Spain. *Quaternary Research (United States), 83*, 24–40. https://doi.org/10.1016/j.yqres.2014.08.005

Rozenberg, J., & Hallegatte, S. (2015). The impacts of climate change on poverty in 2030 and the potential from rapid, inclusive, and climate-informed development. Policy Res. Work. Pap. - World Bank.

Ruiz, M. O., Chaves, L. F., Hamer, G. L., Sun, T., Brown, W. M., Walker, E. D., Haramis, L., Goldberg, T. L., & Kitron, U. D. (2010). Local impact of temperature and precipitation on West Nile virus infection in Culex species mosquitoes in Northeast Illinois, USA. *Parasites and Vectors, 3*, 1–16. https://doi.org/10.1186/1756-3305-3-19

Saatchi, S. S., Harris, N. L., Brown, S., Lefsky, M., Mitchard, E. T. A., Salas, W., Zutta, B. R., Buermann, W., Lewis, S. L., Hagen, S., Petrova, S., White, L., Silman, M., & Morel, A. (2011). Benchmark map of forest carbon stocks in tropical regions across three continents. *Proceedings of the National Academy of Sciences of the United States of America, 108*, 9899–9904. https://doi.org/10.1073/pnas.1019576108

Saboori, B., Sapri, M., & Bin Baba, M. (2014). Economic growth, energy consumption and CO2 emissions in OECD (Organization for Economic Co-operation and Development)'s transport sector: A fully modified bi-directional relationship approach. *Energy, 66*, 150–161. https://doi.org/10.1016/j.energy.2013.12.048

Salimi, M., & Al-Ghamdi, S. G. (2020). Climate change impacts on critical urban infrastructure and urban resiliency strategies for the Middle East. *Sustainable Cities and Society*. https://doi.org/10.1016/j.scs.2019.101948

Santamouris, M., Cartalis, C., Synnefa, A., & Kolokotsa, D. (2015). On the impact of urban heat Island and global warming on the power demand and electricity consumption of buildings—A review. *Energy and Buildings, 98*, 119–124. https://doi.org/10.1016/j.enbuild.2014.09.052

Shahmohamadi, P., Che-Ani, A. I., Etessam, I., Maulud, K. N. A., & Tawil, N. M. (2011). Healthy environment: The need to mitigate urban Heat Island effects on human health. *Procedia Engineering, 20*, 61–70. https://doi.org/10.1016/j.proeng.2011.11.139

Shayegh, S., Manoussi, V., & Dasgupta, S. (2021). Climate change and development in South Africa: The impact of rising temperatures on economic productivity and labour availability. *Climate and Development, 13*, 725–735. https://doi.org/10.1080/17565529.2020.1857675

Sheffield, J., Wood, E. F., & Roderick, M. L. (2012). Little change in global drought over the past 60 years. *Nature, 491*, 435–438. https://doi.org/10.1038/nature11575

Silva, C. H. L., Aragão, L. E. O. C., Anderson, L. O., Fonseca, M. G., Shimabukuro, Y. E., Vancutsem, C., Achard, F., Beuchle, R., Numata, I., Silva, C. A., Maeda, E. E., Longo, M.,

& Saatchi, S. S. (2020a). Persistent collapse of biomass in Amazonian forest edges following deforestation leads to unaccounted carbon losses. *Science Advances, 6*. https://doi.org/10.1126/sciadv.aaz8360

Silva, C. V. J., Aragão, L. E. O. C., Young, P. J., Espirito-Santo, F., Berenguer, E., Anderson, L. O., Brasil, I., Pontes-Lopes, A., Ferreira, J., Withey, K., França, F., Graça, P. M. L. A., Kirsten, L., Xaud, H., Salimon, C., Scaranello, M. A., Castro, B., Seixas, M., Farias, R., & Barlow, J. (2020b). Estimating the multi-decadal carbon deficit of burned Amazonian forests. *Environmental Research Letters, 15*, 114023. https://doi.org/10.1088/1748-9326/abb62c

Smoyer Tomic, K. E. (2003). Heat wave: A social autopsy of disaster in Chicago. JAMA J. Am. Med. Assoc. https://doi.org/10.1001/jama.289.12.1573.

Society, N.G. (NGS) (2022). How climate change impacts water access [WWW document]. Natl. Geogr. Soc.

Solaymani, S. (2019). CO2 emissions patterns in 7 top carbon emitter economies: The case of transport sector. *Energy, 168*, 989–1001. https://doi.org/10.1016/j.energy.2018.11.145

Stagl, J., Mayr, E., Koch, H., Hattermann, F. F., & Huang, S. (2014). Effects of climate change on the hydrological cycle in central and Eastern Europe. In *Advances in global change research* (pp. 31–43). Springer International Publishing. https://doi.org/10.1007/978-94-007-7960-0_3

Tan, J., Zheng, Y., Tang, X., Guo, C., Li, L., Song, G., Zhen, X., Yuan, D., Kalkstein, A. J., Li, F., & Chen, H. (2010). The urban heat Island and its impact on heat waves and human health in Shanghai. *International Journal of Biometeorology, 54*, 75–84. https://doi.org/10.1007/s00484-009-0256-x

Teixidó, N., Gambi, M. C., Parravacini, V., Kroeker, K., Micheli, F., Villéger, S., & Ballesteros, E. (2018). Functional biodiversity loss along natural CO2 gradients. *Nature Communications, 9*, 1–9. https://doi.org/10.1038/s41467-018-07592-1

Thomas, C. D., Cameron, A., Green, R. E., Bakkenes, M., Beaumont, L. J., Collingham, Y. C., Erasmus, B. F. N., Ferreira De Siqueira, M., Grainger, A., Hannah, L., Hughes, L., Huntley, B., Van Jaarsveld, A. S., Midgley, G. F., Miles, L., Ortega-Huerta, M. A., Peterson, A. T., Phillips, O. L., & Williams, S. E. (2004). Extinction risk from climate change. *Nature, 427*, 145–148. https://doi.org/10.1038/nature02121

UN (2015). Global sustainable development report: 2015 edition. Glob. Sustain. Dev. Rep. 2015 Ed.

UNDESA. (2021). *Global population growth and sustainable development*. United Nations.

UNDP. (2004). *Reducing disaster risk—a challenge for development*. United Nations Development Programme.

UNEP (2021). Adaptation gap report 2021 the gathering storm—Adapting to climate change in a post-pandemic world.

United Nations (2015). United Nations Department of Economic and Social Affairs, Population Division Department of Economic and Social Affairs, Population Division. World Popul. Ageing United Nat.

Ürge-Vorsatz, D., Harvey, L. D. D., Mirasgedis, S., & Levine, M. D. (2007). Mitigating CO2 emissions from energy use in the world's buildings. *Building Research and Information, 35*, 379–398. https://doi.org/10.1080/09613210701325883

U.S. Global Change Research Program (2009). Global Climate Change Impacts in the United States, Thomas R. Karl, Jerry M. Melillo, and Thomas C. Peterson (Eds.). Cambridge University Press.

Uttara, S., Bhuvandas, N., & Aggarwal, V. (2012). Impacts of urbanization on environment. *International Journal of Research in Engineering and Applied Sciences*.

Van Der Werf, G. R., Morton, D. C., Defries, R. S., Olivier, J. G. J., Kasibhatla, P. S., Jackson, R. B., Collatz, G. J., & Randerson, J. T. (2009). CO2 emissions from forest loss. *Nature Geoscience, 2*, 737. https://doi.org/10.1038/ngeo671

Van Dijk, A. I. J. M., Beck, H. E., Crosbie, R. S., De Jeu, R. A. M., Liu, Y. Y., Podger, G. M., Timbal, B., & Viney, N. R. (2013). The millennium drought in Southeast Australia (2001-2009):

Natural and human causes and implications for water resources, ecosystems, economy, and society. *Water Resources Research, 49*, 1040–1057. https://doi.org/10.1002/wrcr.20123

Van Gelder, P., De Ronde, J., Neykov, N. M., & Neytchev, P. (2000). Regional frequency analysis of extreme wave heights: Trading space for time. In *Coastal engineering 2000—Proceedings of the 27th international conference on coastal engineering, ICCE 2000* (pp. 1099–1112). American Society of Civil Engineers. https://doi.org/10.1061/40549(276)85

Vasiliades, L., Loukas, A., & Liberis, N. (2011). A water balance derived drought index for Pinios River Basin, Greece. *Water Resource Management, 25*, 1087–1101. https://doi.org/10.1007/s11269-010-9665-1

Vermeulen, S. J., Campbell, B. M., & Ingram, J. S. I. (2012). Climate change and food systems. *Annual Review of Environment and Resources, 37*, 195. https://doi.org/10.1146/annurev-environ-020411-130608

Vetter, S. H., Sapkota, T. B., Hillier, J., Stirling, C. M., Macdiarmid, J. I., Aleksandrowicz, L., Green, R., Joy, E. J. M., Dangour, A. D., & Smith, P. (2017). Greenhouse gas emissions from agricultural food production to supply Indian diets: Implications for climate change mitigation. *Agriculture, Ecosystems and Environment, 237*, 234–241. https://doi.org/10.1016/j.agee.2016.12.024

Villoria-Sáez, P., Tam, V. W. Y., del Río Merino, M., Viñas Arrebola, C., & Wang, X. (2016). Effectiveness of greenhouse-gas emission trading schemes implementation: A review on legislations. *Journal of Cleaner Production, 127*, 49. https://doi.org/10.1016/j.jclepro.2016.03.148

Voelkel, J., Hellman, D., Sakuma, R., & Shandas, V. (2018). Assessing vulnerability to urban heat: A study of disproportionate heat exposure and access to refuge by socio-demographic status in Portland, Oregon. *International Journal of Environmental Research and Public Health, 15*, 640. https://doi.org/10.3390/ijerph15040640

Voogt, J., & Oke, T. (2003). Thermal remote sensing of urban climates. *Remote Sensing of Environment, 86*, 370–384. https://doi.org/10.1016/S0034-4257(03)00079-8

Waltz, E. (2014). Beating the heat. *Nature Biotechnology, 32*, 610–613. https://doi.org/10.1038/nbt.2948

Wang, Q., & Su, M. (2019). The effects of urbanization and industrialization on decoupling economic growth from carbon emission—A case study of China. *Sustainable Cities and Society, 51*, 101758. https://doi.org/10.1016/j.scs.2019.101758

Wang, J., Song, C., & Yuan, R. (2021). CO2 emissions from electricity generation in China during 1997–2040: The roles of energy transition and thermal power generation efficiency. *The Science of the Total Environment, 773*, 145026. https://doi.org/10.1016/j.scitotenv.2021.145026

Wang, J., Chen, Y., Nie, J., Yan, Z., Zhai, P., & Feng, J. (2022). On the role of anthropogenic warming and wetting in the July 2021 Henan record-shattering rainfall. *Scientific Bulletin, 67*, 2055–2059. https://doi.org/10.1016/j.scib.2022.09.011

Wege, M., Salas, L., & LaRue, M. (2021). Ice matters: Life-history strategies of two Antarctic seals dictate climate change eventualities in the Weddell Sea. *Global Change Biology, 27*, 6252–6262. https://doi.org/10.1111/gcb.15828

Welander, A., Lyttkens, C. H., & Nilsson, T. (2015). Globalization, democracy, and child health in developing countries. *Social Science & Medicine, 136-137*, 52. https://doi.org/10.1016/j.socscimed.2015.05.006

Weyrich, P. (2016). Barriers to climate change adaptation in urban areas in Germany (Doctoral dissertation, Climate Service Center Germany (GERICS)).

Winsemius, H. C., Jongman, B., Veldkamp, T. I. E., Hallegatte, S., Bangalore, M., & Ward, P. J. (2018). Disaster risk, climate change, and poverty: Assessing the global exposure of poor people to floods and droughts. *Environment and Development Economics, 23*, 328–348. https://doi.org/10.1017/S1355770X17000444

WMO (1979). Proceedings of the world climate conference: A conference of experts on climate and mankind, World Meteorological Organization. Environ. Conserv. https://doi.org/10.1017/s0376892900003702.

World Meteorological Organization (2021). State of the global climate 2020.

Wright, V. P. (1986). World Energy Outlook. pp. 23–28.
Xu, Y., Ramanathan, V., & Victor, D. G. (2018). Global warming will happen faster than we think. *Nature, 564*, 30. https://doi.org/10.1038/d41586-018-07586-5
Xue, R., Ai, B., Lin, Y., Pang, B., & Shang, H. (2019). Spatial and temporal distribution of aerosol optical depth and its relationship with urbanization in Shandong province. *Atmosphere (Basel), 10*, 110. https://doi.org/10.3390/atmos10030110
Yang, J., Cheng, J., & Huang, S. (2020). CO2 emissions performance and reduction potential in China's manufacturing industry: A multi-hierarchy meta-frontier approach. *Journal of Cleaner Production, 255*, 120226. https://doi.org/10.1016/j.jclepro.2020.120226
Zander, K. K., Cadag, J. R., Escarcha, J., & Garnett, S. T. (2018). Perceived heat stress increases with population density in urban Philippines. *Environmental Research Letters, 13*, 084009. https://doi.org/10.1088/1748-9326/aad2e5

Chapter 2
Research Progress of Urban Thermal Environment Studies

2.1 Concept

2.1.1 Definition

Urban thermal environment (UTE) is defined as the heat-relevant physical environment that determines the human body's feeling of cold-hot extent and health. Especially, it refers to a physical environment that takes the urban land surface temperature and canopy air temperature as the core. It comprises anthropogenic-modified urban atmospheric attributes such as air humidity, wind speed, and atmospheric turbidity, underlying surface properties such as land use and land cover type, thermal inertia, albedo, and emissivity, and solar radiation to influence human bodies and activities (Yao et al., 2018). This term was first used by Jendritzky and Nübler (1981), where the bioclimatic aspects, beside temperature, of urban heat island were concerned. UTE as an important part of the urban environment not only directly links to the quality of urban settlement and the health of urban residents but also indirectly influences the consumption of energy and water resources, vegetation phenology, and ultimately the urban sustainable development.

Urban heat island (UHI) refers to the positive temperature contrast between the city and its surrounding rural area (Fig. 2.1). UHI is caused by encroachment of impervious surface on the pre-urban natural coverage, which modifies the thermal, radiative, and aerodynamic properties of the underlying surface, thereby the energy balance is alternated (RIZWAN et al., 2008). As a climatic phenomenon, the UHI effect can occur in different layers of cities, such as the boundary layer, canopy layer, surface layer, and subsurface layer (Table 2.1) (Oke et al., 2017). Thus, we define the UHI effect at these layers correspondingly as boundary layer UHI, canopy UHI, surface UHI, and subsurface UHI. Due to the difference in the generative mechanism and the unique properties of corresponding layers, these UHI effects occur in different spatiotemporal scales and thus require differential observation

L. Zhou et al., *Urban Surface Thermal Environments*, The Urban Book Series,
https://doi.org/10.1007/978-3-031-94499-4_2

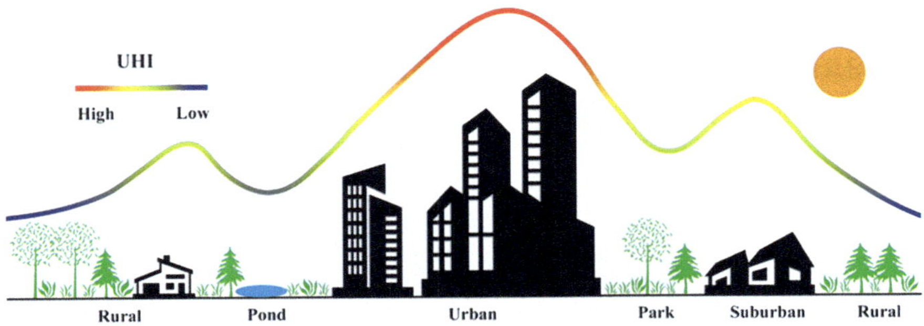

Fig. 2.1 Schematic description of urban heat island

Table 2.1 Four types of urban heat island

UHI type	Scale	Research scope	Processes
Surface UHI	Micro	Interface of the outdoor atmosphere with the solid material, not an actual plane, but the projection and mix of 3D surface temperature	Surface EB
Canopy UHI	Local	The lowest layer of urban boundary layer, between urban surface and the mean building height	Surface EB and EB of UCL air volume
Boundary layer UHI	Local/ meso	Layer between the top of the UCL and the top of the urban boundary layer	EB at top of RSL and BL EB
Subsurface UHI	Local	Layer under urban face, including urban soil and subterranean built fabric	Subsurface EB

Note: *UHI* urban heat island, *EB* energy balance, *RSL* roughness layer

and modeling methods. Moreover, these UHI effects at different layers interact with others. For example, the surface UHI can influence the canopy UHI by modifying sensible heat flux and longwave radiation in urban area. It also influences the subsurface UHI via changing the ground heat flux (Oke, 1982).

Difference Between These Two Concepts Compared to UHI focusing on only one climatic variable (temperature) and is considered as the climatic phenomenon (Table 2.2), UTE emphasizes the heat stress directly felt by human bodies which are comprehensively influenced by multiple climatic variables, e.g., temperature, relative humidity, wind speed, and short- and longwave radiation, although temperature is considered as the core among them. Moreover, the UHI paradigm underlines the urban–rural contrast of temperature, while the UTE paradigm more emphasizes the intra-urban variation of heat stress. In practice, these two terms are used interchangeably in most cases. On the one hand, the local variation of temperature is also concerned in UHI research, and the concept of local urban heat island is proposed (Clinton & Gong, 2013), considering that the current UHI mitigation methods are mostly localized. On the other hand, the land surface temperature or modeled air temperature is the only proxy variable for UTE in many studies, due to the unavailability of other abovementioned meteorological variables at the intra-urban scale.

Table 2.2 Comparison between UTE and UHI

Difference	UTE	UHI
Connotation	Heat-relevant environment in urban areas for human and other biotic elements as well as abiotic urban infrastructure	Positive temperature contrast between urban and surrounding rural areas, especially at the canopy level
Spatial scale	Micro- (building and street) and local- (neighborhood scale)	Local and meso- (city) scale
Involved climatic variable	Temperature, humidity, wind speed, and radiation (on human body)	Temperature
Effects	The perception (comfort) and health of humans	The perception (comfort) and health of human and behavior of other biotic elements

In addition, in practice, the intensity of SUHI greatly depends on the selection of reference temperature. Here we encounter two issues: (1) Rural areas are not stagnant, and changes in rural areas can have an impact on SUHII. (Yao et al., 2019) and (2) the natural elements and their living condition may be not representative in urban areas. For example, in some urban areas, the surrounding coverage is mostly cropland, but in cities it is trees and grass with differential irrigation management. Thus, the reference rural should be not viewed as a "paradigm" of the thermal environment, and the temperature gap cannot be filled and, in some cases, it is unnecessary to do this (Martilli et al., 2020). It is necessary to fully consider in order to avoid these two problems as much as possible.

2.1.2 *Dynamics*

As mentioned above, the root cause of the UHI is the modification of the energy balance (EB) of pre-urban sites by urban development (Oke, 1988). Temporally, UHI arises from differences in urban and rural cooling and warming rates, which could be altered by changing the surface energy balance. Considering the high relevance of surface and canopy UHI to human health, we mainly discussed the general dynamics and their genesis in these two layers.

2.1.2.1 Surface UHI

Diurnal Dynamics of SUHI As mentioned above, UHI arises from the differential rate of temperature dynamics between urban and rural areas (Fig. 2.2), and thus, a discussion about the diurnal dynamic of temperature is needed (Oke et al., 2017): *Shortly after sunrise*, the open rural surface warms rapidly due to increasing direct solar radiation. The roof in urban areas also increased rapidly for the same reason.

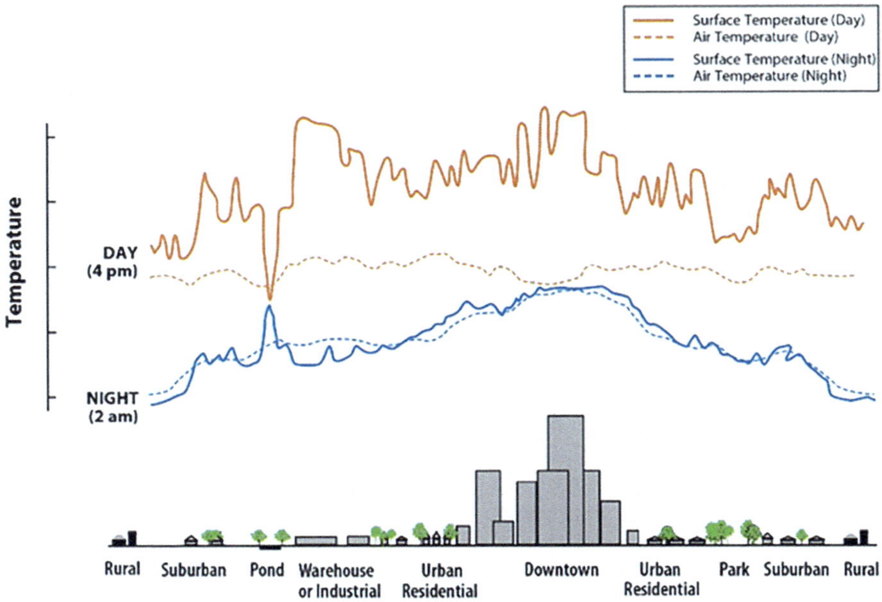

Fig. 2.2 The temperature profile in different areas, with the formation of the UHI. (Source: Earth Resources Observation and Science Center)

However, the complete urban surface warms more slowly, due to the presence of intra-canyon facets with lower surface temperature (ST), which leads to a decrease in SUHI. *During the morning*, the warming rate in both rural and urban areas reaches their peak and then decreases in a similar manner, although the peak time-point for urban areas is later, and thus, the urban cooling rate remains lower than that of the rural after the timepoint, i.e., the surface UHI starts to increase. *From the midday*, ST in both rural and urban areas continues to decrease with increasing cooling rate and the cooling rate reaches the peak just before sunset. The cooling rate of complete ST in urban area is significantly lower than that of rural area, which causes a large gap between ST in urban and rural areas, i.e., SUHI. *During night-time*, the cooling rate in the urban area is lower slightly relative to the rural area, which leads to a gradual increase in the SUHI for both plan and complete surface representations, i.e., even for plan roof, its thermal admittance is also larger than that of rural area.

2.1.2.2 Canopy UHI

Diurnal Dynamics of CUHI *After sunrise*, AT growth is rapid in rural areas because the effects of rapid surface warming on air warming are concentrated in the lowest air layer until morning, while AT warming is slower in urban areas because of slower warming of the inner canyon surface and slower mixing of the cross-canyon circulation. As a result, the UHI gradually decreases. *From midday to early afternoon*: At midday, rural AT catches up to urban AT, and thus, the overnight canopy UHI is rapidly eroded; meanwhile, the changing rate of AT in both areas declines to 0. After the zero point of changing rate, AT in both areas begins to decrease, while the changing rate of AT in rural area is significantly higher than that of urban area. *From the later afternoon to early evening*, the urban cooling rate is much smaller than the rural one, due to the greater heat gain from the larger sensible heat flux from the slowly cooling intra-canyon surface and larger anthropogenic heat source as well as the limited heat loss from the advection and longwave radiation. During this time, the increase of canopy UHI is most rapid. *Later in the night*, the rural cooling rate declines as the dominant net longwave flux decreases continuously due to the lowering AT, while the vertical mixing is limited and heat storage from the substrate has been drained out. In contrast, the surface cooling in the urban area is relatively weak in the previous stage, and thus, the lower layer of the atmosphere has neutral stability and gentle mixing. As a result, the canopy UHI usually reaches its peak at midnight.

2.2 History

The study of the UTE/UHI is as old as the study of urban climate itself, in which the UTE/UHI has always been the core issue. The research history could be divided into four periods (Table 2.3), namely early period (1818–1900), accumulation period (1900–1970), modern foundation (1970–2000), and redevelopment period (2000–present), in terms of the evolution of research topics and techniques.

2.2.1 Early Period (1818–1900)

The scientific documentation and research of UHI could be traced back to the early nineteenth century, which is almost consistent with the establishment of urban climatology, after the advancement in meteorological instruments, such as the thermometer, barometer, and anemometer. The first to do this scientifically was Luke Howard (a chemist and gentleman scientist), and he was also the pioneer in urban climatology with the scientific documentation of London climate for 27 years. With what we would now consider to be unrepresentative exposure of thermometers, he observed, and correctly recognized, the anomalous warmth of the city (both diurnally and seasonally), i.e., the UHI effect. Further, he correctly hypothesized almost all the causes we now study: Anthropogenic heating; multiple reflection; lack of evaporation and the retardation of airflow, etc. These findings and speculation and documentation of other meteorological parameters were together reported in two early volumes entitled *The Climate of London* (Howard, 1833). During the latter half of the nineteenth century, the meteorological observation networks expanded rapidly around the world, especially in Europe. The increasing available data

Table 2.3 Major achievements in each period

Period	Key results	Work
Early period	Anomalous warmth of the city compared to the rural site	Howard (1818)
	Comparison of UHI across multiple European cities	Hann (1885)
Accumulation period	Proposing the mobile measurement method	Schmidt (1927)
	A synthesis of UHI study from the perspective of the landscape	Kratzer (1937)
	First urban climatography	Balchin and Pye (1947)
	Vertical structure of UHI	Duckworth and Sandberg (1954)
Modern foundation	First numerical model of UHI	Summers (1964)
	First study of UHI via satellite thermal imagery	Rao (1972)
	Identifying the vertical structure (sublayers) of UHI	Oke (1976)
	The first large-scale climatic flux observation	Changnon and Ackerman (1977)
	Landsat imagery was used to study SUHI	Carnahan and Larson (1990)
Redeveloping period	Neighborhood-wide CFD modeling of UTE	Bruse and Fleer (1998)
	Proposing the first urban canopy model	Masson (2000)
	MODIS data were first used in the SUHI study	Zhang et al. (2004a)
	Proposing a 3D model to calculate radiant fluxes and mean radiant temperature	Lindberg et al. (2008)
	The first proposal to use local climate zones for studying SUHI	Stewart and Oke (2012)
	Application of ECOSTRESS data to land surface temperature studies	Xue et al. (2020)

promoted the studies of thermal climate in several of the great cities of Europe, including Munich by Wittwer (1960) and Berlin by Kremser (1886).

2.2.2 Accumulation Period (1900–1970)

During the early twentieth century, a large part of research was conducted in Central Europe, which framed the urban effect within the context of landscape studies and microscale climatology. For example, Geiger's first edition of the climate near the ground appeared in 1927, while Kratzer (1937) documented their survey work on the climates of cities in 1937. German-Austrian workers of this period also began studies of basic urban climate processes, e.g., the effect of building/street geometry on the exchange of longwave radiation (Lauscher, 1934) and winds circulation (Albrecht 1933). Moreover, some basic methods for examining local climates were also established during this period. For meteorological variables with relatively low changing rate, their measurements in urban areas were greatly aided by the invention of the mobile station by Schmidt (1927).

The pace accelerated after World War II, in response to military interest in research into dispersion of gases in urban areas and needs for climate information from the urban reconstruction. This period produced some of the most detailed urban climatographic of specific cities, e.g., Bath (Balchin & Pye, 1947), San Francisco (Duckworth & Sandberg, 1954), Uppsala (Sundborg, 1950), etc. Moreover, Duckworth and Sandberg (Fig. 2.3) were the first to fully study the vertical structure of the heat island (Duckworth & Sandberg, 1954). The statistical relationship between measured UHI and urban feature was also explored with the available data from different research conducted on single cities (Oke, 1973). Research during this period was characterized by more detailed descriptions of urban climates within the urban canopy layer, efforts to statistically relate these patterns to urban landscape features and meteorological controls, and initial attempts to investigate atmospheric conditions above roof level. There was a considerable body of empirical data that had been accumulated during this period that supported, complemented, and extended the findings of Howard.

2.2.3 Modern Foundations (1970–2000)

Early efforts toward modeling the urban climate include the numerical heat island models by Summers (1964) and Myrup (1969); the scaled hardware model of Davis and Pearson (1970); the wind tunnel (Fig. 2.3) works on turbulence around

Fig. 2.3 Wind tunnel model. (Tönseth & Welchermill, 2014)

buildings and the urban wind profile by Davenport (2000); and numerical models of radiation transfer in the polluted urban boundary layer (Atwater, 1971). However, attempts to measure the budgets of energy, mass, and momentum remained weak because the knowledge about the role of scale and landscape heterogeneity was lacking and instruments were often deployed with poor exposure and used inappropriate methods, e.g., a typical approach is recording the vertical gradients in wind speed, temperature, and humidity near the ground and estimating the energy exchanges from them.

Heterogeneity and Scale Issues A major challenge to studying the urban climate was how to cope with both the heterogeneity of the urban landscape and varying scales of urban effects. The absence of such a framework meant that observational data and numerical simulations were often incompatible and the fundamental controls that regulate urban climate effects could not be isolated. Oke (1976) conducted the first study to make explicit the distinction between the CUHI and boundary layer UHI and identify the urban roughness sub-layer. Meanwhile, proposing a model to represent types of urban landscapes that would allow representative observations above the urban canopy layer was more difficult to resolve. While boundary-layer theory had been developed for extensive homogeneous surfaces, it was not clear whether these concepts could be applied to highly heterogeneous urban landscapes. Work by Roth and Oke (1993) and Schmid and Bünzli (1995) provided essential steps to solve the issue and showed that urban turbulence is very similar to that over more homogeneous terrain.

During the 1990s, the UC research grew substantially: Both the number of researchers and interest in integrating UC into boundary-layer meteorology discipline increased. For example, much of the observational work shifted from the UCL into the atmosphere above roof level, where the microscale effects of the underlying urban surface might be assumed to be integrated into local-scale (neighborhood-scale) signals. Although the notion of the urban atmosphere as a developing boundary layer was early introduced by Summers (1964), related research has developed significantly through field projects utilizing aerial surveys and balloons to probe the vertical structure of temperature and air pollutants. Large-scale observational campaigns were initiated, including the Urban Air Pollution Dynamics Program in New York (Davidson, 1967), METROMEX in St. Louis (Changnon & Ackerman, 1977), and the Complete Atmospheric Energetics Experiment (CAENEX) in Zaporozhye, U.S.S.R. (Kondrat′ev, 1973).

2.2.4 Redeveloping Period (2000–Present)

Improvement of Local- and Micro-scale Models As computer power has improved, local- and micro-scale models with increasing sophistication and detail were developed. The simple descriptions of the urban surface (e.g., concrete slabs) used in earlier models were replaced by more realistic descriptions of land cover, material properties (e.g., asphalt and concrete), geometry (e.g., street dimensions), and functions (e.g., traffic) that define different types of urban neighborhoods. Advances in computing power have also made it possible to apply computational fluid dynamics (CFD) models (Fig. 2.4), traditionally used to simulate flows around individual buildings, to entire neighborhoods (Hanna et al., 2006). Moreover, there are available modeling tools running on inexpensive personal computers to explore urban micro-climatic variations (e.g., SOLWEIG (Lindberg et al., 2008) and ENVI-met (Bruse & Fleer, 1998)). Cooperation among research teams now allows for more comprehensive research projects (e.g., BUBBLE—an urban boundary-layer meteorology project) (Rotach et al., 2005).

Application of Remote Sensing Another achievement is the development of SUHI research based on remotely sensed thermal imageries. Although Rao (1972) first observed the SUHI phenomenon in 1972, the field remained relatively inactive between 1972 and 2005, with fewer than five studies published per year. However, since 2005, there has been an exponential growth in SUHI publications each year. This increase can be attributed to the growing interest driven by rapid urbanization and global warming, along with the advancements of remote sensing (RS) techniques and computing power, which have greatly enhanced the quantity and quality of RS data and the capability of data processing. *RS data sources*: P. Krishna Rao began the thermal RS research of UHI with ITOS-1 (TIROS-M, meteorological satellite), and he found the evident urban signatures in the eastern United States. In 1990, Landsat data were first applied in SUHI research (Carnahan & Larson, 1990) and gradually became the major satellite record for the SUHI analyses due to its comparatively high spatial resolution (60–120 m), long time series, and high data quality, especially after the free release of the entire Landsat archive in 2008. Conversely, MODIS data, available since 2000, offer high temporal resolution, medium spatial resolution (1 km), global coverage, and free access, making it pos-

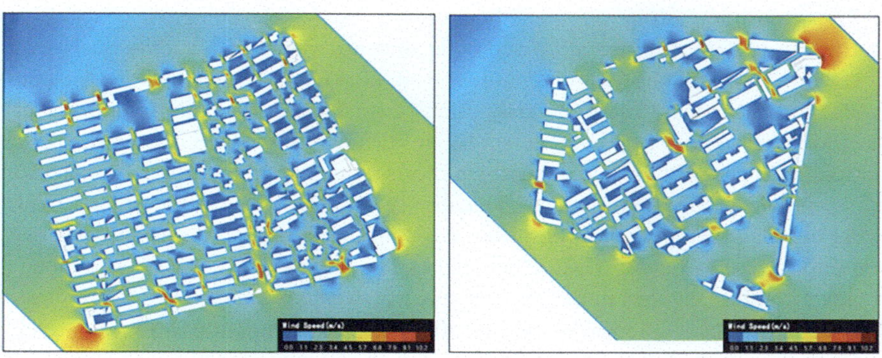

(a) (b)

Fig. 2.4 Wind environment simulation diagram of a low floor area ratio scenario: Low floor area ratio–high building density plot (**a**) and low floor area ratio–low building density plot (**b**). (Li et al., 2024) sibletostudythediurnal,inter-annual,andintra-annualvariabilityoftheSUHI.MODIS-based SUHI research was first published in English in 2004 (Zhang et al., 2004b). ASTER was also on board the Terra satellite in 2000 and the corresponding LST product was released from 2001 until 2004, when Nichol published the first day and night study (Nichol, 2005).

2.3 Data Sources

To characterize the UHI, the study mainly used three types of methods: Field observation, remote sensing observation, and model prediction to obtain temperature as well as humidity, wind speed, radiation, and other related data. The field observation method refers to the field observation of the environmental response to UHI using meteorological and radiation monitoring instruments at the corresponding stations (Grimmond, 2006), and the remote sensing observation method refers to the surface temperature obtained by the satellite-based or airborne sensors (Zhou et al., 2019). The model prediction approach refers to the simulation of the environmental response to the cooling effect of UHI through models and corresponding scenarios (Krayenhoff et al., 2021). A detailed introduction follows:

2.3.1 In Situ Monitoring

In situ strategies require researchers to use meteorological observation instruments and radiometric instruments to make field observations of the cooling effect of urban green infrastructure (UGI). Specifically, the observation methods can be divided into two types: Fixed-point and mobile observations.

The fixed-point observation method uses a fixed meteorological observation instrument to collect temperature and other meteorological data continuously over time (Fig. 2.5). The observation could be from either weather stations or research-grade. However, data from previously established stations, typically setup for standard weather monitoring, are generally insufficient for SUHI research. These stations are generally located at nearby airports, which do not adequately represent the complete urban environment. It generally requires observations to be made under clear and windy conditions, thus avoiding the influence of unstable weather conditions on data quality (Oke et al., 2017), but some studies also choose cloudy weather conditions to examine the influence of weather conditions (Wang et al., 2015b). Such observations also require that the observation site is in a relatively homogeneous part of the local climate zone to ensure homogeneity of the observed source area, but this restriction is relaxed in some studies to understand the effect of land cover/land use in the vicinity of the observation site on the temperature at the observation site (Giridharan et al., 2008).

Fig. 2.5 Fixed station. (Jin et al., 2018)

Although the spatial data points that can be obtained using the fixed-point observation method are fewer and often limited to the microscale, fixed-point observation allows for continuous observations over longer time periods (within a day to several days), which allows for a detailed examination of the vegetation cooling effect in the daily cycle, which is difficult to achieve with several other approaches, such as mobile and remote sensing observations. To explore the spatial characteristics of CUHI, sensor networks require significant logistical support and funding, as many of the former urban meteorological networks are no longer operational. In addition, due to the availability of wireless sensor technology, it has become cost-acceptable to use a large number of fixed sensors in the study to obtain canopy temperatures at the urban scale (Muller et al., 2013; Ziter et al., 2019).

Mobile observation method focuses on the mobile observation of canopy temperature by handheld or vehicle-mounted sensors (automobile and bicycle) (Fig. 2.6). It generally uses a stop-and-go strategy, where observation sites are selected in advance and brief observations are completed by arriving at each site one by one along a route (Ziter et al., 2019). For mobile campaigns conducted at the urban scale, post-hoc data processing is required to correct for the effects of elevation changes and daily cycling and to avoid observation sites with congested traffic during the observation process to avoid data interference from excessive transient anthropogenic heat release (Oke et al., 2017). In many cases, correction for the daily cycle is not straightforward because the warming and cooling rates are not consistent across different local climate zones. Using a single temperature variation

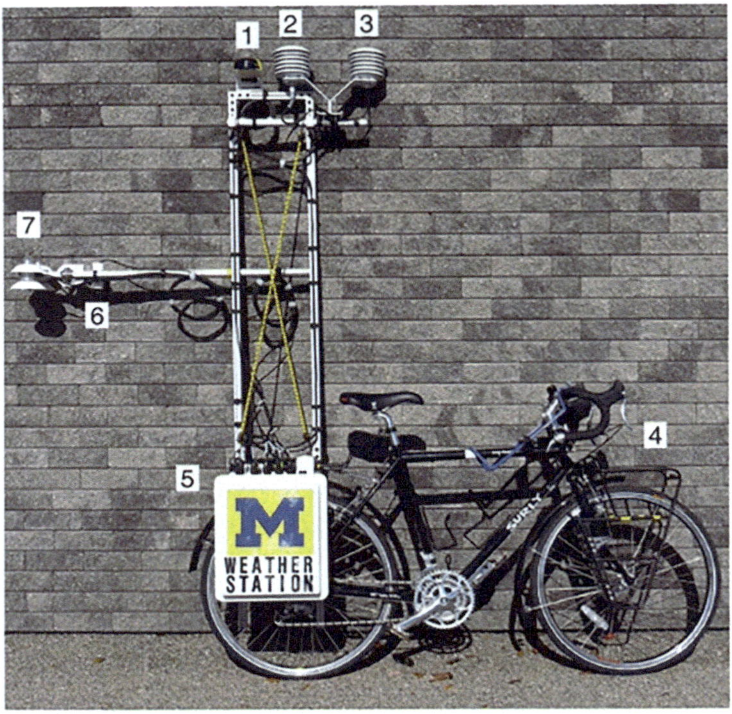

Fig. 2.6 Image of the bicycle-based mobile measurement system. (Rajkovich & Larsen, 2016)

coefficient may make the results significantly biased, but this issue is generally not considered for neighborhood-scale data obtained from pedestrian handheld sensors.

2.3.2 Remote Sensing

LST is determined by a combination of surface thermal radiation and thermodynamic properties and is influenced by ground moisture, surface reflectance, solar and atmospheric downward radiation, and near-surface AT. In turn, AT is mainly influenced by the combination of sensible heat flow emanating from the surface, human activities, and the background temperature of surrounding landscape elements (Cao et al., 2021). In terms of data sources, the thermal environment and UHI effects in the urban surface layer are mainly studied using LST data, which is retrieved from thermal infrared remote sensing images (Tomlinson et al., 2011).

Compared with the urban canopy and boundary layer, the urban LST is closely related to human health and warm and cold perceptions, and it is both an important parameter for describing the urban thermal environment conditions and explaining the formation of the urban thermal environment, as well as one of the core elements of the current urban spatial thermal environment research. Meanwhile, for the study of the thermal environment on urban and regional scale, the retrieved LST from high-resolution remote sensing image offers the advantage of quickly and accurately obtaining continuous large-scale regional temperature data. This contrasts with air temperature (AT) monitored by limited and scattered meteorological detection sites. Several open-access remote sensing images with spatial resolution exist in current studies for retrieving LST (Table 2.4). Among them, at the regional and urban scales, researchers mainly use MODIS sensors carried on Terra and Aqua satellites (Clinton & Gong, 2013). At the local scale, the study uses mainly TM, ETM+, and TIRS sensors on board the Landsat series of satellites and ASTER images on board the Terra satellite (Estoque & Murayama, 2017).

Table 2.4 Main thermal remote sensing data sources for urban studies

Platform/sensor	Time range	Spatial resolution	Revisit period	Advantages	Disadvantages
Airborne/MASTER	1998–present	7 m	–	The highest spatial resolution	No fixed revisit period, U.S. region only
Landsat/TM	1984–2013	120 m	16 days	The longest time series	Low spatial resolution
Landsat/ETM+	1999–2005	60 m	16 days	Medium spatial resolution, Long time series	Poor data quality because of stripes
Landsat/TIRS	2013–present	100 m	16 days	The highest spatial resolution in open-access data obtained from satellite	Data access from 2013 to now
Terra/ASTER	1999–present	90 m	16 days	The sub-highest spatial resolution in open-access data, nighttime data	On-demand access; unbalanced spatiotemporal image
Aqua and Terra/MODIS	1999–present	1000 m	0.5 days	Four times observation in one day with two satellites; nighttime data; the highest time resolution	Low spatial resolution; only satisfied the study at large scale and region
ISS/ECOSTRESS	2018–present	70 m	–	The sub-highest spatial resolution in open-access data, diurnal variation	No fixed revisit period
GF-5/VIMS	2018–present	40 m	–	The highest spatial resolution obtained from satellite	High acquisition cost; no fixed revisit period
HJ-1B/IRS	2008–present	300 m	4 days	High revisit period	Low spatial resolution

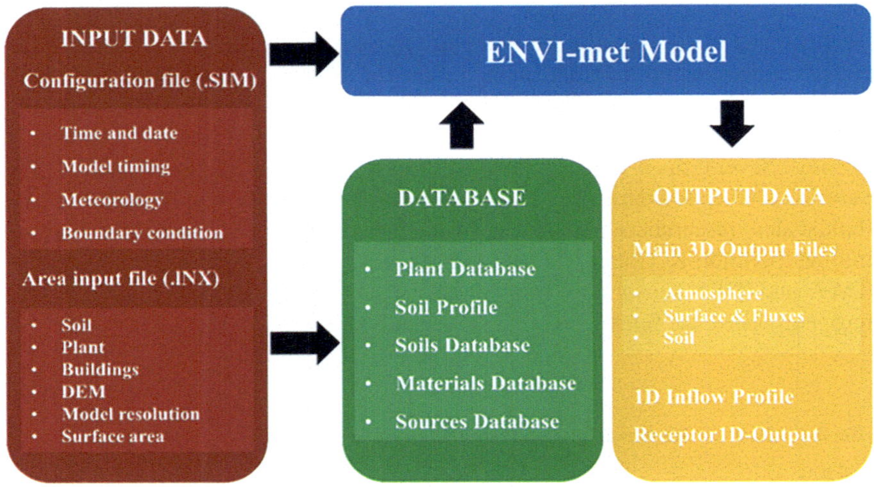

Fig. 2.7 ENVI-met flowcharts. (Maleki & Mahdavi, 2016)

2.3.3 Numerical Simulation

Models at Micro and Local Scales Relevant models can be divided into those based on CFD and on surface energy balance, according to the dominant principle of modeling. These two principles are not used independently; in fact, model designers, after determining the dominant principle of the model, often incorporate the other principle into the model by means of parameterization. In the former type of models (usually at microscales), the building and its facets were explicitly represented in models, so that the air drag effects of buildings and other affiliated objects, especially trees, on canyon ventilation and the three-dimensional transport of heat, momentum, and mass in the canyon could be modeled, while the surface energy balance of each facet and the cubic volumes of trees could also be described. In the latter model (usually at local scales), almost only the transport of heat, momentum, and mass along the vertical direction was considered and parameterized (multi-layer urban climate model), while these elements are averaged at the horizon plane (Krayenhoff, 2014). Currently, at the micro and local scales, the most frequently used tool is the ENVI-met model (Fig. 2.7) where many parameters about the properties of building facets and vegetation could be set, and a series of key physical processes could be effectively represented (Tsoka et al., 2018).

Models at the Mesoscale The models used in mesoscale cooling studies are generally coupled with a mesoscale climate model, of which the weather research and forecasting (WRF) model (Fig. 2.8) is the most commonly used, and an urban canopy model (usually a single-layer canopy model), where the mesoscale climate model provides the upper atmospheric boundary conditions for the urban canopy

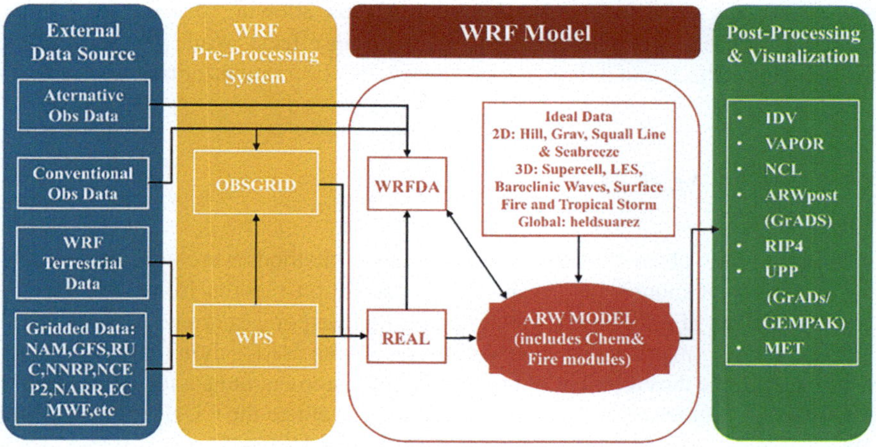

Fig. 2.8 WRF modeling system flowchart. (Source: WRF UsersGuide_v4.2)

mode (Skamarock et al., 2019). The tile-based weighting approach is mostly used in this type of coupling, which decomposes the mesoscale climate model grid cell on the urban surface into separated tiles composed of vegetation, artificial surfaces, water bodies, etc. and obtains their area proportions in the cell. For natural cover tiles such as vegetation and water bodies, they are still modeled by the land surface model included in the mesoscale climate model, while for artificial surfaces, they are modeled by the urban canopy model, and finally the surface fluxes obtained from the different modeling approaches are weighted according to the area proportions within the tuple to obtain some boundary conditions of the mesoscale climate model.

2.3.4 *Comparison*

Comparing the above three types of acquisition methods, in terms of the scales to which they are applied, in situ observation methods are the smallest and are mainly used at microscales such as stations and streets; remote sensing observations are the largest and are mainly used at local scales such as green spaces and neighborhoods and at urban scales. Model prediction methods can achieve simulations at different scales, although their simulation resolution and the cooling mechanisms that can be considered are limited by the scale. In terms of spatial resolution, model prediction methods can achieve the largest resolution (e.g., up to 0.5 m for the ENVI-met model). This is followed by in situ observation methods, which can achieve spatial resolution at the site scale, but the spatial continuity of the resulting data is poor. Remote sensing observation methods usually yield the lowest spatial resolution. In terms of temporal resolution, both in situ and model simulation methods can achieve near-continuous observations or simulations, thus allowing the study of daily dynamics and inter-day variability of UHI cooling effects, while remote sensing observation methods can only make instantaneous observations at transit times.

2.3.5 *Other Methods*

In addition to the above methods, other innovative methods were used in research. For the fix-point monitoring, besides the observation records from the standard weather stations (Fig. 2.9) and research-specific temporal stations, crowdsourced measurements from private weather stations have been used to monitor the CUHI. In addition, with the development of microscale measurement units with the capability of wireless communication, they are also used to evaluate the CUHI or the cooling effects of vegetation (e.g., the i-Tree projects led by USDA Forest Service). They could also be carried by pedestrians to monitor the ambient environment as the smart devices (phones or wearables). Moreover, some cities also deployed a large

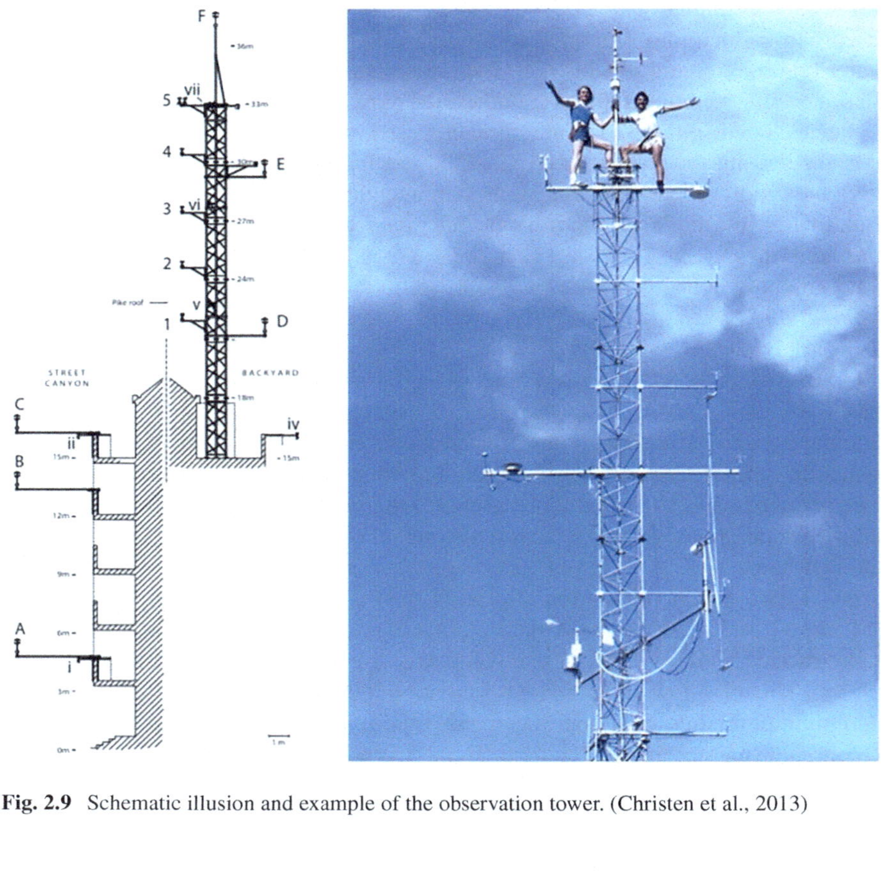

Fig. 2.9 Schematic illusion and example of the observation tower. (Christen et al., 2013)

number of automatic meteorological stations within cities; however, they are hardly accessed by outsiders.

2.4 Methodology

2.4.1 Gap-Filling

Gap-filling could be divided into five strategies in terms of information usage, i.e., spatial interpolation, temporal interpolation, spatiotemporal interpolation, fusion-based filling, and surface energy balance-based filling, here as follows:

2.4.1.1 Spatial Interpolation (With/Without Considering Kernels)

In this strategy, the missing pixel is reconstructed with spatially neighborhood valid pixels and additional covariates sometimes (Fig. 2.10a). The widely employed simple methods include spline curve interpolation, ordinary kriging, and inverse distance weighting (IDW) methods, and they are effective when only a small number of missing pixels occur. However, they cannot address areas with large-scale missing or high spatial heterogeneity. Therefore, auxiliary data as covariate need to be introduced. The commonly used covariates include elevation (Ke et al., 2012), location (longitude and latitude) (Zhou et al., 2012), and NDVI (Ke et al., 2013). However, these former two covariates are more suitable for large-scale areas, while NDVI is suitable for urban areas. Although the introduction of covariates could improve the performance of the spatial interpolation method, under conditions with concentrated clouds or extensive clouds, these methods may fail to achieve satisfying results. Thus, information in other dimensions needs to be considered.

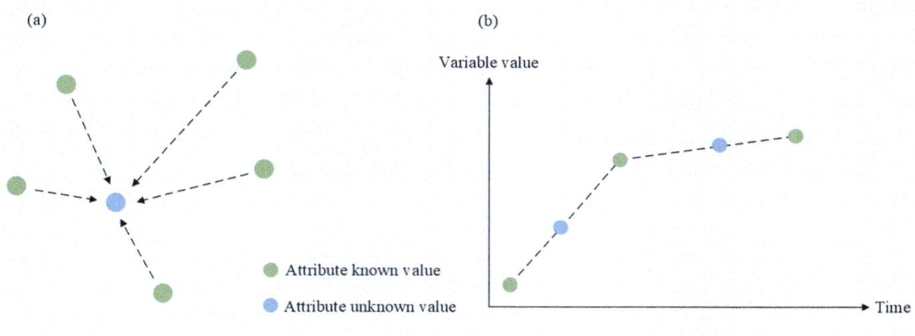

Fig. 2.10 (**a**) Spatial interpolation principle, (**b**) time linear interpolation

2.4.1.2 Temporal Interpolation (Data-Driven/Physical Models)

In this strategy, linear interpolation (Zhang et al., 2015) is the most simple and feasible method (Fig. 2.10b); however, it may produce unreasonable estimates with large errors, especially during periods with dramatically changing weather conditions.

Theory foundations of DTC/ATC: The temporal dynamics of land surface is controlled by surface energy balance, especially the solar radiation during daytime and longwave emitted radiation during nighttime, which both comply with simple physical laws approximately, i.e., the former changes with sun elevation angle and the latter decay in an exponential way (Göttsche & Olesen, 2001). Therefore, the diurnal or annual dynamics of LST, i.e., diurnal time cycle (DTC) or annual time cycle (ATC) could be fitted by the semi-physical model with harmonic terms as well as additional exponential decaying terms with different assumptions or constraints on surface energy balance terms (Duan et al., 2012; Huang et al., 2014). *Application of DTC/ATC*: The fitted ATC/DTC could be used to temporally interpolate the missed pixels. DTC and ATC have been used in some studies (Van De Kerchove et al., 2012; Van Den Bergh et al., 2007). Improvement for the original form has also been made, e.g., Xu and Shen (2013) developed the harmonic analysis of time series (HANTS) algorithm to produce annual 8-day MODIS LST series, while Liu et al. (2019) proposed the hybrid ATC (ATCH), where multiple harmonics are combined with linear functions of scaling factors.

Trade-offs: It should be borne in mind that there is a trade-off between accuracy and generalizability in model forms. For example, sophisticated ATC models generally have high accuracies but low generalization abilities due to their requirements for large numbers of parameters. Conversely, simple ATC models with one or two harmonics have good generalization abilities but relatively poor fitting performances. *Application of ML*: In addition to physical models, machine learning (ML) methods with the ability to model complex and interrelated relationships have also been considered. For example, Arslan and Sekertekin (2019) used the long short-term memory (LSTM) neural networks with the ability to alleviate the vanishing gradient problem for daily MODIS LST time series with RMSEs ranging from 2 K to 9 K and ranging from 1 K to 5 K for daytime and nighttime. *Ad hoc methods*: In addition to these general methods, some methods are developed for special data characteristics. For example, considering the interrelationship between Terra and Aqua LST, Crosson et al. (2012) developed a mutual-filling method for these two datasets, where the interconversion is realized by a mean difference per pixel or more complicatedly, a conversion coefficient derived by multiple regression with auxiliary variables, such as latitude, longitude, and elevation.

2.4.1.3 Spatiotemporal Interpolation

In this strategy, both spatially and temporally adjacent valid pixels or all available spatiotemporally neighboring valid pixels (i.e., within a spatiotemporal moving window) are considered to interpolate the missed pixels, although in various forms (Fig. 2.11).

Basic sub-strategy is to use the multitemporal method first and then apply a spatial reconstruction method if the first step fails to recover all missing pixels or identify abnormal pixels (Zeng et al., 2015). *Spatiotemporal moving window* as the other sub-strategy is also considered (Li et al., 2018) using the valid pixels (distribution) within a spatiotemporal window to construct a quantile regression to obtain the target value, after a daily merging of four diurnal MODIS LST imageries into two imageries in the reference time with mutual filling abovementioned. Weiss et al. (2014) considered both the spatial neighboring pixels in the waited-filled imagery and those in the imagery on the same day but adjacent years (including the pixel in the target location) and used ICW method to combine them.

Beyond spatiotemporal proximity: Based on these abovementioned methods, some studies also considered the property similarity, i.e., using similar pixels in terms of factors such as terrain factors (DEM, slope, aspect, NDVI), solar factors (solar radiation), time, NDVI, and geography to correct the estimated hypothetical value (Tan et al., 2021; Yu et al., 2015). *ML methods*: Under large-extent and long-term cloud cover conditions, the reconstruction performance worsens due to the complicated and nonlinear relationships that occur across space and time between LSTs. Therefore, ML methods with a stronger ability to handle complicated nonlinear relationships have been proposed in recent studies. Sarafanov et al. (2020)

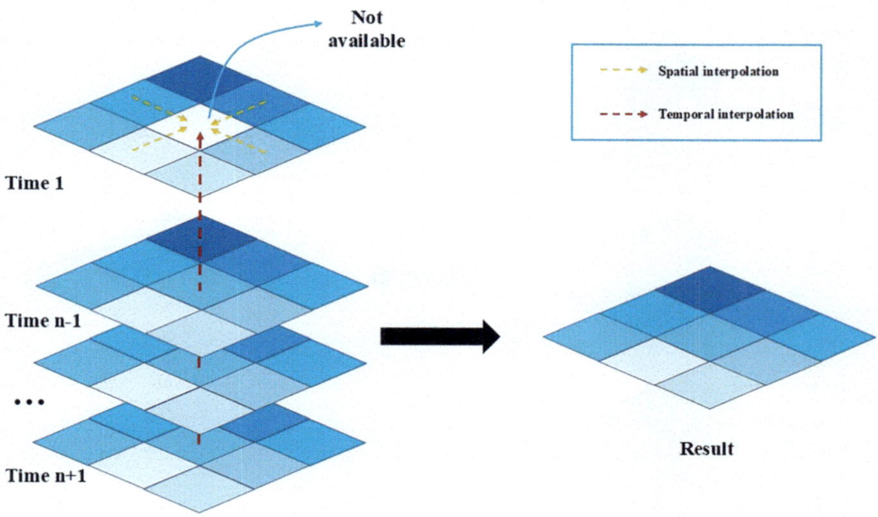

Fig. 2.11 Spatiotemporal interpolation

compared lasso regression, random forest (RF), extra trees, support vector regression, and k-nearest neighbors regression methods to gap-fill the Sentinel-3 LST and MODIS LST data.

2.4.1.4 Fusion-Based Filling

In this strategy, the LST from other sources is used for filling (Fig. 2.12). Especially, to reconstruct the cloudy-contaminated pixel, most studies used the passive microwave (PMW) or meteorological reanalysis data. Therefore, the gap-filling could be viewed as a downscaling of these two coarse resolution in specific locations, i.e., missed pixels.

PMW-based method: Kou et al. (2016) introduced the Bayesian maximum entropy method (spatiotemporal geostatistical scheme) to merge MODIS LST and AMSR-E LST. Soil temperature measurements, with R^2 of 0.80 but a high RMSE of 11.2 K, can be explained by the systematic bias between the LST and soil temperature. Therefore, Duan et al. (2017) first converted the subsurface temperature into LST and then merged with MODIS LST, while the orbital gaps in the AMSR-E data were filled by temporal linear interpolation and IDW. *Reanalysis data-based methods*: Long et al. (2020) fused the MODIS LST and China Land Data Assimilation System (CLDAS) LST using enhanced STARFRM (ESTARFM) and then used a bias correction approach to minimize the differences between the fused LST and MODIS LST. Zhang et al. (2021) proposed an RTM method to merge daily MODIS LST and GLDAS LST based on the model of decomposing LST time series. *Caveats*: For PMW-based methods, physical differences between the two

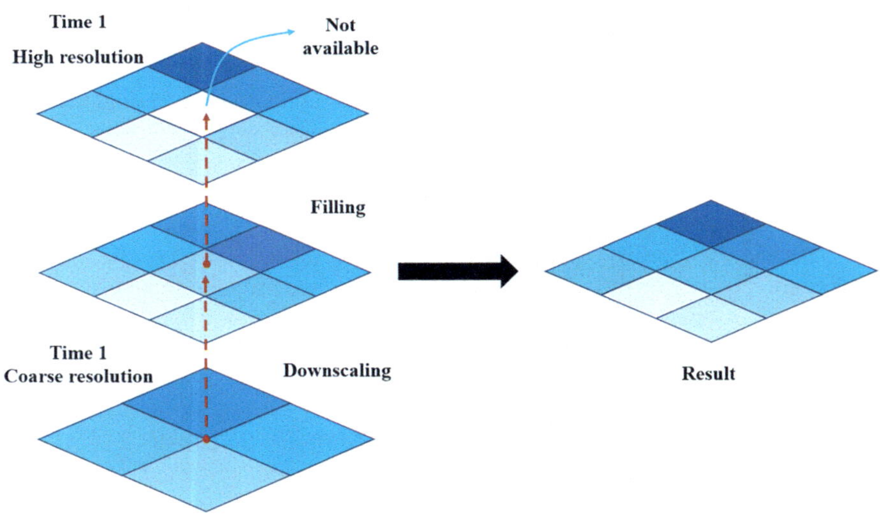

Fig. 2.12 Fusion-based filling

temperature measurements: (i) The orbital gaps and the coarse spatial resolutions of PMW data and (ii) lack of practical, accurate spatial downscaling methods for PMW LST over highly heterogeneous surfaces, especially in urban areas.

2.4.1.5 Surface Energy Balance-Based Gap-Filling

In this strategy, a hypothetical LST was first spatiotemporally interpolated, and then the estimated LST was corrected by the LST differences between cloudy pixels and neighboring valid pixels derived from the surface energy budget (SEB) model (Fig. 2.13).

Basic Methods Jin (1999) first proposed a SEB-based neighboring-pixel (NP) interpolation method, where LST differences between cloudy pixels and neighboring valid pixels derived from SEB model and spatially interpolated hypothetical LST were combined. The validation based on field experiments and climate community model/biosphere-atmosphere transfer scheme simulation data showed that the general accuracy ranged from 1 K to 2 K on a monthly basis. Lu et al. (2011) developed a temporal SEB-based NP approach to calculate the LST differences between cloudy pixels and temporally neighboring cloud-free pixels for MSG/SEVIRI LST (3 km 15 min). In situ measurements showed an RMSE of 5.55 K and a bias of −3.68 K in Kenya and an RMSE of 5.11 K and a bias of 0.37 K in Burkina Faso. Yu et al. (2014) developed a spatiotemporal NP interpolation for daily MODIS

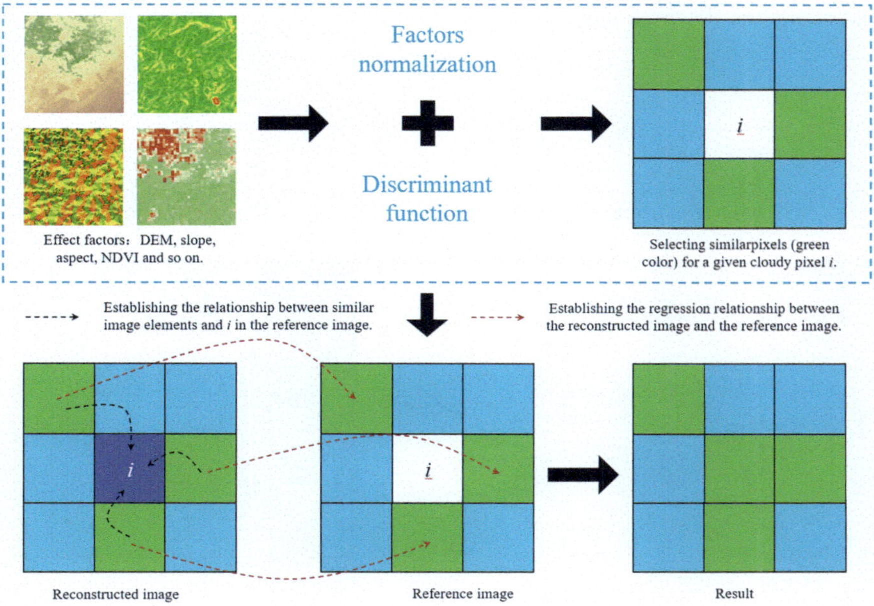

Fig. 2.13 Surface energy balance-based gap-filling

LST data. In situ measurements showed an RMSE of 4.12 K and a bias of 0.25 K during the daytime and an RMSE of 2.90 K and a bias of −0.13 K during the nighttime.

Introduction of Auxiliary Variables To account for terrain effects and others, auxiliary data were introduced. Yu et al. (2019) made some modifications, where auxiliary factors including DEM, slope, aspect, NDVI, and theoretical solar radiation are introduced to help find a similar pixel with cloudy pixel. In situ measurements showed an average error less than 1.00 K and an RMSE less than 3.20 K. Zeng et al. (2018) first found out the pixel similar with cloudy pixel in terms of VI, which is used to estimate hypothetical LST via a weighted linear regression, and then the LST difference derived from SEB using MODIS albedo, MODIS emissivity, and Global Land Surface Satellite (GLASS) downward shortwave radiation was used to correct it. The validation based on surface radiation budget observing network (SURFRAD) data showed R^2 ranging from 0.72 to 0.93 and MAEs ranging from 3.65 K to 6.70 K.

2.4.2 *Spatiotemporal Downscaling*

The downscaling or the resolution enhancement could occur in the spatial (Fig. 2.14), temporal, or simultaneously spatiotemporal dimension. Therefore, the section could be divided into three parts, i.e., spatial downscaling, temporal downscaling, and spatiotemporal downscaling.

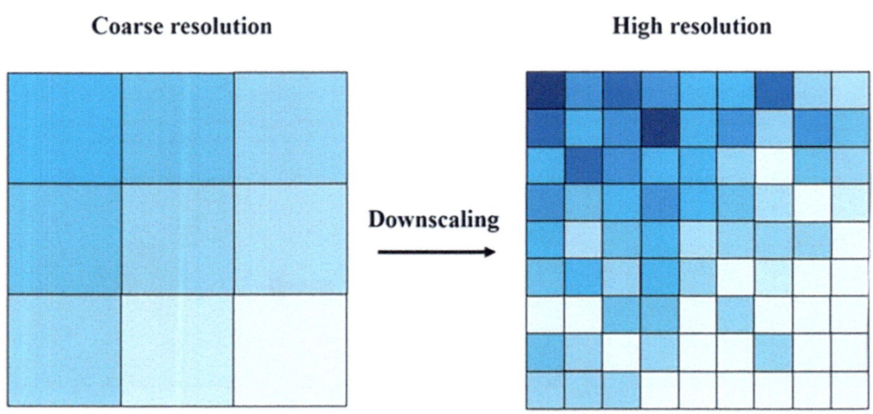

Fig. 2.14 Spatial downscaling

2.4.2.1 Spatial Downscaling

Requirement for Scaling Factor The enhancement of low-spatial-resolution LST images usually relies on high-spatial-resolution auxiliary data that are statistically or physically correlated to LST, which is a key assumption indicated by Zhan et al. (2013). Otherwise, spatial interpolation methods mentioned in the previous section can also be used to enhance the spatial resolution of LST data (Metz et al., 2017). However, spatially interpolated LSTs are merely a smooth expansion based on the coarse LST and are unable to reconstruct the rich details of the actual LST. The auxiliary parameters used to construct the relationship are often called "kernels" (or "scaling factors").

Regression Kernels The mainly used kernels are derived from the VNIR sensor within the same platform which usually has a higher SR, including spectral indices, albedo, emissivity, and vegetation fraction. Besides, terrain factors, solar incidence angle, and land cover fraction are also considered. Especially in urban areas and other less vegetated areas, the NDBI can be used to portray the distribution pattern for LST (Zawadzka et al., 2020). Two trends are present in the selection of kernels: (1) From single to multivariate: Recent studies, e.g., Dong et al. (2020) used more than 20 kernels to achieve high global operability. (2) From one-class to hybrid: Recent studies tend to employ regression kernels from multiple data sources (Amazirh et al., 2019; Zhang et al., 2019).

Regression Scale In the context of LST spatial enhancement methods, scale issues are represented as whether the kernel–LST relationships stay invariant at different SR (scale invariance) and what shape a suitable regression window should be (i.e., localization strategy) (Zhan et al., 2013). The *scale invariance* assumption may be insufficient and many studies found the scale dependence. Aniruddha even found that the scale dependence is a function of the study area characteristics and regression tools (Table 2.5).

To determine the *window shape*, Jeganathan et al. (2011) found that 7 × 7 covered too many multiple land cover types, while 3 × 3 produced unstable results, so a 5 × 5 window was the most appropriate as it balanced the trade-off between sample size and landscape heterogeneity (thus reducing the representativeness of the corresponding regression relationships), the size is also recommended by other studies (Chen et al., 2014). In essence, the key issue is how to extract a

Table 2.5 Regression tools

Simple tools	ML tools
Multiple linear regression	ANN
Quadratic regression	SOFM
Piecewise linear regression	SVMR
Principal component regression	RF
Geographically weighted regression	CNN

homogeneous local area where kernel–LST relationships are stable. Therefore, future studies should focus on selecting similar samples for regression by identifying useful spectral, spatial, or temporal features from remote sensing images and auxiliary data while balancing both enhancement accuracy and computational complexity (Mao et al., 2021).

2.4.2.2 Temporal Downscaling Interpolation-Based

This strategy utilizes physical or statistical models (Fig. 2.15), such as DTC/ATC models to fit temporally discrete LST observations and then produce a denser LST time series. The same procedure is also applied in the temporal interpolation-based gap-filling. Compared to the limitation of the spatial interpolation method in spatial downscaling, the temporal interpolation methods are suitable for temporal downscaling, because the temporal variation of LST has a greater smoothness than the spatial one.

In practice, some studies used the low-SR but high-TR satellite data (e.g., geostationary satellite) to fit the model and applied it to the high-SR polar-orbiting observations to enable frequent time series estimates at a diurnal LST scale (Jin & Dickinson, 1999; Sun & Pinker, 2005). Furthermore, the model forms were modified and terms related to the dynamics of surface property and atmospheric status were introduced (Duan et al., 2013; Inamdar et al., 2008). This simplification that LST series are cyclically stationary is hardly feasible in reality, which causes unexpected uncertainties because of multiple factors, such as changes in land cover or

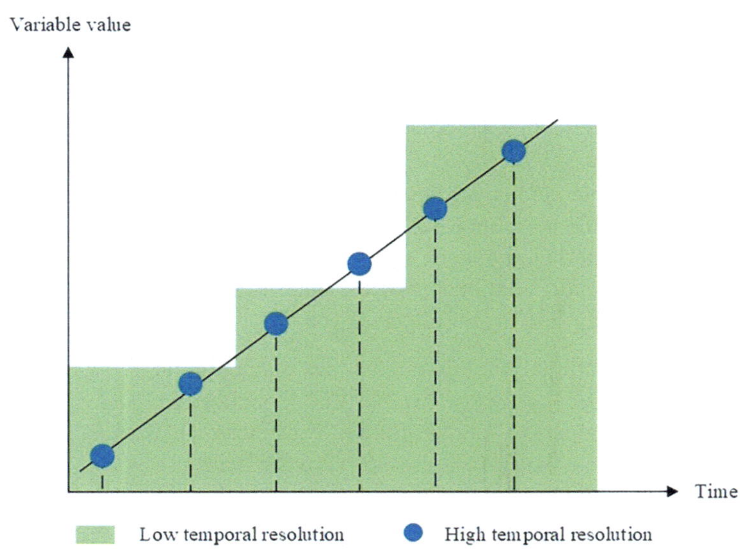

Fig. 2.15 Temporal downscaling

human disturbances (heterogeneity/mixed pixel) (Quan et al., 2014; Stine et al., 2009). To address the mixed pixel issue, Quan et al. (2014) combined a DTC model with a linear mixing model using multitemporal and multi-angular LST observations and their corresponding component fractions retrieved by VNIR reflectance.

2.4.2.3 Fusion-Based

This strategy utilizes the spatiotemporal data fusion methods to merge frequent coarse-resolution images and sparse fine-resolution images into synthesized images with both high spatial and temporal resolutions.

STARFM method was applied (Liu & Weng, 2012). It estimates daily Landsat-like reflectance using a weighted sum of reflectance difference derived from coarse pixels and fine-resolution reflectance in similar neighboring pixels. Wu et al. (2015b) extended the STARFM and proposed the spatiotemporal integrated temperature fusion model (STITFM), which is used to integrate LSTs from multiscale polar-orbiting and geostationary satellite observations to generate diurnal Landsat-like LSTs in their study. *Adaptation of STARFRM for LST*: However, the STARFM and its variants suffer from low accuracy and weak robustness in complex and heterogeneous landscapes. Some modifications have been proposed in other fields to address the issue over landscapes with mixed and changing land cover types, which have also been applied in LST fusion. For example, both enhanced STARFM (ESTARFM) (Fig. 2.16) proposed by Zhu et al. (2010) and unmixing-based spatiotemporal image fusion method (spatial and temporal data fusion approach (STDFA) have been applied in Wu et al. (2015a), respectively.

Physical models, such as temperature mixing models and temperature cycle models, were also introduced in fusion-based methods. Weng et al. (2014) proposed the spatiotemporal adaptive data fusion algorithm for temperature mapping (SADFAT) algorithm (Fig. 2.17), where a linear spectral mixture model and an ATC model were introduced and a conversion coefficient was used to relate the thermal radiance change in a mixed pixel at a coarse resolution to that of a fine-resolution pixel, with MODIS and Landsat data. Quan et al. (2018) proposed BLEnd spatiotemporal temperatures (BLEST) by integrating a linear temperature mixing model,

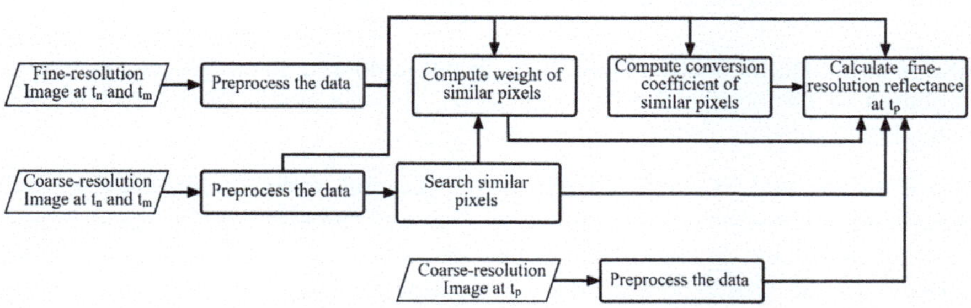

Fig. 2.16 The flowchart of the ESTARFM algorithm

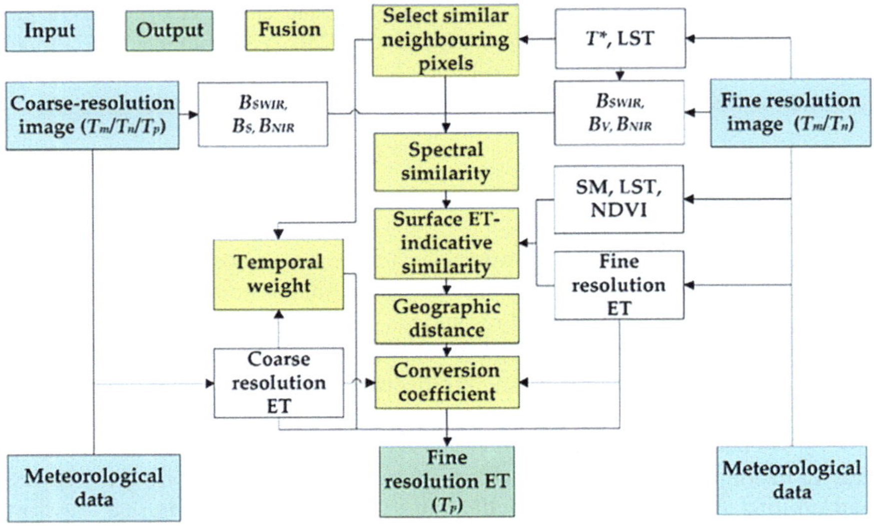

Fig. 2.17 Schematic diagram of the spatiotemporal adaptive data fusion algorithm for evapotranspiration mapping (SADFAET). (Wang et al., 2019)

a DTC model, and an ATC model, where besides the conversion coefficients derived from temperature cycle models, residuals of the weight function were also interpolated using the thin plate spline technique to reduce the biases from land cover type change.

ML methods are also introduced (learning-based fusion). For example, Moosavi et al. (2015) proposed a hybrid wavelet-artificial intelligence fusion approach (WAIFA), which coupled wavelet transformations with ANN to fuse MODIS and Landsat LST data, making a better performance in terms of accuracy. This implied that incorporating machine ML methods could help the model deal with the non-stationary properties of LST and capture the complex relationships between observations from different sensors.

2.4.2.4 Simultaneous Spatiotemporal Downscaling

Spatial and temporal enhancement methods are limited by the temporal and spatial resolution of the original LSTs, respectively. Thus, a combination of these two methods is needed to furthermore improve the resolution of LST (Fig. 2.18).

Two Fundamental Strategies Here are two general strategies for this aim, i.e., "Regression-Then-Fusion" (first spatial and then temporal downscaling (using the fusion method)) and "Fusion-Then-Regression" with a reverse sequence (Xia et al., 2018). Bai et al. (2015) adopted the former strategy. They first enhanced Landsat ETM+ LST from 60 m to 30 m using the extreme learning machine and VNIR bands and then fused them with MODIS LST observations by SADFAT. Zhan et al. (2016)

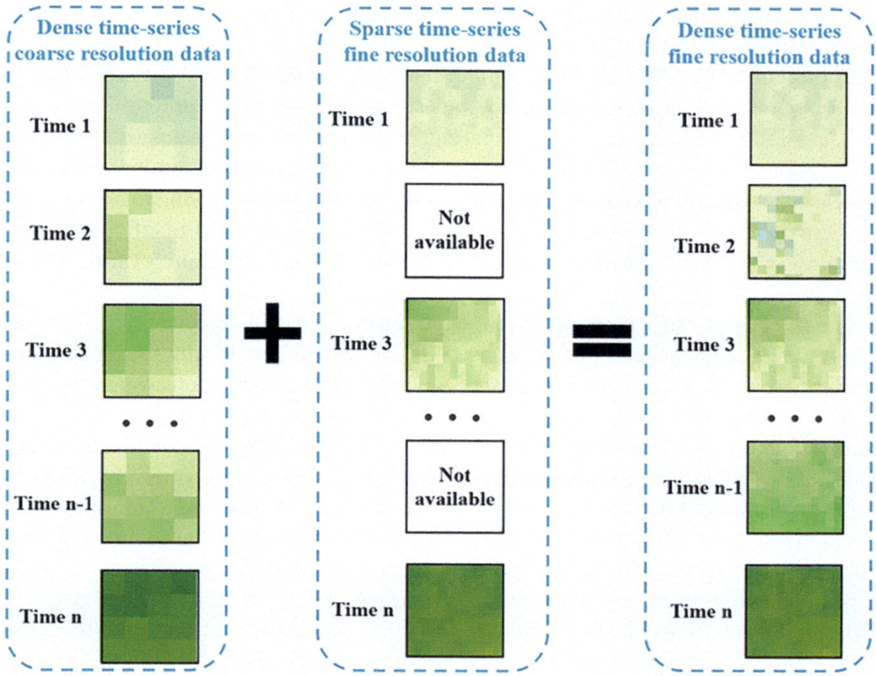

Fig. 2.18 Spatiotemporal downscaling

adopted the latter strategy. They established temperature cycle models at the coarse-resolution scale and then spatially downscaled the parameters with a scaling factor, which was used to produce temporally continuous LSTs at the SR of scaling factors.

Comparison and Combination Xia et al. (2018) compared these two strategies with data. They found that the choice of strategies depended on the performance of spatial downscaling at the start or end time. On this basis, in the following study, Xia et al. proposed the weighted combination of kernel-driven and fusion-based method (CKFM), which takes two kinds of downscaling simultaneously and then combines these two results via weights calculated from error estimation. In the first part, the regression is first realized with Landsat VNIR bands and then fused with MODIS LST. In the second part, STARFM and thin plate spline interpolation are used.

2.4.3 Characterization of Urban Heat Islands

UHI Effect at the Urban Scale At the urban scale, there are a great number of methods used to measure the intensity (Schwarz et al., 2011). They could be classified into three types: difference-based, heat island footprint, and ISA-based. In the first type, the UHI was measured as the temperature difference between urban area and its surrounding rural area, or a specific land use type, e.g., forest, water, and local climate zone (LCZ) nature. Among them, the urban–rural difference in average temperature is mostly used in previous research. However, due to the existence of heat island footprint, i.e., UHI effect outside urban boundary via advection, the magnitude of UHI may be underestimated as the rural area's temperature was increased. When there is a difference in temperature between urban areas and other land cover, a series of criteria will be added, such as the selected pixel, and should be far away from the urban area and water bodies, but with similar elevation.

UHI Effect at the Intra-urban Scale (Distribution) Local-scale methods for portraying the spatial structure of the urban surface thermal environment can be summarized into three categories: (1) landscape-based methods, such as source–sink landscape, thermal landscape, and heat landscape (Meng et al., 2010). This type of method spatially partitions the thermal images based on subsurface landscape, temperature classification, or clustering and analyzes the landscape pattern by measuring the thermal contribution of each class of patches. (2) Methods based on local cold and hot spots: First, spatial segmentation of thermal images based on the spatial similarity of temperature fields and then clarification of the distribution of cold hotspots through local cold hotspot analysis (Guo et al., 2015). (3) Gradient-based approach (Sun, 2018): The variation of surface temperature with distance from the urban center is analyzed by buffer zone circles or profile lines. Moreover, in urban climatology, the urban canopy thermal environment is usually expressed as a series of isotherms by analogy with topography, which is contained layer by layer between them, forming a composite spatial structure containing different layers of heat islands and heat centers.

2.4.4 Analysis of Impacting Factors

The influencing factor (Fig. 2.19) analysis process is generally designed to quantitatively assess the relationship between cooling effect measures and their influencing factors and to identify the dominant factors among them. Although in some parametric simulation studies, the effects of each influencing factor can be obtained directly by varying individual model parameters (Ryu & Baik, 2012), for observational studies, the effects of influence factors still need to be assessed by statistical methods. Current studies still mainly use traditional statistical methods, such as correlation analysis and multiple regression analysis; conversely, machine learning methods are gradually being applied (Gage & Cooper, 2017). Machine learning methods are more advantageous in portraying the nonlinear responses of dependent variables to explanatory variables and interactions among explanatory variables than the additive and linear effects assumptions embedded in multiple regression analysis. However, machine learning methods are still inferior to multiple regression methods in their ability to explain and attribute response relationships.

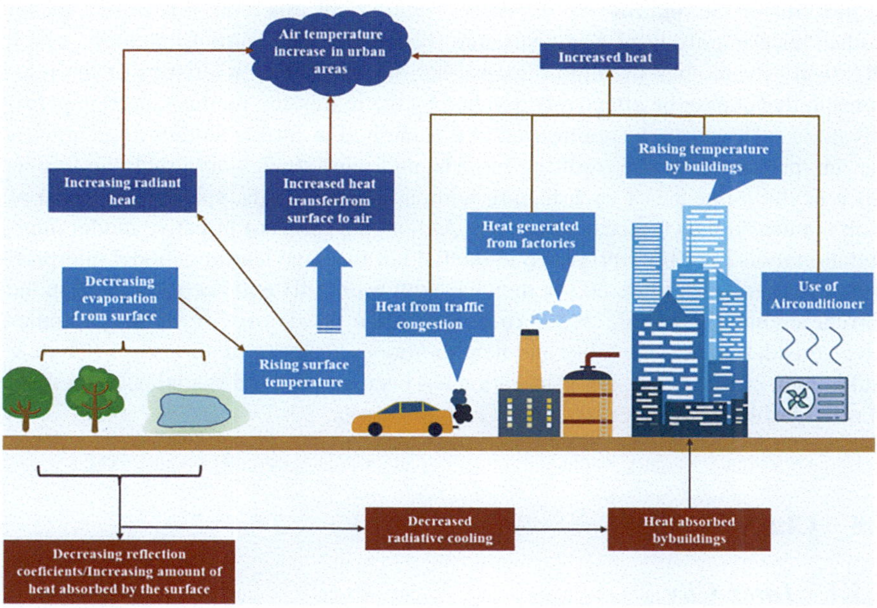

Fig. 2.19 The causes of the urban heat island effect. (Anjum et al., 2019)

Meanwhile, the machine learning method requires a high sample size, and many site-scale studies can provide only tens of observed sample points; thus, the method is more suitable for remote sensing data studies that can easily provide a large number of samples.

In addition, to identify the key factors influencing the cooling effect of UHI, previous studies have mostly quantified the relative importance of factors by different methods. Among them, the standardized coefficient method based on the linear regression model is still the main method to examine the relative importance, while the hierarchical decomposition method, random forest, machine learning method, and geographic detector method are also gradually applied. In addition to dominant factor identification, nonlinear features such as threshold effects and moderating effects in relationships have also received attention from researchers. The threshold effect refers to the significant change in the response of the dependent variable before and after a certain threshold of the independent variable. Ziter et al. (2019) found that the maximum decrease in temperature due to an increase in the percentage of vegetation cover was reached when the percentage of vegetation cover was greater than 40. The moderating effect, however, refers to the fact that the moderating variable can influence the strength and direction of the relationship between another explanatory variable and the target variable. Although the effects of moderating variables as categorical variables on the SUHI or LST such as climate zone type (Yu et al., 2018) and season (Chen et al., 2014) are of interest to researchers.

In terms of the *relationship attribution analysis* of this type of research, recent studies have mainly used a tree-structure-based machine learning method, which effectively overcomes the high-dimensional, nonlinear characteristics of the relationship that cannot be effectively handled by the regression methods used in previous studies. However, the application of this method in current studies often ignores the interpretability of the resulting models, and most studies only rank the importance of the influence of each feature without examining the specific influence of each feature on the surface temperature. Although a series of effective model interpretation tools have been proposed in the field of machine learning model interpretability research (Abusitta, 2024), they have not been fully and deeply applied in the current studies (Sun et al., 2020; Yu et al., 2020). In addition, few current studies have considered the scalability and neighborhood effects of relationships (Chun & Guldmann, 2014), and less consideration has been given to the spatial heterogeneity of relationships (Berger et al., 2017; Gage & Cooper, 2017).

2.5 Characteristics

2.5.1 Inter-City

2.5.1.1 Global Scale

Spatial Variation On a global scale, the SUHI phenomenon was observed widely by many studies. For example, as reported by Chakraborty and Lee (2019), 87% and 93% of the urban clusters showed positive daytime and nighttime SUHI, respectively, during 2000–2019. However, the magnitude is not consistent among different studies, given the difference in the observed period and the delineation of urban and suburban extents, but they all observed that the SUHII of cities presented significant variations along the latitude or across the biomes/climate zones. For example, Zhang et al. (2010) found an average of 3.8 °C UHI in cities built in biomes dominated by forests; 1.9 °C UHI in cities embedded in grass-shrubs biomes; and only a weak UHI or sometimes an urban heat sink (UHS) in cities in arid and semi-arid biomes. Chakraborty and Lee (2019) found that there are pronounced positive surface UHI intensities around the equator, at 20° south, and between 40° and 60° north, while the nighttime UHI shows lesser latitudinal variability: There is a pronounced nighttime UHI at 30° north and around 10° south. They also found that for daytime, the highest surface UHI is for the equatorial urban clusters, followed by snow, warm temperate, and arid zone, especially, SUHI in the last zone has nearly zero value. In contrast, during nighttime, arid urban clusters have the highest SUHII, while urban clusters in other climate zones have a similar SUHI value.

Surface Heat Sink Moreover, the significant negative SUHII, i.e., the surface heat sink, was also observed (Clinton & Gong, 2013). In contrast to the general expectation, the heat sink was more observed in the daytime rather than the nighttime with a less significant SUHI. As reported by Clinton and Gong (2013), at the pixel scale (size of 1 km), the heat sink during daytime has greater maximum intensity and more persistent exposure than the SUHI during nighttime. Moreover, the phenomena were more significant in snow and deciduous needleleaf forest biomes in Central Asia. In terms of overall exposure on cooling degree days to SUHI and SUHS, "barren," "grassland," "open shrub," "closed shrub," and "savanna" experience the largest average values of cumulative heat. This study also found that both islands and sinks are possible in the same location, though the timing may differ.

Diurnal Dynamics In addition to the average magnitude, the diurnal and seasonal patterns of SUHII were also a concern for many studies (Fig. 2.20). In these studies, a general diurnal pattern of SUHI was observed, which was analogous to that of LST, i.e., with the increase and decrease in solar altitude, the SUHI first strengthens and then declines during daytime, while with the urban-rural difference in the cooling rate, the SUHI slowly declines during nighttime. However, a distinct pattern was also observed in the arid zone, i.e., the highest surface UHI intensity is at 2230 LT and the lowest is at 1030 LT, which may be attributed to the low thermal admittance, low roughness, or the low evapotranspiration rate of the surface around the urban cluster. Furthermore, the diurnal range of SUHI was also considered, and Chakraborty and Lee (2019) found that it was the highest for the equatorial urban clusters (0.88 °C), followed by 0.79 °C for snow, 0.56 °C for warm temperate, and 0.53 °C for arid.

Seasonal Variation SUHI represented significant seasonal dynamics which were also controlled by the variation of solar radiation (zenith and sunshine duration) as in the diurnal case. In general, the daytime surface UHI shows higher values during

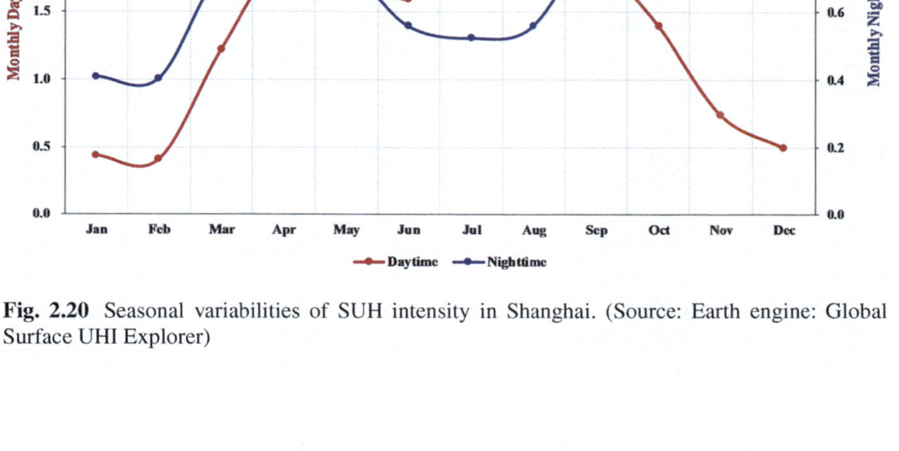

Fig. 2.20 Seasonal variabilities of SUH intensity in Shanghai. (Source: Earth engine: Global Surface UHI Explorer)

the summer, with the highest intensities in July and the lowest values in December. In terms of the climate zone, as reported by Chakraborty and Lee (2019), the warm temperate and snow urban clusters show comparable patterns with low values during the boreal winter and high values during the boreal summer. The daytime surface UHI in the equatorial climate zone shows the opposite pattern to the warm temperate and snow climate zones, with the highest values during boreal autumn and the lowest values during boreal spring. The seasonality of the global nighttime surface UHI is akin to the daytime pattern with June highs and November lows. The nighttime surface UHI of the warm temperate climate zone shows high values during boreal summer (July) and low values in boreal autumn. The arid zone again shows two peaks during the year: One in March and the other in October. The minimum surface UHI in the arid urban clusters is in July. The equatorial zone shows the highest surface UHI intensity in boreal winter (January) and the minimum in boreal summer. For the snow climate zone, urban clusters also show two peaks during the year: One in February and the other in July.

2.5.1.2 Continental Scale

Spatial Variation of SUHI in Europe At the continental scale, the SUHI presents more detailed variation. For example, in Europe with diverse local climate types and numerous cities, many studies have examined the spatiotemporal characteristics of European cities' SUHII and found the SUHI varied considerably due to the difference in latitude, distance to coast, background climate and biome type, altitude, and topography. Ward et al. (2016) investigated the SUHI of 70 cities during both the heat wave (HW) and non-heat wave summer periods, and they found that the SUHI varied among cities during both HW and non-HW periods and were stratified by urban size and climate types. Moreover, they found that compared to the non-HW period, the prevalent HW event in July 2006 generally increased the SUHI, while the difference also varied considerably with urban size and climate type, especially, the difference was more significant in the cold climate than in hot climate. Nastran et al. (2019) also considered cities with a larger and evener distribution (403 cities) and a longer period (10 years) to examine the SUHI in Europe and its response to different greening planning families (PF). They found although the SUHII was generally higher in the northern cities, the longitude had a stronger influence on SUHI than the latitude: SUHI increased with longitude, especially for cities in oceanic planning families.

Temporal Variation of SUHI in Europe Compared to these two previous studies, Schwarz et al. (2011) additionally considered the diurnal variation of SUHI. Moreover, they considered multiple SUHI indicators, such as the urban–rural difference, urban–other types difference, urban–other single land cover, and SUHI footprint parameter. They found a significant and generally consistent diurnal pattern for different SUHI indicators, i.e., following four-time points (01:30, 10:30, 13:30, 22:30), the indicator value first increases then decreases (maximum in 13:30 and minimum in 01:30), which is similar to what Chakraborty and Lee (2019) found at the global scale. It is surprising that only a few studies concerned the temporal evolution of SUHI for a long-term period. For example, Benas et al. (2017) examined the trend of urban LST and SUHI in the 17 large Mediterranean cities during 2001–2012. They found a positive trend for both LST and SUHI for most of the cities. Furthermore, a 0.1 °C decade^{-1} increase in urban LST corresponded to an increase in SUHI by about 0.04 °C decade^{-1}. Parida et al. (2021) looked at the change of SUHI in more recent times to examine the variation of SUHI between the COVID-19 lockdown period and the previous usual period. They found a larger decrease in nighttime LST was also seen in urban areas (by ~1 to 2 °C) compared to rural landscapes.

Africa Compared to studies for European cities, the counterpart in undeveloped Africa and Latin American continents was less documented, especially. To the best of our knowledge, only two recent studies, i.e., Li et al. (2021) and Simwanda et al. (2019) investigated the spatiotemporal characteristics of SUHI in Africa at a large scale, although the concerned cities were still scarce. Li et al. (2021) found that for five most populated cities in East Africa (namely Khartoum, Addis Ababa, Nairobi, Kampala, and Dar es Salaam) during 2003–2017, mean annual daytime SUHI intensities were higher than the nighttime counterpart in most cities. Moreover, Kampala in Uganda with humid climate has the most significant day and night time SUHI (2.21 °C and 0.82 °C in 2017) and Khartoum in Sudan with arid zones even had the negative SUHII value. They also found that from 2003 to 2017, the annual SUHII generally increased.

Latin America We also found only a few studies concerning the SUHI in Latin American cities at a large scale. To the best of our knowledge, only one study, i.e., Wu et al. (2019), addressed this issue. This study investigated the UHI of 44 cities in South America with MODIS LST during 2003–2016, which were mainly located near the sea and had a population in 2014. Wu et al. (2019) found that the mean daytime SUHII was higher than the mean nighttime SUHII in all areas except the arid climate zone. In the daytime, cities in the northwestern (Columbia) and south-eastern (Brazil) parts present larger SUHII values, especially in summer reaching 6 °C. Moreover, in the warm temperate climate zone, the daytime SUHII is larger in summer than in the other seasons. In contrast, in the arid zone, the daytime SUHII was negative across for seasons, especially in spring with an average SUHI reaching −2.84 °C. The nighttime SUHII showed less obvious seasonal variations. Considering the temporal evolution of SUHII, most cities showed an insignificant temporal trend in the SUHII from 2003 to 2016.

2.5.1.3 National Scale

China There are a great number of studies concerning the spatiotemporal characteristics of SUHI in China. For example, Zhou et al. (2014a) investigated the diurnal and seasonal dynamics of 32 major cities in China during 2003–2011. This study found that annual mean SUHII varied from 0.01 to 1.87 °C in the day and 0.35 to

1.95 °C at night. Diurnally, higher SUHIIs for the day and night were observed in the southeastern and northern regions, respectively. Particularly, Northeast China, with a humid-cold climate, exhibited the greatest SUHII in both day (1.43 °C) and night (1.90 °C). Seasonally, they found a higher SUHI in summer than in winter during the day, and the opposite during the night for most cities. Temporally, Yao et al. (2017; 2018) investigated the SUHI in 31 Chinese major cities during 2001–2015, although the former considers three types of indicators of SUHI, including SUHI intensity, area of SUHI, and percentage of area with increasing SUHII (PAISUHII). Wang et al. (2015a) considered a larger number of major cities, and they found that some authors also shifted focus from major cities to prefecture-level cities. They found that spatially, southern China experienced more intense daytime SUHIIs, while the opposite was true for nighttime SUHIIs. Nighttime SUHIIs in northern China were greater than those in southern China.

United States Compared to China, there are only a small number of studies investigating the SUHI phenomena at a national scale (Chakraborty et al., 2020; Imhoff et al., 2010; Li et al., 2017). Among these studies, Imhoff et al. (2010) explored the SUHI of the 38 most populous cities during 2003–2005 across eight biomes. This study found that the annual SUHI effect was significant with an average SUHII of 2.9 °C, except for urban areas in biomes with arid and semi-arid climates. Among biomes, cities with the largest (8 °C on average) were built in biomes dominated by temperate broadleaf and mixed forests. Chakraborty investigated the inter-urban and intra-urban variability of SUHI in 497 urbanized areas composed of 55,871 census tracts during 2013–2017. They found that, generally, the annual average SUHI intensity is 1.38 ± 0.66 °C during daytime and 0.40 ± 0.28 °C for nighttime. Seasonally, daytime SUHII was the largest during summer (1.91 ± 0.97 °C) and the lowest during winter for most cities (418 and 439 out of the 497 cities). Li et al. (2017) considered a substantially larger sample, including approximately 5000 urban clusters larger than 10 km^2. They found that SUHI significantly varies across cities with a range of approximately −5.0 °C to 7.9 °C in 2010. Temporally, they observed a prevalent pattern that SUHII was highest in summer and lowest in winter and lowest in midnight and highest in early afternoon. Spatially, the Northern cities have higher SUHI than the urban areas in Southern regions.

India Akin to the United States, there are also only scarce studies about the UHI in India at national scale (Kumar & Mishra, 2019; Kumar et al., 2017; Mohammad & Goswami, 2021; Raj et al., 2020; Shastri et al., 2017). For example, Raj et al. (2020) investigated the SUHI of 44 major cities in 2000–2017. They found that most cities exhibited SUHI effect at an annual scale with SUHII up to 2 °C. Temporally, although statistically insignificant, most cities exhibit a positive trend in daytime SUHII during monsoon and post-monsoon periods, but a negative trend during the winter and summer seasons, while nighttime SUHII was increased in all seasons for most cities. Mohammad and Goswami further explored variations of SUHI across climate zones with 150 major cities during 2003–2018. They observed that urban cool islands occurred over the hot desert, hot steppe, and tropical monsoon climatic zone during daytime in both summer (−0.25 to −0.17 °C) and winter (−0.33 to 0.17 °C) seasons. During nighttime, SUHII was positive irrespective of climatic region and seasons, with a range of 0.48–1 °C in summer and 0.46–1.32 °C in winter. Both Shastri et al. (2017) and Kumar et al. (2017) noticed the unexpectedly prevalent daytime urban cool island effect during the pre-monsoon summer, even during the heat wave period (March–May, the hottest period in India), while the nighttime UHI effect is almost all significant, due to the lack of vegetation and moisture in non-urban areas relative to cities.

2.5.2 *Intra-city*

In terms of thermal landscape, Meng et al. (2010) found that in Beijing, the patch density in the main urban area was larger than that of the suburban area, while the pattern reversed for the mean patch size, indicating a more continuous thermal landscape pattern in the suburban area. Moreover, most landscape metrics were sensitive to the grid size used to aggregate surface temperature. *In terms of gradient perspective*, the spatial structure of UTE in the Shanghai metropolitan region showed an obvious trend of spatial aggregation phenomenon. From the urban center to surrounding areas, the mean LST was gradually decreasing in all directions, from coast to inland, the extent of high-temperature zones in daytime is generally larger than that in nighttime, and the former is about 1.5 times the latter. *In terms of landscape-differentiation methods*, recently, local climate zones proposed by Stewart and Oke (2012) have been gradually used to investigate the intra-urban variation of UTE. Bechtel et al. (2019) investigated the intra-urban SUHI pattern based LCZ via the MODIS and Landsat LST data and LCZs mapping deriving from WUDAPT L0 data in 50 global cities. They found that compact types (LCZ 1–3) and commercial/industrial types (LCZ 8 and LCZ 10) generally have stronger SUHI intensities than the open types (LCZ 4–5 and especially LCZ 6), while LCZ 9 (scattered buildings) has a considerably lower SUHI intensity than all other built classes.

2.6 Drivers at Multiscales

Here, we are mainly concerned with two topics: The drivers of spatial pattern of temperature within cities, i.e., UTE, and the urban–rural contrast of temperature, i.e., UHI intensity (UHII). It is well known that the temporal thermal behavior (e.g., cooling and heating rates) of the surface and canopy layers of land, regardless of whether the land is of urban or rural type, depends on the surface energy budget (SEB) that occurs on it (Chen et al., 2020). Therefore, the difference in SEB of different lands exactly shapes the variance of thermal behavior and thus the spatial heterogeneity of temperature, which should focus on the analysis of drivers of UTE and UHI (Fig. 2.21).

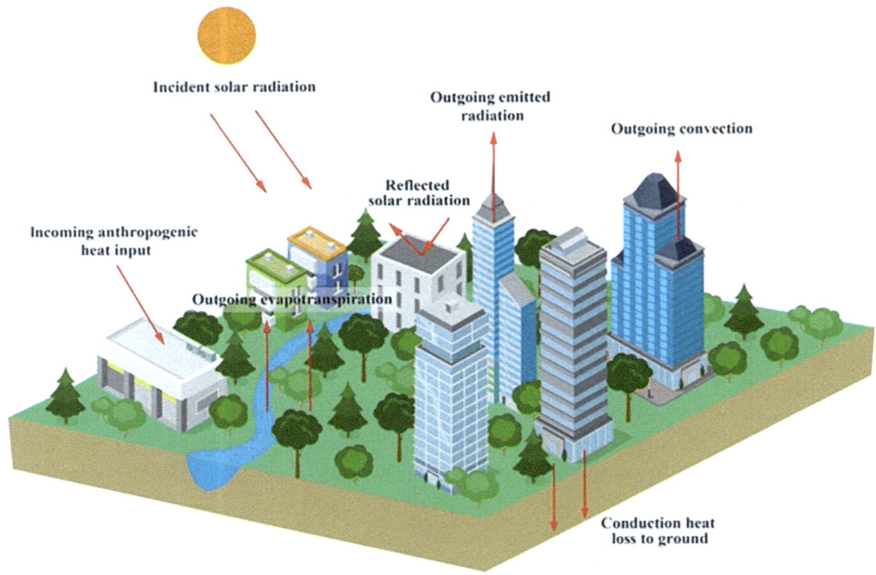

Fig. 2.21 Surface energy budget in urban areas

From *the perspective of SEB*, the drivers of UTE and UHI should be related to the SEB terms, namely the radiation-relevant terms (upward and downward short- and longwave radiation), turbulent terms (sensible, latent, and advective heat flux), anthropogenic heat flux, and heat storage flux (also called ground heat flux in some research). The wind speed and direction influence all three turbulent heat fluxes. Based on surface properties and weather conditions, researchers also investigated the drivers at the higher level, such as land cover and use, socioeconomic factors, and urban form, which have a relatively indirect link to UHI and UTE. In terms of UHII, the previous research always concerns the effects of urban–rural contrast in surface properties, weather conditions, macroscale urban form, and socioeconomic factors, while in terms of UTE, the previous research mainly concerns the effect of landscape pattern and three-dimensional local urban form as well as human activity.

From the above *comparison*, we can observe a gap of scale in the drivers. For the macroscale UHII, its drivers are also macroscale, while for the UTE, its drivers are mainly local-scale. Thus, in addition to the dichotomy of UHI or UTE, the driver could be also grouped in terms of spatiotemporal scales. In terms of spatial scale, the drivers could be classified as meso-(urban), local, and microscale, while in terms of temporal scales, the drivers could be grouped into inter-annual, seasonal, monthly, day-to-day, and diurnal scales. Here, we mainly used the spatiotemporal scale framework to elaborate the drivers of UTE and UHI. In each scale, the importance of factors was sorted by the level at which they linked to SEB, while difference in temporal scale will be also emphasized.

2.6.1 Synoptic and Meso-scales

In these scales, the drivers of UHII at urban scale could be grouped into temporal factors (sunshine duration and sun elevation), background macro-climate, meteorological factors and types, urban–rural contrast in surface properties, anthropogenic heat emission, urban form feathering by the size, boundary characteristics (overall shape and complexity (fractal dimension) and polycentricity), density and arrangement of built-up area, landscape composition and fragmentation of land cover and use, building height, and socioeconomic factors. Among these drivers, the background macro-climate and meteorological factors generally could be viewed at regional and synoptic scales, while the others aggregated from urban and rural areas could be viewed as mesoscale (Fig. 2.22a).

U-R contrasts in surface properties have a primary control on the SEB and UHI. The concerned surface properties in previous research include the thermal ones (thermal capacity and thermal conduction), the radiative ones (albedo and emissivity), and the turbulent-exchange ones (roughness, canopy conductance, and moisture availability). Note that the moisture also influences the thermal and radiative properties. The plant coverage or vegetation indices have been also seen as a surface property in some research, given its relationship with thermal, radiative, and turbulent properties. Thermal capacity and thermal conduction influence the storage heat flux and the changing rate of temperature. Generally, the urban area has a

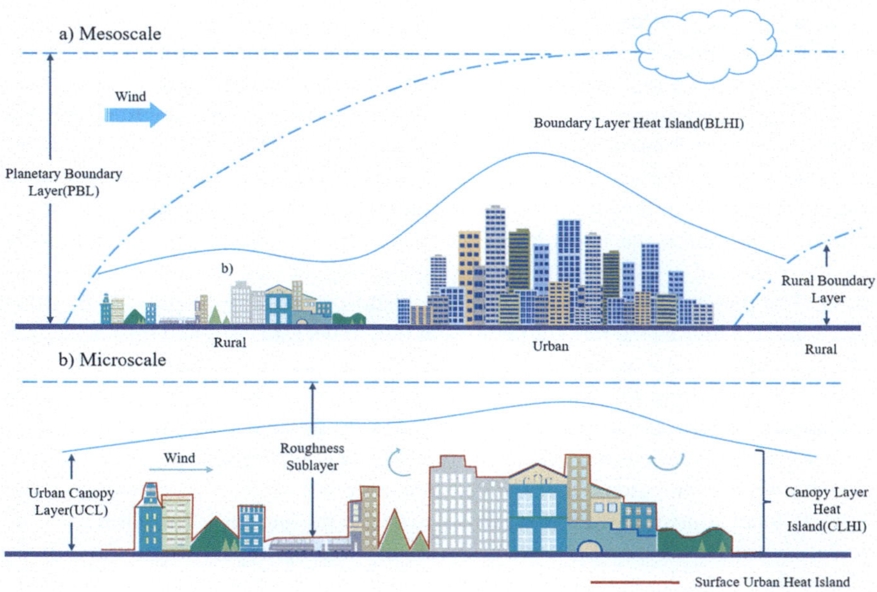

Fig. 2.22 The main components of the urban atmosphere. (**a**) Mesoscale structure showing the Planetary Boundary Layer (PBL) and Boundary Layer Heat Island (BLHI); (**b**) Microscale structure illustrating the Urban Canopy Layer (UCL), Roughness Sublayer, and Canopy Layer Heat Island (CLHI), with the Surface Urban Heat Island indicated by a red line. (Redraw from Oke, 1997)

greater thermal capacity and conduction, and thus greater heat storage flux, which has a detrimental effect on the U-R contrast in the cooling effect during nighttime. Albedo and emissivity affect the short- and longwave radiation budget, respectively, and thus the net radiation flux at the surface. The turbulent fluxes depend on the contrast in the vertical air-surface status (temperature and moisture) or horizontal gradient (temperature and moisture) and resistance to turbulent exchange. The first and third factors are partially determined by the canopy conductance (moisture availability) and roughness.

Metrological and synoptic weather factors investigated in the previous study could be divided into quantitative and qualitative factors. The former factors include known near-surface meteorological factors, such as wind speed and direction, cloud coverage (cloudiness) and insolation, ambient humidity, and precipitation (PBL height), while the latter factors include multiple kinds of factors at the mesoscale and synoptic scale, such as the mesoscale land-sea breeze UHI circulation and synoptic-scale weather types/pattern as well as the large-scale weather system. In addition, some quantitative factors rely on a few meteorological elements, such as sky condition, atmospheric stability, geostrophic wind, or the passage of cold/warm front. Generally, the quantitative factors influence or reflect the quantitative factors, i.e., the former's effects on SEB and UHI rely on the latter.

For the *quantitative factors*, previous research found that the increase in the wind speed will facilitate the advection and reduce the UHI, but some studies also found that when the U-R contrast in available energy (Li and Bou-Zeid, 2013) is negative during daytime, the increasing wind speed favors the increase in the sensible heat flux in urban area instead of advection, and thus increases the CUHI. The wind direction determines the wind source and the temperature of advection air mass, thus different wind directions may have a contrast effect in UHI. The cloud coverage and insolation determined the insolation arriving surface during daytime and the loss of energy via longwave radiation and thus influenced the net radiation, especially due to the difference in the surface properties. The urban and rural areas have different responses to the change in meteorological factors. The increase in relative humidity may inhibit the latent heat flux in the moist conditions or promote it by decreasing the soil moisture depletion and facilitating the dew formation in dry conditions. The precipitation will increase the moisture availability, especially in urban areas, which increases the ratio of latent heat flux to net radiation and thus decreases UHI, in addition to the high cloud coverage associated with precipitation.

For the *qualitative factors*, we first look at the weather type/pattern. From a circulation-to-environment perspective, it could be divided into two classes, i.e., weather typing classifications (WTs) (also called air mass classifications) and circulation pattern classifications (CPCs) (Kassomenos et al., 2022). The former used the relatively local atmospheric characteristics to define the air mass. The latter used the circulation pattern on a larger scale to define the weather types, where the core is, and in addition the direction of the airflow and the location of the study area to the center of the high/low are also important. Generally, for the air mass classification, the dry and warm air mass are generally conducive to the development of UHI, especially for the nocturnal one (Khan et al., 2021; Hardin et al., 2018), while the moist air mass is disadvantageous. Khan et al. (2021) even found that the humid air mass during daytime is in favor of UCI effect. The reason could be that nocturnal dry air mass during nighttime is related to low relative humidity and low cloud coverage, which facilitate the rapid radiative cooling in rural areas, while the daytime humid air mass decreases the insolation and inhibits the evapotranspiration cooling in urban areas in some cases (Khan et al., 2021).

2.6.2 *Mesoscale Urban Form*

Types of Mesoscale Urban Form In previous research, the examined macroscale urban form could be generally divided into two types, i.e., boundary geometries, such as urban area and boundary compactness (Liang et al., 2020b; Liu et al., 2021), and intra-urban landscape patterns of built-up areas and other land covers (Yue et al., 2019), while the first type of urban form could be also viewed as the landscape metrics at patch level to a large extent, which obtained more concerns than the second type at class and landscape level. In addition, some distinct aspects of urban form were also considered in studies, such as the sprawling/compact development modes (Yang et al., 2016), polycentricity (Debbage & Shepherd, 2015), as well as spatial interactions between cells (Li et al., 2020) or the centrality of cells (Lu et al., 2022).

Boundary Geometry The boundary geometry factors examined in previous studies include urban area, shape compactness, shape complexity (fractal dimension or shape index), contiguity, elongation, and shape isometry (Liang et al., 2020a, b; Zhou et al., 2017). Among these factors, the urban area was considered in almost all studies and it was observed to have a positive effect on UHII, except Li et al. (2020) found a negative effect in urban areas based on a parameterized numerical simulation experiment in Berlin. From a causal perspective, the urban area itself, however, has an indirect effect on urban–rural difference in surface properties and SEB terms, rather than it associates with urban scale (and thus socioeconomic factors) and other urban form factors (e.g., shape complexity), as Li et al. (2017) stated, "urban area size is not a major direct driving factor of SUHI, but a useful surrogate of those factors." Meanwhile, other boundary geometry factors were also found to have either a positive or negative effect in different studies or at different times. For example, Liu et al. (2021) found that the shape complexity (characterized by fractal dimension) and shape isometry (characterized by Boyce-Clark shape index) have a negative effect on SUHII based on 266 Chinese cities, while Zhou et al. (2017) found contrasting results for these two factors based on over 5000 European cities.

Mechanisms The inconsistent effect sign and moderating effect of urban development gradient and climate zone on urban form factors can be attributed to the multiple pathways through which boundary geometry influences SUHII or average LST. Generally, boundary-relevant urban form is related to intra-urban landscape patterns, energy consumption, air pollutant emission, and ventilation. In a larger picture, the boundary-relevant urban form is related to the urban development modes (sprawling/compact urbanization) and socioeconomic factors (urban scale). For example, Liu et al. (2021) explained the effect of compactness on mean LST as its influence on or association with urban density, urban ventilation, and transport demand, since higher boundary shape compactness is associated with compact development, implying higher urban density (although it means higher SUHII) and thus greater resistance for ventilation to reach urban cores, while the opportunity for contact with the cooler rural surface was also reduced under a more compact shape. In addition, the compact development mode is related to less traffic demand, less congestion, and less building energy consumption (infrastructure), which is related to lower emissions of air pollutants and waste heat. Here, only the impedance to ventilation is conducive to UHI, while other related changes will help mitigate UHI.

2.6.3 Neighborhood Scale

Two-Dimensional Factors This study focuses on the local influences that control the internal spatial differentiation of the urban surface thermal environment. As mentioned above, the local influencing factors include four main categories of land cover and land use, urban location characteristics, landscape pattern indicators, and three-dimensional building and vegetation morphological characteristics. Land use and land cover are considered to be the most dominant influencing factors and have also been most extensively studied (Deilami et al., 2018). Among land covers, mainly impervious surfaces, vegetation, bare ground, and water bodies have a controlling effect on thermal environmental partitioning. Impervious surfaces reduce water evaporation from urban surfaces and absorb and store more solar radiation and heat, thus having a warming effect (Zhou et al., 2014b). Impervious surfaces are considered to have the most significant effect on LST, and Imhoff et al. (Imhoff et al., 2010) even found that they explain up to 70% of the LST variance. The relationship between vegetation and LST is slightly weaker than that of impervious surfaces, which can reduce their own temperature and that of the surrounding area through shading effects, evapotranspiration, and lower thermal storage.

Three-Dimensional Urban Morphology Due to the growing conflict between the dramatic increase in urban population and the shortage of building land, the rapid expansion of cities in the two-dimensional direction has been accompanied by the extension of urban building stock in height (Gong et al., 2015). The three-dimensional structure and morphology of cities (building scale, block height to width ratio, etc.) become important factors influencing urban micrometeorology (Oke, 1982). However, due to the difficulties in obtaining overall urban 3D height data, scholars have only begun to examine the effects of urban 3D building and vegetation morphological features on the surface thermal environment at the overall urban scale in the last decade, whereas previously, studies considering 3D factors were limited to urban localities and often utilized only sky view factor (SVF), which reflects building shadow information and has not been considered for other 2D features. Among the currently considered 3D factors, in addition to SVF, factors characterizing urban geometry such as average building height and standard deviation, building vegetation height difference, and building exterior surface area, volume, surface roughness, volume ratio, and complete surface ratio (complete aspect ratio) have been included in the relationship study (Alavipanah et al., 2018).

2.6.4 Microscale

Geometric properties include orientation (slope angle and azimuth) and openness to sun and sky, which influence the net radiation. By day, the highest T_0 occurs on facets that maximize the local irradiance (incident solar radiation). *Radiative properties* include albedo α and emissivity ε, which influence the net radiation. If a surface is warmer than its surroundings and it has high ε, it will radiate heat more effectively which promotes its cooling. *Thermal properties* include thermal conductivity and heat capacity, which directly influence the heat storage flux and indirectly influence other terms. The highest T_0 by day occurs on facets made of materials with low heat conductivity and capacity, which resist heat transfer into the substrate and instead concentrate it in a thin surface layer that gets very hot. At night, these properties make the surface cool rapidly. *Moisture properties* determine the water available for evaporation and thus the latent heat flux. The highest T_0 is found generally over dry facets that do not channel much available energy into evaporating water. In contrast, surfaces with access to water experience lower T_0 by day due to evaporative cooling and a lower diurnal temperature range.

2.7 Effects

The high temperatures caused by the UHI have various impacts on the environment and quality of life in communities (Fig. 2.23). UHI drives higher demand for air conditioning to cool buildings. Case studies from various countries and regions reveal that as temperatures rise, electricity usage for air conditioning increases significantly, contributing to overall energy consumption and peak power demand. Reliance on fossil fuel power generation to meet these needs also leads to increased air pollution and greenhouse gas emissions. These pollutants are harmful to human health and contribute to complex air quality issues such as ground-level ozone formation (smog), fine particulate matter, and acid rain, posing risks to human wellbeing. Furthermore, UHI leads to higher daytime temperatures and reduced nighttime cooling, resulting in heat-related deaths and illnesses such as general discomfort, respiratory difficulties, heat cramps, heat exhaustion, and non-fatal heatstroke. Water temperature affects various aspects of aquatic organisms, particularly

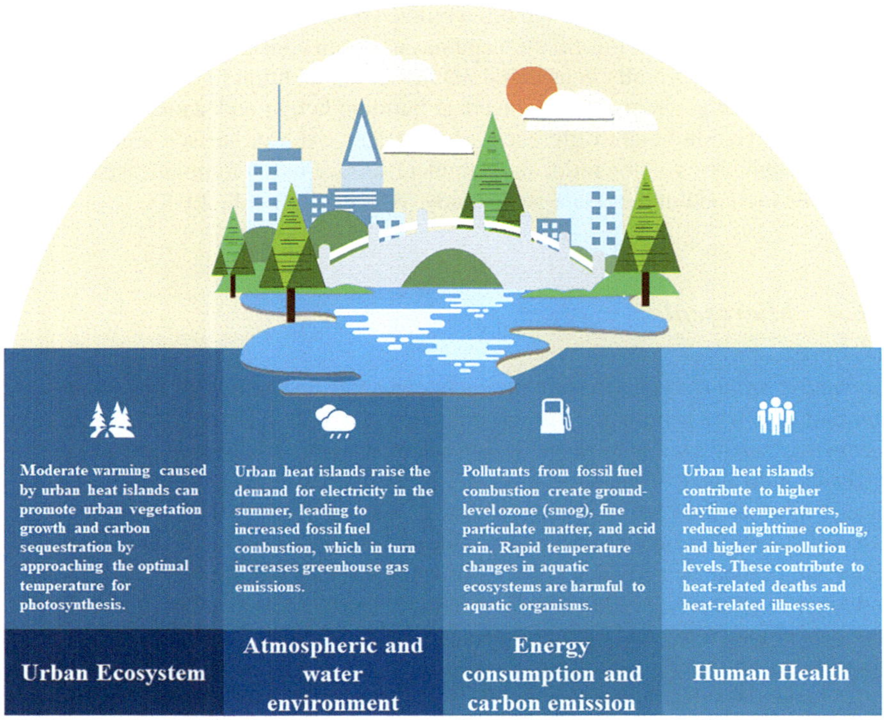

Fig. 2.23 Effects of urban heat island

the metabolism and reproduction of many aquatic species. UHI causes rainwater runoff to become warmer, leading to rapid temperature fluctuations in aquatic ecosystems that exert pressure on aquatic life, even posing lethal risks.

2.7.1 Urban Ecosystems

2.7.1.1 Urban Warming (Long-Term Effects)

Phenology UHI has an impact on the phenology, especially the start and end (the timing of bud break and leaf fall or discoloration) of the growing season (Fig. 2.24). Samuel et al. (2016) investigated the impacts of satellite data in the city of Madison during 2012–2014. They found that the growing season length (GSL) is about 5d longer than surrounding rural areas. However, parks within urban areas experience a subdued expression of GSL lengthening resulting from interactions between the UHI and a park cool island effect. In addition, along the urban–rural gradient, there is a very strong linear relationship between the LSP and the LST along the urban–rural gradient, especially within 6 km of the urban edge. Ren et al. (2018) also found

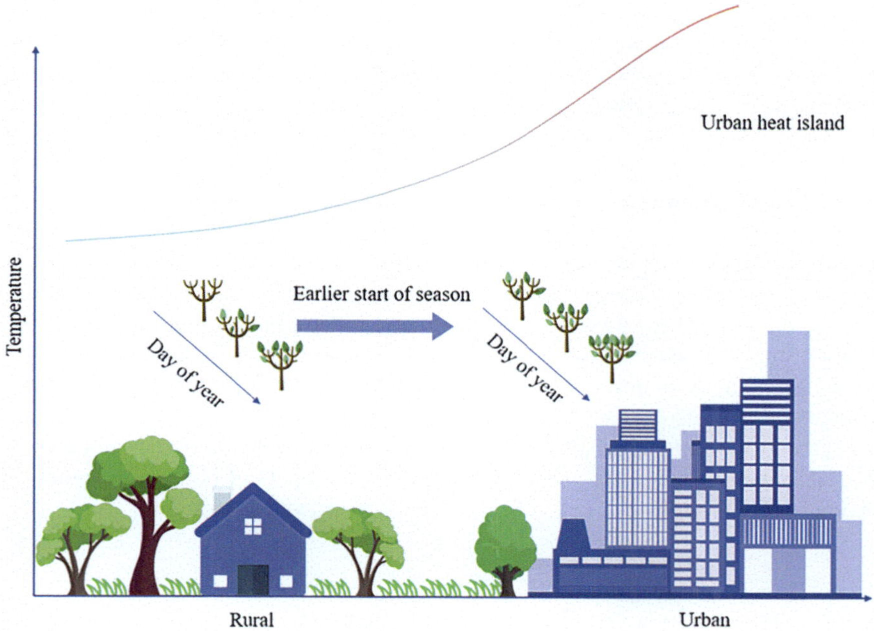

Fig. 2.24 Effect of UHI on phenology

that SOS was negatively correlated with land surface temperature (LST), with a correlation coefficient of −0.24 ($P < 0.01$), and EOS and GSL were positively correlated with LST, with correlation coefficients of 0.56 and 0.44 ($P < 0.01$). However, some studies found contrasting results. For example, Kabano et al. (2021) found that in Kampala, Uganda (a city with tropical climate), growing season length increases along the urban–rural gradient, and vegetation in the most built-up LCZ class had the shortest growing seasons.

Growth The effect of UHI on growth of urban vegetation has been demonstrated and most studies found that the moderate warming induced by UHI could promote the growth and carbon sequestration of urban vegetation via approaching the optimal temperature for photosynthesis, while extreme warming will have a reverse effect. For example, Wei et al. (2021) found that the indirect influences of urbanization on annual net primary productivity (NPP) are related to temperature. Hardiman et al. (2017) used vegetation photosynthesis and respiration model to investigate the urban biogenic fluxes and the consequences of UHI in Petersham, Worcester, and Boston, USA. They found that although gross ecosystem exchange in Boston was enhanced in the spring and fall, it was offset by the reduction in summer associated with heat stress. Moreover, the ecosystem respiration was intensified throughout the year, and thus, the net influence of UHI reduced Boston's estimated annual storage by 22.8%. It also suggests that the effects of additional warming depend on the background temperature, as there is an optimal temperature for each species (Pau et al., 2018). Some studies even found negative effects of UHI across whole growing season in some arid regions, such as Raleigh, USA (Meineke et al., 2016) and Melbourne, Australia (Nitschke et al., 2017).

2.7.1.2 High Temperature (Short-Term)

High temperature in short-term heat wave periods has a significant negative effect on vegetation's functioning, growth, and survival (mortality). In the urban context, it interacts with other stresses, such as air and soil pollutants, and especially the drought stress, which usually occurs with heat waves and is enlarged by heat waves, given the increased atmospheric moisture demand under higher temperature. Conversely, we also notice that management measures such as irrigation could strengthen the ability of urban vegetation to avoid heat stress via transpiration cooling. Here we review recent progress and group the effects of high temperature on vegetation into four levels, i.e., cell, leaf, whole plant, and community.

2.7.2 Atmospheric and Water Environment

2.7.2.1 Air Quality

Impacts on Pollution Generation One of the most cited research results was for Los Angeles, where it was found that for every 1 °C rise in temperature above 22 °C, the incidence of smog increased by 5%. Wilby (2008) utilized statistical downscaling models in lieu of general circulation models for London's nocturnal UHI implications, which projected increased ozone concentrations through the 2050s relative to present conditions. Otanicar et al. (2010) investigated further transportation air quality by quantifying the impact of the UHI on light-duty gasoline vehicles using the US Environmental Protection MOBILE6 modeling techniques. Those findings indicated that the major impact is on evaporative emissions, i.e., vapors escaping from the vehicle's fuel system. Starting emissions, however, decrease throughout the year, as Otanicar et al. predicted, given start emissions are highly dependent on temperature for operating and catalytic converter efficiencies.

2.7.2.2 Water Quality

Increased water temperatures resulting from a rise in pavement temperature and stormwater runoff can impact chemical and biological processes that support water quality, and nonpoint source runoff is deemed a major source of pollutants in urban areas (Fig. 2.25). The percentage of impervious surface area helps determine stream water quality as defined by ecological indicators such as benthic macroinvertebrate

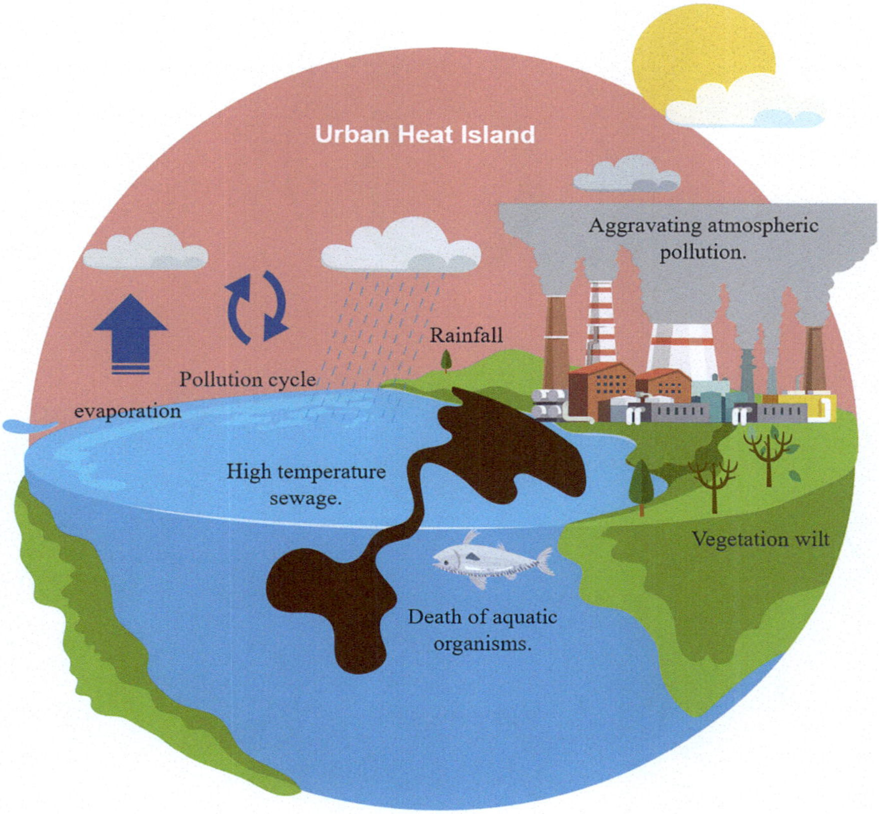

Fig. 2.25 Effect of UHI on atmospheric and water environment

community composition and fish density and abundance. Biofouling of urban waterways can limit runoff transport, and small increases in water temperature typically lead to notable changes in algal growth rates (Butterwick, 2005), promoting fouling and potentially competing with or harming other species, particularly fish. Muller et al. (2014) examined the role of UHI on subsurface water infrastructure in Oberhausen, Germany, and the potential for impacts on water quality. They identified an average subsurface UHI at 70 cm below ground surface of 9 °C between the city center and a rural control location.

2.7.3 *Energy Consumption and Carbon Emissions*

Energy demand impacts resulting from UHI include possible increased demand resulting from mechanical cooling as well as reductions resulting from reduced winter heating (Fig. 2.26). *Impacts of UHI on cooling and heating load*: As an

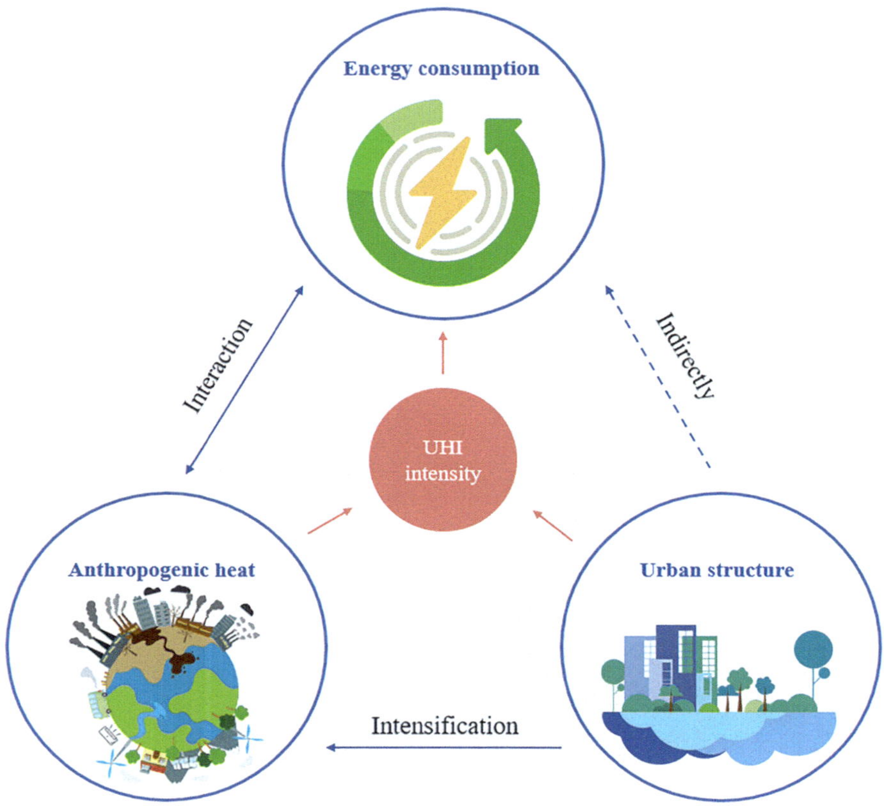

Fig. 2.26 The impact of UHI on energy consumption

example, Rong (2006) examined the role of the UHI in the state of Texas and documented that residential cooling load increased by approximately 6%, whereas the heating load decreased by approximately 16%, and an overall energy decrease of approximately 1% was observed. Giridharan and Kolokotroni (2009) examined the winter UHI in London and found that although it was observable, it was not as predictable as the summer UHI. Kolokotroni et al. (2012) further examined the role of the UHI in London, where many office buildings are not mechanically cooled. As such, there were only marginal increases in energy consumption and CO_2 emissions between urban office buildings and their rural counterparts. However, in modeling in the future with mechanically cooled buildings in the London urban environment, they found that CO_2 emissions as a proxy of energy consumption increased between 480% and 670% in the city center.

2.7.4 *Human Health*

Impacts on Thermal Comfort Radiation reflected solar radiation may increase potential health risks to humans. Though reflective pavements mainly increase their reflectivity to visible light, some reflective materials such as white clay can increase the intensity of reflected ultraviolet (UV) radiation to people. UV radiation is harmful to live cells and can result in sunburn, increased rates of aging of the skin, and skin cancer, with its damage accumulating over years. Childhood sun exposure may play an important role in the development of skin cancer later in adult life (Boumans et al., 2014). Therefore, the amount of reflected radiation should be considered when planning ground and building paving, especially in schoolyards and playgrounds. Moreover, light-colored reflective pavement surfaces can cause glare and visual pollution, which can harm eyesight after a long period of exposure. Reflection from light-colored surfaces can disturb occupants of taller neighboring buildings when applied to roofs; make pedestrians on nearby sidewalks suffer when applied to walls; and provide less lane demarcation due to the poor visibility of white lines when applied to light-colored roads, potentially increasing driving risks (Fig. 2.27).

Impacts on Heat-Related Diseases Others are expected to be manifested in the form of heightened incidence of allergic respiratory diseases such as asthma resulting from increased air pollutants (D'Amato et al., 2013); the complication of existing medical morbidity among vulnerable populations, including those with mental

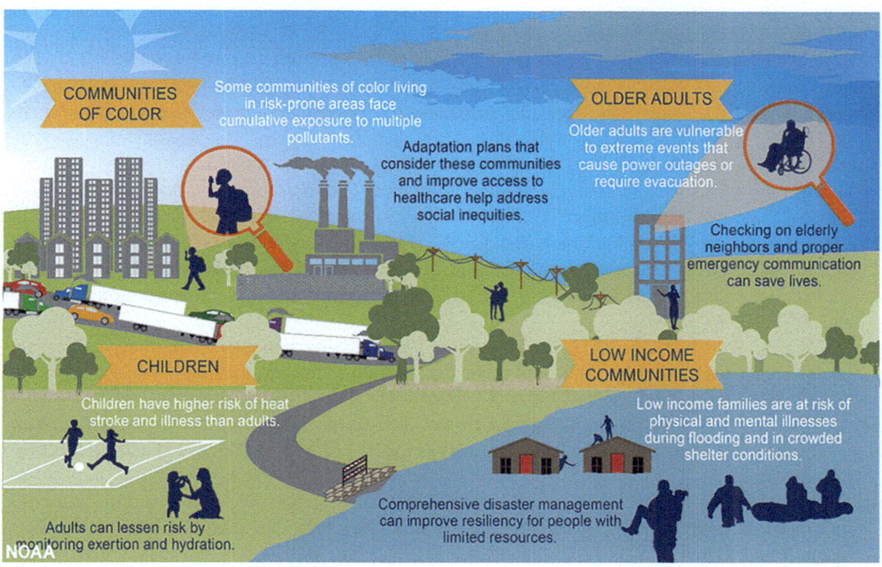

Fig. 2.27 The impact of global warming on human health. (Source: EPA (National Climate Assessment))

illnesses (Chand & Murthy, 2008); or even an expected change in the reach of tropical diseases such as malaria, dengue, West Nile virus, and others. Lin et al. (2012) estimated that excessive heat would be two to six times higher in 2080–2099 than in 1991–2004. The study used a baseline estimate for respiratory disease burden attributable to extreme heat in New York at 100 hospital admissions or $644,069 in direct hospitalization costs and 616 days of hospitalization per year.

Physiological and Socioeconomic Condition Realizing better health outcomes from extreme heat exposure relies heavily on the adaptive capacity of the individual, which in turn is influenced by the demographic, socioeconomic status, and geographic location of the individual. Characterization of vulnerable populations at the community level is essential in developing programmatic interventions or downstream policy changes. Findings from an ecological study designed on Phoenix and Philadelphia neighborhoods suggested that the most important heat risk factors are place-specific. Other risk factors found by the study include a higher proportion of Black residents, low housing values, and linguistically and socially isolated residents (Patz et al., 2008).

References

Abusitta, A., Li, M. Q., & Fung, B. C. (2024). Survey on Explainable AI: Techniques, challenges and open issues. *Expert Systems with Applications, 255*, 124710. https://doi.org/10.1016/j.eswa.2024.124710

Alavipanah, S., Schreyer, J., Haase, D., Lakes, T., & Qureshi, S. (2018). The effect of multi-dimensional indicators on urban thermal conditions. *Journal of Cleaner Production, 177*, 115. https://doi.org/10.1016/j.jclepro.2017.12.187

Albrecht, F. (1933). Untersuchungen der vertikalen Luftzirkulation in der Grossstadt. *Met. Zt, 50*, 93–98.

Amazirh, A., Merlin, O., & Er-Raki, S. (2019). Including Sentinel-1 radar data to improve the disaggregation of MODIS land surface temperature data. *ISPRS Journal of Photogrammetry and Remote Sensing, 150*, 11. https://doi.org/10.1016/j.isprsjprs.2019.02.004

Anjum, S. S., Noor, R. M., Aghamohammadi, N., Ahmedy, I., Mat Kiah, L., Hussin, N., Anisi, M. H., & Qureshi, M. A. (2019). Modeling traffic congestion based on air quality for greener environment: An empirical study. *IEEE Access, 7*, 57100–57119. https://doi.org/10.1109/ACCESS.2019.2914672

Arslan, N., & Sekertekin, A. (2019). Application of long short-term memory neural network model for the reconstruction of MODIS land surface temperature images. *Journal of Atmospheric and Solar - Terrestrial Physics, 194*, 105100. https://doi.org/10.1016/j.jastp.2019.105100

Atwater, M. A. (1971). The radiation budget for polluted layers of the urban environment. *Journal of Applied Meteorology, 10*, 205. https://doi.org/10.1175/1520-0450(1971)010<0205:trbfpl>2.0.co;2

Bai, Y., Wong, M. S., Shi, W. Z., Wu, L. X., & Qin, K. (2015). Advancing of land surface temperature retrieval using extreme learning machine and spatiotemporal adaptive data fusion algorithm. *Remote Sensing, 7*, 4424. https://doi.org/10.3390/rs70404424

Balchin, W. G. V., & Pye, N. (1947). A micro-climatological investigation of bath and the surrounding district. *Quarterly Journal of the Royal Meteorological Society, 73*, 297–323.

Bechtel, B., Demuzere, M., Mills, G., Zhan, W., Sismanidis, P., Small, C., & Voogt, J. (2019). SUHI analysis using local climate zones—A comparison of 50 cities. *Urban Climate, 28*, 100451. https://doi.org/10.1016/j.uclim.2019.01.005

Benas, N., Chrysoulakis, N., & Cartalis, C. (2017). Trends of urban surface temperature and heat island characteristics in the Mediterranean. *Theoretical and applied climatology, 130*, 807–816. https://doi.org/10.1007/s00704-016-1905-8

Berger, C., Rosentreter, J., Voltersen, M., Baumgart, C., Schmullius, C., & Hese, S. (2017). Spatio-temporal analysis of the relationship between 2D/3D urban site characteristics and land surface temperature. *Remote sensing of environment, 193*, 225–243.

Boumans, R. J., Phillips, D. L., Victery, W., & Fontaine, T. D. (2014). Developing a model for effects of climate change on human health and health–environment interactions: Heat stress in Austin, Texas. *Urban Climate, 8*, 78–99. https://doi.org/10.1016/j.uclim.2014.03.001

Bruse, M., & Fleer, H. (1998). Simulating surface-plant-air interactions inside urban environments with a three dimensional numerical model. *Environmental Modelling and Software, 13*, 373. https://doi.org/10.1016/S1364-8152(98)00042-5

Butterwick, C., Heaney, S. I., & Talling, J. F. (2005). Diversity in the influence of temperature on the growth rates of freshwater algae, and its ecological relevance. *Freshwater Biology, 50*(2), 291–300. https://doi.org/10.1111/j.1365-2427.2004.01317.x

Cao, J., Zhou, W., Zheng, Z., Ren, T., & Wang, W. (2021). Within-city spatial and temporal heterogeneity of air temperature and its relationship with land surface temperature. *Landscape and Urban Planning, 206*, 103979. https://doi.org/10.1016/j.landurbplan.2020.103979

Carnahan, W. H., & Larson, R. C. (1990). An analysis of an urban heat sink. *Remote Sensing of Environment, 33*, 65. https://doi.org/10.1016/0034-4257(90)90056-R

Chakraborty, T., & Lee, X. (2019). A simplified urban-extent algorithm to characterize surface urban heat islands on a global scale and examine vegetation control on their spatiotemporal variability. *International Journal of Applied Earth Observation and Geoinformation, 74*, 269–280. https://doi.org/10.1016/j.jag.2018.09.015

Chakraborty, T., Hsu, A., Manya, D., & Sheriff, G. (2020). A spatially explicit surface urban heat island database for the United States: Characterization, uncertainties, and possible applications. *ISPRS Journal of Photogrammetry and Remote Sensing, 168*, 74. https://doi.org/10.1016/j.isprsjprs.2020.07.021

Chand, P. K., & Murthy, P. (2008). Climate change and mental health. *Regional Health Forum.*, 43–48.

Changnon, S. A., & Ackerman, B. (1977). Summary of METROMEX, volume 1: Weather anomalies and impacts. *Bulletin (Illinois State Water Survey) no. 62.*

Chen, X., Li, W., Chen, J., Rao, Y., & Yamaguchi, Y. (2014). A combination of TsHARP and thin plate spline interpolation for spatial sharpening of thermal imagery. *Remote Sensing, 6*, 2845. https://doi.org/10.3390/rs6042845

Chen, C., Wang, L., Myneni, R. B., & Li, D. (2020). Attribution of land-use/land-cover change induced surface temperature anomaly: How accurate is the first-order Taylor series expansion? *Journal of Geophysical Research—Biogeosciences, 125*. https://doi.org/10.1029/2020JG005787

Christen, A., Oke, T. R., Grimmond, S., Steyn, D. G., & Roth, M. (2013). 35 years of urban climate research at the "Vancouver-sunset" flux tower. https://api.semanticscholar.org/CorpusID:132588374.

Chun, B., & Guldmann, J. M. (2014). Spatial statistical analysis and simulation of the urban heat island in high-density central cities. *Landscape and urban planning, 125*, 76–88. https://doi.org/10.1016/j.landurbplan.2014.01.016

Clinton, N., & Gong, P. (2013). MODIS detected surface urban heat islands and sinks: Global locations and controls. *Remote Sensing of Environment, 134*, 294. https://doi.org/10.1016/j.rse.2013.03.008

Crosson, W. L., Al-Hamdan, M. Z., Hemmings, S. N. J., & Wade, G. M. (2012). A daily merged MODIS Aqua-Terra land surface temperature data set for the conterminous United States. *Remote Sensing of Environment, 119*, 315. https://doi.org/10.1016/j.rse.2011.12.019

D'Amato, G., Baena-Cagnani, C. E., Cecchi, L., Annesi-Maesano, I., Nunes, C., Ansotegui, I., ... & Canonica, W. G. (2013). Climate change, air pollution and extreme events leading to increasing prevalence of allergic respiratory diseases. *Multidisciplinary respiratory medicine, 8*, 1–9. https://doi.org/10.1186/2049-6958-8-12

Davenport, A.G., 2000. Estimating the roughness of cities and sheltered country.

Davidson, B. (1967). A summary of the New York urban air pollution dynamics research program. *Journal of the Air Pollution Control Association, 17*, 154. https://doi.org/10.1080/00022470.1967.10468961

Davis, M. L., & Pearson, J. E. (1970). Modeling urban atmospheric temperature profiles. *Atmospheric Environment, 1967*, 4. https://doi.org/10.1016/0004-6981(70)90062-4

Debbage, N., & Shepherd, J. M. (2015). The urban heat island effect and city contiguity. *Computers, Environment and Urban Systems, 54*, 181. https://doi.org/10.1016/j.compenvurbsys.2015.08.002

Deilami, K., Kamruzzaman, M., & Liu, Y. (2018). Urban heat island effect: A systematic review of spatiotemporal factors, data, methods, and mitigation measures. *International Journal of Applied Earth Observation and Geoinformation, 67*, 30–42.

Dong, P., Gao, L., Zhan, W., Liu, Z., Li, J., Lai, J., Li, H., Huang, F., Tamang, S. K., & Zhao, L. (2020). Global comparison of diverse scaling factors and regression models for downscaling Landsat-8 thermal data. *ISPRS Journal of Photogrammetry and Remote Sensing, 169*, 44. https://doi.org/10.1016/j.isprsjprs.2020.08.018

Duan, S. B., Li, Z. L., Wang, N., Wu, H., & Tang, B. H. (2012). Evaluation of six land-surface diurnal temperature cycle models using clear-sky in situ and satellite data. *Remote Sensing of Environment, 124*, 15. https://doi.org/10.1016/j.rse.2012.04.016

Duan, S. B., Li, Z. L., Wu, H., Tang, B. H., Jiang, X., & Zhou, G. (2013). Modeling of day-to-day temporal progression of clear-sky land surface temperature. *IEEE Geoscience and Remote Sensing Letters, 10*, 1050. https://doi.org/10.1109/LGRS.2012.2228465

Duan, S. B., Li, Z. L., & Leng, P. (2017). A framework for the retrieval of all-weather land surface temperature at a high spatial resolution from polar-orbiting thermal infrared and passive microwave data. *Remote Sensing of Environment, 195*, 107. https://doi.org/10.1016/j.rse.2017.04.008

Duckworth, F. S., & Sandberg, J. S. (1954). The effect of cities upon horizontal and vertical temperature gradients. *Bulletin of the American Meteorological Society, 35*, 198. https://doi.org/10.1175/1520-0477-35.5.198

Estoque, R. C., & Murayama, Y. (2017). Monitoring surface urban heat island formation in a tropical mountain city using Landsat data (1987–2015). *ISPRS Journal of Photogrammetry and Remote Sensing, 133*, 18. https://doi.org/10.1016/j.isprsjprs.2017.09.008

Gage, E. A., & Cooper, D. J. (2017). Urban forest structure and land cover composition effects on land surface temperature in a semi-arid suburban area. *Urban Forestry & Urban Greening, 28*, 28. https://doi.org/10.1016/j.ufug.2017.10.003

Giridharan, R., & Kolokotroni, M. (2009). Urban heat island characteristics in London during winter. *Solar Energy, 83*(9), 1668–1682. https://doi.org/10.1016/j.solener.2009.06.007

Giridharan, R., Lau, S. S. Y., Ganesan, S., & Givoni, B. (2008). Lowering the outdoor temperature in high-rise high-density residential developments of coastal Hong Kong: The vegetation influence. *Building and Environment, 43*, 1583. https://doi.org/10.1016/j.buildenv.2007.10.003

Gong, J., Hu, Y., Liu, M., Bu, R., Chang, Y., Li, C., & Wu, W. (2015). Characterization of air pollution index and its affecting factors in industrial urban areas in northeastern China. *Polish Journal of Environmental Studies, 24*, 1579.

Göttsche, F. M., & Olesen, F. S. (2001). Modelling of diurnal cycles of brightness temperature extracted from meteosat data. *Remote Sensing of Environment, 76*, 337. https://doi.org/10.1016/S0034-4257(00)00214-5

Grimmond, C. S. B. (2006). Progress in measuring and observing the urban atmosphere. *Theoretical and Applied Climatology, 84*, 3. https://doi.org/10.1007/s00704-005-0140-5

Guo, W., Liu, S., Xu, J., Wu, L., Shangguan, D., Yao, X., Wei, J., Bao, W., Yu, P., Liu, Q., & Jiang, Z. (2015). The second Chinese glacier inventory: Data, methods and results. *Journal of Glaciology, 61*, 357. https://doi.org/10.3189/2015JoG14J209

Hann, J. (1885). On the reduction of temperature means from short series of observations to the equivalents of longer periods. *Quarterly Journal of the Royal Meteorological Society, 11*, 28–38. https://doi.org/10.1002/QJ.4970115305

Hanna, S. R., Brown, M. J., Camelli, F. E., Chan, S. T., Coirier, W. J., Hansen, O. R., Huber, A. H., Kim, S., & Reynolds, R. M. (2006). Detailed simulations of atmospheric flow and dispersion in downtown Manhattan: An applications of five computational fluid dynamics models. *Bulletin of the American Meteorological Society, 87*, 1713. https://doi.org/10.1175/BAMS-87-12-1713

Hardiman, B. S., Wang, J. A., Hutyra, L. R., Gately, C. K., Getson, J. M., & Friedl, M. A. (2017). Accounting for urban biogenic fluxes in regional carbon budgets. *Science of the Total Environment, 592*, 366. https://doi.org/10.1016/j.scitotenv.2017.03.028

Hardin, A. W., Liu, Y., Cao, G., & Vanos, J. K. (2018). Urban heat island intensity and spatial variability by synoptic weather type in the northeast U.S. *Urban Climate, 24*, 747. https://doi.org/10.1016/j.uclim.2017.09.001

Howard, L., 1818. The climate of London: Deduced from meteorological observations, made at different places in the neighbourhood of the metropolis. W. Phillips, George Yard, Lombard Street, sold also by J. and A. Arch~....

Howard, L. (1833). *The climate of London* (Vol. I). Harvey and Darton.

Huang, F., Zhan, W., Duan, S. B., Ju, W., & Quan, J. (2014). A generic framework for modeling diurnal land surface temperatures with remotely sensed thermal observations under clear sky. *Remote Sensing of Environment, 150*, 140. https://doi.org/10.1016/j.rse.2014.04.022

Imhoff, M. L., Zhang, P., Wolfe, R. E., & Bounoua, L. (2010). Remote sensing of the urban heat island effect across biomes in the continental USA. *Remote Sensing of Environment, 114*, 504. https://doi.org/10.1016/j.rse.2009.10.008

Inamdar, A. K., French, A., Hook, S., Vaughan, G., & Luckett, W. (2008). Land surface temperature retrieval at high spatial and temporal resolutions over the southwestern United States. *Journal of Geophysical Research Atmospheres, 113*. https://doi.org/10.1029/2007JD009048

Jeganathan, C., Hamm, N. A. S., Mukherjee, S., Atkinson, P. M., Raju, P. L. N., & Dadhwal, V. K. (2011). Evaluating a thermal image sharpening model over a mixed agricultural landscape in India. *International Journal of Applied Earth Observation and Geoinformation, 13*, 178–191.

Jendritzky, G., & Nübler, W. (1981). A model analysing the urban thermal environment in physiologically significant terms. *Archives for Meteorology, Geophysics, and Bioclimatology Series B, 29*, 313. https://doi.org/10.1007/BF02263308

Jin, M., & Dickinson, R. E. (1999). Interpolation of surface radiative temperature measured from polar orbiting satellites to a diurnal cycle 1. Without clouds. *Journal of Geophysical Research Atmospheres, 104*, 2105. https://doi.org/10.1029/1998JD200005

Jin, H., Cui, P., Wong, N., & Ignatius, M. (2018). Assessing the effects of urban morphology parameters on microclimate in Singapore to control the urban heat island effect. *Sustainability, 10*(1), 206. https://doi.org/10.3390/su10010206

Kabano, P., Lindley, S., & Harris, A. (2021). Evidence of urban heat island impacts on the vegetation growing season length in a tropical city. *Landscape and Urban Planning, 206*, 103989. https://doi.org/10.1016/j.landurbplan.2020.103989

Kassomenos, P., Kissas, G., Petrou, I., Begou, P., Khan, H. S., & Santamouris, M. (2022). The influence of daily weather types on the development and intensity of the urban heat island in two Mediterranean coastal metropolises. *Science of the Total Environment, 819*, 153071. https://doi.org/10.1016/j.scitotenv.2022.153071

Ke, L., Song, C., & Ding, X. (2012). *Reconstructing complete MODIS LST based on temperature gradients in northeastern Qinghai-Tibet plateau, in: International geoscience and remote sensing symposium (IGARSS)*. https://doi.org/10.1109/IGARSS.2012.6350664

Ke, L., Ding, X., & Song, C. (2013). Reconstruction of time-series MODIS LST in Central Qinghai-Tibet plateau using geostatistical approach. *IEEE Geoscience and Remote Sensing Letters, 10*, 1602. https://doi.org/10.1109/LGRS.2013.2263553

Khan, H. S., Santamouris, M., Kassomenos, P., Paolini, R., Caccetta, P., & Petrou, I. (2021). Spatiotemporal variation in urban overheating magnitude and its association with synoptic air-masses in a coastal city. *Scientific Reports, 11*, 6762. https://doi.org/10.1038/s41598-021-86089-2

Kondrat′ev, K. I. (1973). *The complete atmospheric energetics experiment*. International Council of Scientific Unions, World Meteorological Organization.

Kolokotroni, M., Ren, X., Davies, M., & Mavrogianni, A. (2012). London's urban heat island: Impact on current and future energy consumption in office buildings. *Energy and buildings, 47*, 302–311. https://doi.org/10.1016/j.enbuild.2011.12.019

Kou, X., Jiang, L., Bo, Y., Yan, S., & Chai, L. (2016). Estimation of land surface temperature through blending MODIS and AMSR-E data with the Bayesian maximum entropy method. *Remote Sensing, 8*. https://doi.org/10.3390/rs8020105

Kratzer, A. (1937). *Das Stadtklima [City climate]*. Friedr. Vieweg & Sohn.

Krayenhoff, E. S. (2014). A multi-layer urban canopy model for neighbourhoods with trees.

Krayenhoff, E. S., Broadbent, A. M., Zhao, L., Georgescu, M., Middel, A., Voogt, J. A., Martilli, A., Sailor, D. J., & Erell, E. (2021). Cooling hot cities: A systematic and critical review of the numerical modelling literature. *Environmental Research Letters., 16*. https://doi.org/10.1088/1748-9326/abdcf1

Kremser, V. (1886). Vortrag uber das Klima von Berlin. Festschrift d. 59 Vers. dd Naturf und Artie 37–38.

Kumar, R., & Mishra, V. (2019). Decline in surface urban heat island intensity in India during heatwaves. *Environmental Research Communications, 1*. https://doi.org/10.1088/2515-7620/ab121d

Kumar, R., Mishra, V., Buzan, J., Kumar, R., Shindell, D., & Huber, M. (2017). Dominant control of agriculture and irrigation on urban heat island in India. *Scientific Reports, 7*, 14054. https://doi.org/10.1038/s41598-017-14213-2

Lauscher, F. (1934). Wärmeausstrahlung und Horizonteinengung. Sitzungsberichte der Akademie der Wissenschaft. *Mathematische-Naturwisserschaftliche Klasse (Vienna), 143*, 503–519.

Li, D., Bou-Zeid, E., (2013). "Synergistic interactions between urban heat islands and heat waves: The impact in cities is larger than the sum of its parts." *Journal of applied Meteorology and Climatology 52*(9): 2051–2064. https://doi.org/10.1175/JAMC-D-13-02.s1.

Li, X., Stringer, L. C., & Dallimer, M. (2021). The spatial and temporal characteristics of urban heat island intensity: implications for East Africa's urban development. *Climate, 9*(4), 51. https://doi.org/10.3390/cli9040051

Li, X., Zhou, Y., Asrar, G. R., Imhoff, M., & Li, X. (2017). The surface urban heat island response to urban expansion: A panel analysis for the conterminous United States. *Science of the Total Environment, 605–606*, 426. https://doi.org/10.1016/j.scitotenv.2017.06.229

Li, X., Zhou, Y., Asrar, G. R., & Zhu, Z. (2018). Creating a seamless 1 km resolution daily land surface temperature dataset for urban and surrounding areas in the conterminous United States. *Remote Sensing of Environment, 206*, 84. https://doi.org/10.1016/j.rse.2017.12.010

Li, Y., Schubert, S., Kropp, J. P., & Rybski, D. (2020). On the influence of density and morphology on the urban heat island intensity. *Nature Communications, 11*. https://doi.org/10.1038/s41467-020-16461-9

Li, J., Yang, Z., Zhao, X., Li, Y., Huang, X., Chen, Y., & Shi, F. (2024). Study of the correlation between the urban wind–heat environment and urban development elements in high-density urban areas: A case study of Central Shanghai. *Buildings, 14*(2), 315. https://doi.org/10.3390/buildings14020315

Lin, S., Hsu, W. H., Van Zutphen, A. R., Saha, S., Luber, G., & Hwang, S. A. (2012). Excessive heat and respiratory hospitalizations in New York State: estimating current and future public health burden related to climate change. *Environmental health perspectives, 120*(11), 1571–1577. https://doi.org/10.1289/ehp.1104728

Liang, Z., Wang, Y., Huang, J., Wei, F., Wu, S., Shen, J., Sun, F., & Li, S. (2020a). Seasonal and diurnal variations in the relationships between urban form and the urban heat island effect. *Energies (Basel), 13*. https://doi.org/10.3390/en13225909

Liang, Z., Wu, S., Wang, Y., Wei, F., Huang, J., Shen, J., & Li, S. (2020b). The relationship between urban form and heat island intensity along the urban development gradients. *Science of the Total Environment, 708*, 135011. https://doi.org/10.1016/j.scitotenv.2019.135011

Lindberg, F., Holmer, B., & Thorsson, S. (2008). SOLWEIG 1.0—Modelling spatial variations of 3D radiant fluxes and mean radiant temperature in complex urban settings. *International Journal of Biometeorology, 52*, 697. https://doi.org/10.1007/s00484-008-0162-7

Liu, H., & Weng, Q. (2012). Enhancing temporal resolution of satellite imagery for public health studies: A case study of West Nile virus outbreak in Los Angeles in 2007. *Remote Sensing of Environment, 117*, 57. https://doi.org/10.1016/j.rse.2011.06.023

Liu, Z., Zhan, W., Lai, J., Hong, F., Quan, J., Bechtel, B., Huang, F., & Zou, Z. (2019). Balancing prediction accuracy and generalization ability: A hybrid framework for modelling the annual dynamics of satellite-derived land surface temperatures. *ISPRS Journal of Photogrammetry and Remote Sensing, 151*, 189. https://doi.org/10.1016/j.isprsjprs.2019.03.013

Liu, H., Huang, B., Zhan, Q., Gao, S., Li, R., & Fan, Z. (2021). The influence of urban form on surface urban heat island and its planning implications: Evidence from 1288 urban clusters in China. *Sustainable Cities and Society, 71*, 102987. https://doi.org/10.1016/j.scs.2021.102987

Long, D., Yan, L., Bai, L., Zhang, C., Li, X., Lei, H., Yang, H., Tian, F., Zeng, C., Meng, X., & Shi, C. (2020). Generation of MODIS-like land surface temperatures under all-weather conditions based on a data fusion approach. *Remote Sensing of Environment, 246*, 111863. https://doi.org/10.1016/j.rse.2020.111863

Lu, L., Venus, V., Skidmore, A., Wang, T., & Luo, G. (2011). Estimating land-surface temperature under clouds using MSG/SEVIRI observations. *International Journal of Applied Earth Observation and Geoinformation, 13*, 265. https://doi.org/10.1016/j.jag.2010.12.007

Lu, Y., Yue, W., He, T., & Shan, Z. (2022). Urban form centrality and thermal environment: An empirical study of Chinese megacities. *Sustainable Cities and Society, 83*, 103955. https://doi.org/10.1016/j.scs.2022.103955

Maleki, A., & Mahdavi, A. (2016). Evaluation of urban heat islands mitigation strategies using 3dimentional urban micro- climate model ENVI-met. https://api.semanticscholar.org/CorpusID:130942054.

Mao, Q., Peng, J., & Wang, Y. (2021). Resolution enhancement of remotely sensed land surface temperature: Current status and perspectives. *Remote Sensing, 13*. https://doi.org/10.3390/rs13071306

Martilli, A., Krayenhoff, E. S., & Nazarian, N. (2020). Is the urban heat island intensity relevant for heat mitigation studies? *Urban Climate, 31*, 100541. https://doi.org/10.1016/j.uclim.2019.100541

Masson, V. (2000). A physically-based scheme for the urban energy budget in atmospheric models. *Boundary-Layer Meteorology, 94*, 357. https://doi.org/10.1023/A:1002463829265

Meineke, E., Youngsteadt, E., Dunn, R. R., & Frank, S. D. (2016). Urban warming reduces aboveground carbon storage. *Proceedings of the Royal Society B: Biological Sciences, 283*, 20161574.

Meng, D., Li, X., Gong, H., et al. (2010). Thermal landscape pattern and thermal environmental effects of typical urban landscapes in Beijing (in Chinese). *Acta Ecologica Sinica, 30*(13), 3491–3500. https://doi.org/10.20103/j.stxb.2010.13.014

Metz, M., Andreo, V., & Neteler, M. (2017). A new fully gap-free time series of land surface temperature from MODIS LST data. *Remote Sensing, 9*. https://doi.org/10.3390/rs9121333

Mohammad, P., & Goswami, A. (2021). Quantifying diurnal and seasonal variation of surface urban heat island intensity and its associated determinants across different climatic zones over Indian cities. *GIScience & Remote Sensing, 58*, 955. https://doi.org/10.1080/15481603.2021.1940739

Moosavi, V., Talebi, A., Mokhtari, M. H., Shamsi, S. R. F., & Niazi, Y. (2015). A wavelet-artificial intelligence fusion approach (WAIFA) for blending Landsat and MODIS surface temperature. *Remote Sensing of Environment, 169*, 243. https://doi.org/10.1016/j.rse.2015.08.015

Muller, C. L., Chapman, L., Grimmond, C. S. B., Young, D. T., & Cai, X. (2013). Sensors and the city: A review of urban meteorological networks. *International Journal of Climatology., 33*, 1585. https://doi.org/10.1002/joc.3678

Müller, N., Kuttler, W., & Barlag, A. B. (2014). Analysis of the subsurface urban heat island in Oberhausen, Germany. *Climate research, 58*(3), 247–256. https://doi.org/10.3354/cr01195

Myrup, L. O. (1969). A numerical model of the urban heat island. *Journal of Applied Meteorology, 8*, 908. https://doi.org/10.1175/1520-0450(1969)008<0908:anmotu>2.0.co;2

Nastran, M., Kobal, M., & Eler, K. (2019). Urban heat islands in relation to green land use in European cities. *Urban Forestry & Urban Greening, 37*, 33–41. https://doi.org/10.1016/j.ufug.2018.01.008

Nichol, J. (2005). Remote sensing of urban heat islands by day and night. *Photogramm Engineering & Remote Sensing, 71*, 613. https://doi.org/10.14358/PERS.71.5.613

Nitschke, C. R., Nichols, S., Allen, K., Dobbs, C., Livesley, S. J., Baker, P. J., & Lynch, Y. (2017). The influence of climate and drought on urban tree growth in Southeast Australia and the implications for future growth under climate change. *Landscape and Urban Planning, 167*, 275. https://doi.org/10.1016/j.landurbplan.2017.06.012

Oke, T. R. (1973). City size and the urban heat island. *Atmospheric Environment, 1967*, 7. https://doi.org/10.1016/0004-6981(73)90140-6

Oke, T. R. (1976). The distinction between canopy and boundary-layer urban heat islands. *Atmosphere (Basel), 14*, 268. https://doi.org/10.1080/00046973.1976.9648422

Oke, T. R., (1997). Urban environments. The surface climates of Canada, Bailey, WG. Oke TR & Rouse WR Eds., Mc Gill-Queen's University Press, *Montréal*, 303-327.

Oke, T. R. (1982). The energetic basis of the urban heat island. *Quarterly Journal of the Royal Meteorological Society, 108*, 1. https://doi.org/10.1002/qj.49710845502

Oke, T. R. (1988). The urban energy balance. *Progress in Physical Geography, 12*, 471. https://doi.org/10.1177/030913338801200401

Oke, T. R., Mills, G., Christen, A., & Voogt, J. A. (2017). Urban climates. *Urban Climates*. https://doi.org/10.1017/9781139016476

Otanicar, T. P., Carlson, J. D., Golden, J. S., Kaloush, K. E., & Phelan, P. E. (2010). Impact of the urban heat island on light duty vehicle emissions for the Phoenix, AZ area. *International Journal of Sustainable Transportation, 4*(1), 1–13. https://doi.org/10.1080/15568310802428552

Parida, B. R., Bar, S., Kaskaoutis, D., Pandey, A. C., Polade, S. D., & Goswami, S. (2021). Impact of COVID-19 induced lockdown on land surface temperature, aerosol, and urban heat in Europe and North America. *Sustainable Cities and Society, 75*, 103336. https://doi.org/10.1016/j.scs.2021.103336

Patz, J. A., Vavrus, S. J., Uejio, C. K., & McLellan, S. L. (2008). Climate change and waterborne disease risk in the Great Lakes region of the U.S. *American Journal of Preventive Medicine, 35*, 451. https://doi.org/10.1016/j.amepre.2008.08.026

Pau, S., Detto, M., Kim, Y., & Still, C. J. (2018). Tropical forest temperature thresholds for gross primary productivity. *Ecosphere, 9*. https://doi.org/10.1002/ecs2.2311

Quan, J., Chen, Y., Zhan, W., Wang, J., Voogt, J., & Li, J. (2014). A hybrid method combining neighborhood information from satellite data with modeled diurnal temperature cycles over consecutive days. *Remote Sensing of Environment, 155*, 257. https://doi.org/10.1016/j.rse.2014.08.034

Quan, J., Zhan, W., Ma, T., Du, Y., Guo, Z., & Qin, B. (2018). An integrated model for generating hourly Landsat-like land surface temperatures over heterogeneous landscapes. *Remote Sensing of Environment, 206*, 403. https://doi.org/10.1016/j.rse.2017.12.003

Raj, S., Paul, S. K., Chakraborty, A., & Kuttippurath, J. (2020). Anthropogenic forcing exacerbating the urban heat islands in India. *Journal of Environmental Management, 257*, 110006. https://doi.org/10.1016/j.jenvman.2019.110006

Rajkovich, N., & Larsen, L. (2016). A bicycle-based field measurement system for the study of thermal exposure in Cuyahoga County, Ohio, USA. *International Journal of Environmental Research and Public Health, 13*(2), 159. https://doi.org/10.3390/ijerph13020159

Rao, P. K. (1972). Remote sensing of urban "heat islands" from an environmental satellite. *The Bulletin of the American Meteorological Society*.

Ren, Q., He, C., Huang, Q., & Zhou, Y. (2018). Urbanization impacts on vegetation phenology in China. *Remote Sensing, 10*. https://doi.org/10.3390/rs10121905

Rizwan, A. M., Dennis, L. Y. C., & Liu, C. (2008). A review on the generation, determination and mitigation of urban heat island. *Journal of Environmental Sciences, 20*, 120. https://doi.org/10.1016/S1001-0742(08)60019-4

Rong, F. (2006). Impact of urban sprawl on US residential energy use. http://hdl.handle.net/1903/3848

Rotach, M. W., Vogt, R., Bernhofer, C., Batchvarova, E., Christen, A., Clappier, A., Feddersen, B., Gryning, S. E., Martucci, G., Mayer, H., Mitev, V., Oke, T. R., Parlow, E., Richner, H., Roth, M., Roulet, Y. A., Ruffieux, D., Salmond, J. A., Schatzmann, M., & Voogt, J. A. (2005). BUBBLE—An urban boundary layer meteorology project. *Theoretical and Applied Climatology, 81*, 231. https://doi.org/10.1007/s00704-004-0117-9

Roth, M., & Oke, T. R. (1993). Turbulent transfer relationships over an urban surface. I. Spectral characteristics. *Quarterly Journal of the Royal Meteorological Society, 119*, 1071. https://doi.org/10.1002/qj.49711951311

Ryu, Y. H., & Baik, J. J. (2012). Quantitative analysis of factors contributing to urban heat island intensity. *Journal of Applied Meteorology and Climatology, 51*(5), 842–854. https://doi.org/10.1175/JAMC-D-11-098.1

Samuel, C. Z., Jason, S., Aditya, S., Christopher, J. K., Philip, A. T., & Steven, P. L., II. (2016). Urban heat island impacts on plant phenology: Intra-urban variability and response to land cover. *Environmental Research Letters, 11*.

Sarafanov, M., Kazakov, E., Nikitin, N. O., & Kalyuzhnaya, A. V. (2020). A machine learning approach for remote sensing data gap-filling with open-source implementation: An example regarding land surface temperature, surface albedo and ndvi. *Remote Sensing, 12*. https://doi.org/10.3390/rs12233865

Schmid, H. P., & Bünzli, B. (1995). The influence of surface texture on the effective roughness length. *Quarterly Journal of the Royal Meteorological Society, 121*, 1. https://doi.org/10.1002/qj.49712152102

Schmidt, W., 1927. Die Verteilung der Minimumtemperaturen in der Frostnacht des 12.5. 1927 im Gemeindegebeit von Wien, Frostschritte der Landwirstschaft 2.

Schwarz, N., Lautenbach, S., & Seppelt, R. (2011). Exploring indicators for quantifying surface urban heat islands of European cities with MODIS land surface temperatures. *Remote Sensing of Environment, 115*, 3175–3186.

Shastri, H., Barik, B., Ghosh, S., Venkataraman, C., & Sadavarte, P. (2017). Flip flop of day-night and summer-winter surface urban heat island intensity in India. *Scientific Reports, 7*. https://doi.org/10.1038/srep40178

Simwanda, M., Ranagalage, M., Estoque, R. C., & Murayama, Y. (2019). Spatial analysis of surface urban heat islands in four rapidly growing African cities. *Remote Sensing, 11*(14), 1645. https://doi.org/10.3390/rs11141645

Skamarock, W. C., Klemp, J. B., Dudhia, J., Gill, D. O., Liu, Z., Berner, J., Wang, W., Powers, J. G., Duda, M. G., & Barker, D. M. (2019). A description of the advanced research WRF model version 4. *National Center for Atmospheric Research, 145*, 550.

Stewart, I. D., & Oke, T. R. (2012). Local climate zones for urban temperature studies. *Bulletin of the American Meteorological Society, 93*, 1879. https://doi.org/10.1175/BAMS-D-11-00019.1

Stine, A. R., Huybers, P., & Fung, I. Y. (2009). Changes in the phase of the annual cycle of surface temperature. *Nature, 457*, 435. https://doi.org/10.1038/nature07675

Summers, P.W., 1964. An urban ventilation model applied to Montreal.

Sun, F., Liu, M., Wang, Y., Wang, H., & Che, Y. (2020). The effects of 3D architectural patterns on the urban surface temperature at a neighborhood scale: Relative contributions and marginal effects. *Journal of Cleaner Production, 258*, 120706. https://doi.org/10.1016/j.jclepro.2020.120706

Sun, Y., Gao, C., Li, J., Li, W., & Ma, R. (2018). Examining urban thermal environment dynamics and relations to biophysical composition and configuration and socio-economic factors: A case study of the Shanghai metropolitan region. *Sustainable cities and society, 40*, 284–295. https://doi.org/10.1016/j.scs.2017.12.004

Sun, D., & Pinker, R. T. (2005). Implementation of GOES-based land surface temperature diurnal cycle to AVHRR. *International Journal of Remote Sensing, 26*, 3975. https://doi.org/10.1080/01431160500117634

Sundborg, Å. (1950). Local climatological studies of the temperature conditions in an urban area. *Tellus, 2*, 222. https://doi.org/10.3402/tellusa.v2i3.8544

Tan, J., Che, T., Wang, J., Liang, J., Zhang, Y., & Ren, Z. (2021). Reconstruction of the daily MODIS land surface temperature product using the two-step improved similar pixels method. *Remote Sensing, 13*. https://doi.org/10.3390/rs13091671

Tomlinson, C. J., Chapman, L., Thornes, J. E., & Baker, C. (2011). Remote sensing land surface temperature for meteorology and climatology: A review. *Meteorological Applications., 18*, 296. https://doi.org/10.1002/met.287

Tönseth, D., & Welchermill, K. (2014). Design of hollow reinforced concrete columns in the tubed mega frame : Dimensionering av ihåliga armerade betongpelare i "tubed mega frame". https://api.semanticscholar.org/CorpusID:107517245.

Tsoka, S., Tsikaloudaki, A., & Theodosiou, T. (2018). Analyzing the ENVI-met microclimate model's performance and assessing cool materials and urban vegetation applications–a review. *Sustainable Cities and Society, 43*, 55. https://doi.org/10.1016/j.scs.2018.08.009

Van De Kerchove, R., Lhermitte, S., Veraverbeke, S., & Goossens, R. (2012). Spatiotemporal variability in remotely sensed land surface temperature, and its relationship with physiographic variables in the Russian Altay Mountains. *International Journal of Applied Earth Observation and Geoinformation, 20*, 4. https://doi.org/10.1016/j.jag.2011.09.007

Van Den Bergh, F., Van Wyk, M. A., Van Wyk, B. J., & Udahemuka, G. (2007). A comparison of data-driven and model-driven approaches to brightness temperature diurnal cycle interpolation. *SAIEE Africa Research Journal., 98*, 81. https://doi.org/10.23919/saiee.2007.9488131

Wang, J., Huang, B., Fu, D., & Atkinson, P. M. (2015a). Spatiotemporal variation in surface urban heat island intensity and associated determinants across major Chinese cities. *Remote Sensing, 7(4)*, 3670. https://doi.org/10.3390/rs70403670

Wang, Y., Bakker, F., de Groot, R., Wortche, H., & Leemans, R. (2015b). Effects of urban trees on local outdoor microclimate: Synthesizing field measurements by numerical modelling. *Urban Ecosystem, 18*, 1305. https://doi.org/10.1007/s11252-015-0447-7

Wang, T., Tang, R., Li, Z.-L., Jiang, Y., Liu, M., & Niu, L. (2019). An improved spatio-temporal adaptive data fusion algorithm for evapotranspiration mapping. *Remote Sensing, 11*(7), 761. https://doi.org/10.3390/rs11070761

Ward, K., Lauf, S., Kleinschmit, B., & Endlicher, W. (2016). Heat waves and urban heat islands in Europe: A review of relevant drivers. *Science of the Total Environment, 569–570*, 527. https://doi.org/10.1016/j.scitotenv.2016.06.119

Wei, S., Chen, Q., Wu, W., & Ma, J. (2021). Quantifying the indirect effects of urbanization on urban vegetation carbon uptake in the megacity of Shanghai, China. *Environmental Research Letters, 16*. https://doi.org/10.1088/1748-9326/ac06fd

Weiss, D. J., Atkinson, P. M., Bhatt, S., Mappin, B., Hay, S. I., & Gething, P. W. (2014). An effective approach for gap-filling continental scale remotely sensed time-series. *ISPRS Journal of Photogrammetry and Remote Sensing, 98*, 106. https://doi.org/10.1016/j.isprsjprs.2014.10.001

Weng, Q., Fu, P., & Gao, F. (2014). Generating daily land surface temperature at Landsat resolution by fusing Landsat and MODIS data. *Remote Sensing of Environment, 145*, 55. https://doi.org/10.1016/j.rse.2014.02.003

Wilby, R. L. (2008). Constructing climate change scenarios of urban heat island intensity and air quality. *Environment and Planning B: Planning and Design, 35*(5), 902–919. https://doi.org/10.1068/b33066t

Wittwer, W. C. (1960). *Grundzuge der Klimatologie von Bayern* (p. 1). Landes-und Volkskunde d.

Wu, M., Li, H., Huang, W., Niu, Z., & Wang, C. (2015a). Generating daily high spatial land surface temperatures by combining ASTER and MODIS land surface temperature products for environmental process monitoring. *Environmental Sciences: Processes and Impacts, 17*, 1396. https://doi.org/10.1039/c5em00254k

Wu, P., Shen, H., Zhang, L., & Göttsche, F. M. (2015b). Integrated fusion of multi-scale polar-orbiting and geostationary satellite observations for the mapping of high spatial and temporal resolution land surface temperature. *Remote Sensing of Environment, 156*, 169. https://doi.org/10.1016/j.rse.2014.09.013

Wu, X., Wang, G., Yao, R., Wang, L., Yu, D., & Gui, X. (2019). Investigating surface urban heat islands in South America based on MODIS data from 2003-2016. *Remote Sensing, 11*. https://doi.org/10.3390/rs11101212

Xia, H., Chen, Y., Zhao, Y., & Chen, Z. (2018). "Regression-then-fusion" or "fusion-then-regression"? A theoretical analysis for generating high spatiotemporal resolution land surface temperatures. *Remote Sensing, 10*, 1382.

Xu, Y., & Shen, Y. (2013). Reconstruction of the land surface temperature time series using harmonic analysis. *Computational Geosciences, 61*, 126. https://doi.org/10.1016/j.cageo.2013.08.009

Xue, J., Anderson, M. C., Gao, F., Hain, C., Sun, L., Yang, Y., Knipper, K. R., Kustas, W. P., Torres-Rua, A., & Schull, M. (2020). Sharpening ECOSTRESS and VIIRS land surface temperature using harmonized Landsat-sentinel surface reflectances. *Remote Sensing of Environment, 251*, 112055. https://doi.org/10.1016/j.rse.2020.112055

Yang, L., Niyogi, D., Tewari, M., Aliaga, D., Chen, F., Tian, F., & Ni, G. (2016). Contrasting impacts of urban forms on the future thermal environment: Example of Beijing metropolitan area. *Environmental Research Letters, 11*. https://doi.org/10.1088/1748-9326/11/3/034018

Yao, R., Wang, L., Huang, X., Niu, Z., Liu, F., & Wang, Q. (2017). Temporal trends of surface urban heat islands and associated determinants in major Chinese cities. *Science of the Total Environment, 609*, 742. https://doi.org/10.1016/j.scitotenv.2017.07.217

Yao, R., Wang, L., Huang, X., Zhang, W., Li, J., & Niu, Z. (2018). Interannual variations in surface urban heat island intensity and associated drivers in China. *Journal of Environmental Management, 222*, 86. https://doi.org/10.1016/j.jenvman.2018.05.024

Yao, R., Wang, L., Huang, X., Gong, W., & Xia, X. (2019). Greening in rural areas increases the surface urban heat island intensity. *Geophysical Research Letters, 46*, 2204. https://doi.org/10.1029/2018GL081816

Yu, S., Chen, Z., Yu, B., Wang, L., Wu, B., Wu, J., & Zhao, F. (2020). Exploring the relationship between 2D/3D landscape pattern and land surface temperature based on explainable eXtreme Gradient Boosting tree: A case study of Shanghai, China. *Science of the Total Environment, 725*, 138229. https://doi.org/10.1016/j.scitotenv.2020.138229

Yu, W., Ma, M., Wang, X., & Tan, J. (2014). Estimating the land-surface temperature of pixels covered by clouds in MODIS products. *Journal of Applied Remote Sensing, 8*, 083525. https://doi.org/10.1117/1.jrs.8.083525

Yu, W., Nan, Z., Wang, Z., Chen, H., Wu, T., & Zhao, L. (2015). An effective interpolation method for MODIS land surface temperature on the Qinghai-Tibet plateau. *IEEE Journal of Selected Topics in Applied Earth Observations and Remote Sensing*. https://doi.org/10.1109/JSTARS.2015.2464094

Yu, W., Tan, J., Ma, M., Li, X., She, X., & Song, Z. (2019). An effective similar-pixel reconstruction of the high-frequency cloud-covered areas of Southwest China. *Remote Sensing, 11*. https://doi.org/10.3390/rs11030336

Yue, W., Liu, X., Zhou, Y., & Liu, Y. (2019). Impacts of urban configuration on urban heat Island: An empirical study in China mega-cities. *Science of the Total Environment, 671*, 1036. https://doi.org/10.1016/j.scitotenv.2019.03.421

Yu, Z., Xu, S., Zhang, Y., Jørgensen, G., & Vejre, H. (2018). Strong contributions of local background climate to the cooling effect of urban green vegetation. *Scientific reports, 8*(1), 6798. https://doi.org/10.1038/s41598-018-25296-w

Zawadzka, J., Corstanje, R., Harris, J., & Truckell, I. (2020). Downscaling Landsat-8 land surface temperature maps in diverse urban landscapes using multivariate adaptive regression splines and very high resolution auxiliary data. *International Journal of Digital Earth, 13*, 899. https://doi.org/10.1080/17538947.2019.1593527

Zeng, C., Shen, H., Zhong, M., Zhang, L., & Wu, P. (2015). Reconstructing MODIS LST based on multitemporal classification and robust regression. *IEEE Geoscience and Remote Sensing Letters, 12*, 512. https://doi.org/10.1109/LGRS.2014.2348651

Zeng, C., Long, D., Shen, H., Wu, P., Cui, Y., & Hong, Y. (2018). A two-step framework for reconstructing remotely sensed land surface temperatures contaminated by cloud. *ISPRS Journal of Photogrammetry and Remote Sensing, 141*, 30. https://doi.org/10.1016/j.isprsjprs.2018.04.005

Zhan, W., Chen, Y., Zhou, J., Wang, J., Liu, W., Voogt, J., Zhu, X., Quan, J., & Li, J. (2013). Disaggregation of remotely sensed land surface temperature: Literature survey, taxonomy, issues, and caveats. *Remote Sensing of Environment, 131*, 119. https://doi.org/10.1016/j.rse.2012.12.014

Zhan, W., Huang, F., Quan, J., Zhu, X., Gao, L., Zhou, J., & Ju, W. (2016). Disaggregation of remotely sensed land surface temperature: A new dynamic methodology. *Journal of Geophysical Research, 121*. https://doi.org/10.1002/2016JD024891

Zhang, X., Friedl, M. A., Schaaf, C. B., & Strahler, A. H. (2004a). Climate controls on vegetation phenological patterns in northern mid- and high latitudes inferred from MODIS data. *Global Change Biology, 10*, 1133. https://doi.org/10.1111/j.1529-8817.2003.00784.x

Zhang, X., Friedl, M. A., Schaaf, C. B., Strahler, A. H., & Schneider, A. (2004b). The footprint of urban climates on vegetation phenology. *Geophysical Research Letters, 31*. https://doi.org/10.1029/2004GL020137

Zhang, P., Imhoff, M. L., Wolfe, R. E., & Bounoua, L. (2010). Characterizing urban heat islands of global settlements using MODIS and nighttime lights products. *Canadian Journal of Remote Sensing, 36*, 185. https://doi.org/10.5589/m10-039

Zhang, G., Xiao, X., Dong, J., Kou, W., Jin, C., Qin, Y., Zhou, Y., Wang, J., Menarguez, M. A., & Biradar, C. (2015). Mapping paddy rice planting areas through time series analysis of MODIS land surface temperature and vegetation index data. *ISPRS Journal of Photogrammetry and Remote Sensing, 106*, 157–171.

Zhang, X., Zhao, H., & Yang, J. (2019). Spatial downscaling of land surface temperature in combination with TVDI and elevation. *International Journal of Remote Sensing, 40*, 1875. https://doi.org/10.1080/01431161.2018.1489164

Zhang, X., Zhou, J., Liang, S., & Wang, D. (2021). A practical reanalysis data and thermal infrared remote sensing data merging (RTM) method for reconstruction of a 1-km all-weather land surface temperature. *Remote Sensing of Environment, 260*, 112437. https://doi.org/10.1016/j.rse.2021.112437

Zhou, Y., Qin, Z., & Bao, G. (2012). Land surface temperature estimation under cloud cover with GIDS. *Yaogan Xuebao- Journal of Remote Sensing, 16*, 492–504.

Zhou, D., Zhao, S., Liu, S., Zhang, L., & Zhu, C. (2014a). Surface urban heat island in China's 32 major cities: Spatial patterns and drivers. *Remote Sensing of Environment, 152*, 51. https://doi.org/10.1016/j.rse.2014.05.017

Zhou, W., Qian, Y., Li, X., Li, W., & Han, L. (2014b). Relationships between land cover and the surface urban heat island: Seasonal variability and effects of spatial and thematic resolution of land cover data on predicting land surface temperatures. *Landscape Ecology, 29*, 153. https://doi.org/10.1007/s10980-013-9950-5

Zhou, B., Rybski, D., & Kropp, J. P. (2017). The role of city size and urban form in the surface urban heat island. *Scientific Reports, 7*. https://doi.org/10.1038/s41598-017-04242-2

Zhou, D., Xiao, J., Bonafoni, S., Berger, C., Deilami, K., Zhou, Y., Frolking, S., Yao, R., Qiao, Z., & Sobrino, J. A. (2019). Satellite remote sensing of surface urban heat islands: Progress, challenges, and perspectives. *Remote Sensing, 11*. https://doi.org/10.3390/rs11010048

Zhu, X., Chen, J., Gao, F., Chen, X., & Masek, J. G. (2010). An enhanced spatial and temporal adaptive reflectance fusion model for complex heterogeneous regions. *Remote Sensing of Environment, 114*, 2610. https://doi.org/10.1016/j.rse.2010.05.032

Ziter, C. D., Pedersen, E. J., Kucharik, C. J., & Turner, M. G. (2019). Scale-dependent interactions between tree canopy cover and impervious surfaces reduce daytime urban heat during summer. *Proceedings of the National Academy of Sciences of the United States of America, 116*, 7575. https://doi.org/10.1073/pnas.1817561116

Chapter 3
Spatiotemporal Patterns of Landscapes and Surface Temperature

3.1 Introduction

Significant impacts of UHI: The deterioration of the UTE has become one of the most obvious problems of global urban climate change, which is spatially manifested in the increasing temperature difference between cities and their surroundings globally, i.e., the intensification of the UHI effect (O'Malley et al., 2015). This phenomenon has significant negative impacts on the health of urban residents (Chen et al., 2021). At the same time, the negative impact of UHI effect is likely to be intensified due to urbanization and global warming (Halder et al., 2021).

Relationship between LULC and thermal environment: China's population is rapidly migrating toward large-and medium-sized cities and urban agglomerations. Previous research has indicated a positive correlation between urban population, urban area, and UHI intensity (Peng et al., 2012). The population migration trend will generate much anthropogenic heat in urban areas, exacerbating UHI effects. Consequently, a growing population will suffer heat-related issues in the future. Previous studies have highlighted the significant role of landscape and land use/land cover (LULC) changes resulting from spatial urbanization and urban renewal in exacerbating the deterioration of the thermal environment (Yang et al., 2022). These changes contribute to the local variability of the thermal environment (Liu & Weng, 2009; Zhou et al., 2011). Therefore, an examination of how LULC changes disrupt regional thermal environments within the context of urbanization, as well as the impact mechanisms of local urban landscape structure on the spatial variations in UTE, can offer valuable insights. This research provides a scientific basis for optimizing regional and urban spatial planning, managing urban expansion, and enhancing the mitigation of regional and UTE. These efforts are crucial for building inclusive, secure, resilient, and sustainable cities and human settlements.

Representativeness of the study area: The Guanzhong Plain urban agglomeration is a crucial economic corridor facilitating exchanges between eastern and

L. Zhou et al., *Urban Surface Thermal Environments*, The Urban Book Series,
https://doi.org/10.1007/978-3-031-94499-4_3

western China. Meanwhile, the conflict between the ecological conservation and urban expansion is growing (Peng et al., 2022). Notably, the region has experienced significant changes in land use, particularly the conversion of farmland into construction land and transformations between farmland and natural ecological land. These shifts in land use structure and land use transitions in the Guanzhong Plain urban agglomeration create an ideal area for examining the influence of LULC changes on regional thermal environments.

3.2 Study Area

The first level of the study area considered in this book is the Guanzhong Plain urban agglomeration. The urban agglomeration is located in the central part of China, belonging to the middle and lower reaches of the Yellow River basin, with the Qinling Mountains in the south and the Loess Plateau in the north, and the Weihe and Fenhe rivers as the main water systems, important for communicating the economic exchanges between the east and the west. The region covers an area of 1.07×10^5 km^2 and involves 11 prefecture-level cities in Shaanxi, Gansu, and Shanxi provinces (Fig. 3.1), including the entire region of Xi'an City, Tongchuan City, Baoji City, Xianyang City, Weinan City, Shangluo City, etc. The Guanzhong Plain is an important birthplace of Chinese civilization and has a unique strategic position in the national modernization. The land use type in the urban agglomeration is mainly farmland, accounting for more than 40%. There are rapid interconversions between farmland and construction land as well as other natural ecological land, thus facilitating the examination of the temperature effects of various land use changes, especially those not directly related to urbanization.

The second level of the study area considered in this book is the Xi'an-Xianyang metropolitan area (Cui et al., 2024), i.e., the major urban areas and adjacent districts and counties of Xi'an and Xianyang cities. Xi'an-Xianyang metropolitan area has a continental monsoon climate with four distinct seasons. In 2020, the annual average temperature of Xi'an and Xianyang is 14.3 °C and 14.1 °C, the annual precipitation is 691.3 mm and 709.5 mm, and the permanent resident population is up to 12.96 million and 3.95 million at the end of the year, respectively. Xi'an and Xianyang are only 25 km away from each other, which historically have been part of the same city. Their economic, trade, and social systems are closely connected; thus, the built-up area at the junction continues to increase, and the SUHI effect is further exacerbated. Therefore, it is necessary to take the heat island of Xi'an–Xianyang metropolitan area as a whole to formulate heat mitigation strategies for urban sustainable development. Its moderate scale also facilitates the extraction of heat island areas and the identification of the spatial hierarchy of their thermal environments.

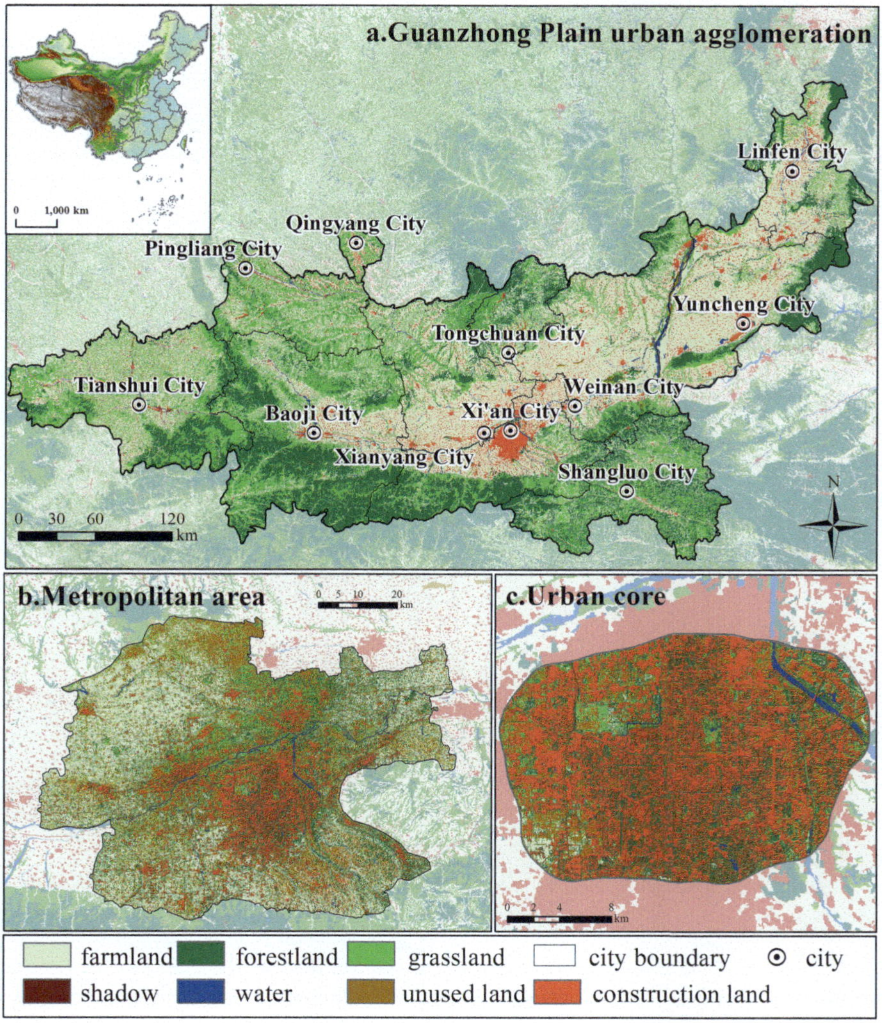

Fig. 3.1 Study area. (**a**) Guanzhong Plain urban agglomeration. (**b**) Metropolitan area. (**c**) Urban core

The third level of the study area of this book is the urban core of Xi'an. Xi'an is not only the capital of Shaanxi Province but also the ancient capital of 13 dynasties in China and the starting point of the Silk Road. As of 2022, Xi'an has a total built-up area of 868.2 km^2 and an urban population of 10,034,400, with an urbanization rate of 79.59% (Li et al., 2019). It is the region with the most significant urbanization and thermal environment deterioration in the study area and thus the region with the most frequent and influential heat wave events.

3.3 Methodology

3.3.1 *Standard Deviation Ellipse*

The standard deviation ellipse method can accurately reveal the spatial distribution law and evolutionary characteristics of the study object (Hoh et al., 2000). Its main parameters include the center of gravity of the ellipse, the azimuth angle, and the long and short axes of the ellipse. The elliptical center of gravity can characterize the center of gravity of geographical elements and their migration trajectory, the azimuth represents the main trend direction of their distribution in two-dimensional space, and the long and short axes of the ellipse reflect the distribution form of geographical features. The calculation of its three parameters is as follows (Milano et al., 2018).

$$\overline{X} = \sum_{i=1}^{n} wx_i / \sum_{i=1}^{n} w_i, \overline{Y} = \sum_{i=1}^{n} wy_i / \sum_{i=1}^{n} w_i \tag{3.1}$$

$$\tan\alpha = \left[\left(\sum_{i=1}^{n} w_i x'^2 - \sum_{i=1}^{n} w_i y'^2 \right) + \sqrt{ \left(\sum_{i=1}^{n} w_i x'^2 - \sum_{i=1}^{n} w_i y'^2 \right)^2 + 4\sum_{i=1}^{n} w_i x'^2 y'^2 } \right] / 2\sum_{i=1}^{n} w_i x' y' \tag{3.2}$$

$$\sigma_x = \sqrt{ \sum_{i=1}^{n} \left(wx'\cos\alpha - wy'\sin\alpha \right)^2 / \sum_{i=1}^{n} w_i^2 } \tag{3.3}$$

$$\sigma_y = \sqrt{ \sum_{i=1}^{n} \left(wx'\sin\alpha - wy'\cos\alpha \right)^2 / \sum_{i=1}^{n} w_i^2 } \tag{3.4}$$

where $\overline{X}$ and $\overline{Y}$ are horizontal and vertical coordinates of the standard deviation elliptic center of gravity, respectively; (x_i, y_i) is the spatial location of the geographical element i; w_i is its corresponding weight; α is the azimuth angle; δx and δy are the standard deviations along the x and y axes, respectively; and x' and y' represent the difference between the coordinates of the center of gravity and the coordinates of the geographic feature.

3.3.2 *Elastic Net Regression*

Elastic net regression is a linear regression model with added regular terms, which can be seen as a compromise between ridge regression and Lasso regression, which considers both L1 and L2 regularization and has a great advantage to deal with the linearly dependent variables. In principle, the method controls the proportion of two types of regularizations through parameter *l1_ratio* and obtains the best estimation result by adjusting the parameter.

$$\underset{W}{\operatorname{argmin}} \frac{1}{2n_{\text{sample}}} Y - XW_2^2 + \alpha l1\_\text{ratio}\, W_1 + 0.5\alpha \left(1 - l1\_\text{ratio}\right) W_2^2 \tag{3.5}$$

where Y is the explanatory variable, X is the explanatory variable, W is the regression coefficient to be estimated, and α is the parameter that controls the size of the regularization. This method has been used in thermal environment studies, especially for estimating the contribution of linearly correlated scale coefficients of several land types to LST (Zhao et al., 2020).

3.4 Spatiotemporal Patterns of Land Use and Urban Thermal Environment

3.4.1 Land Use

Considering the land use structure, from 2000 to 2018, the land use type of the Guanzhong Plain urban agglomeration was mainly farmland, with an average of 44.80%, followed by grassland and forestland, with an average of 27.35% and 21.58%, respectively (Table 3.1). In addition to these three main types of land use, the two types of construction land in rural settlements and construction land occupy an average of 3.56% and 1.15% of the land, respectively, with an average of 1.39% of water and the smallest average of unused land, only 0.17%. Spatially, farmland is mainly distributed in the plains of the central and eastern parts of the urban agglomeration and occupies most of the land in the plains (Fig. 3.2). Correspondingly, forestland and grassland are mainly distributed in mountains and gentle slopes. In particular, forestland is mainly concentrated in the southwest of the urban agglomeration in the steep Qinling Mountains and Liupan Mountains. Grasslands are not only distributed in the abovementioned areas but also widely distributed in Shangluo City, Qiaoshan Mountains (Meridian Ridge) southern foothills, and in the northern

Table 3.1 Changes in land use type area in Guanzhong Plain urban agglomeration from 2000 to 2018 (km^2)

Year	Farmland	Forestland	Grassland	Water	Town land	Rural settlements	Unused land
2000	49,628.61	22,984.49	28,841.19	1240.69	880.32	3515.65	159.89
2005	48,863.99	23,001.31	29,112.23	1279.46	1036.53	3709.94	149.19
2010	47,737.50	23,186.41	29,693.44	1263.4	1279.69	3846.88	153.59
2015	47,041.64	23,156.47	29,653.97	1239.76	1934.88	3960.7	173.08
2018	46,785.63	23,391.97	29,229.76	1143.05	2316.13	4033.38	260.66

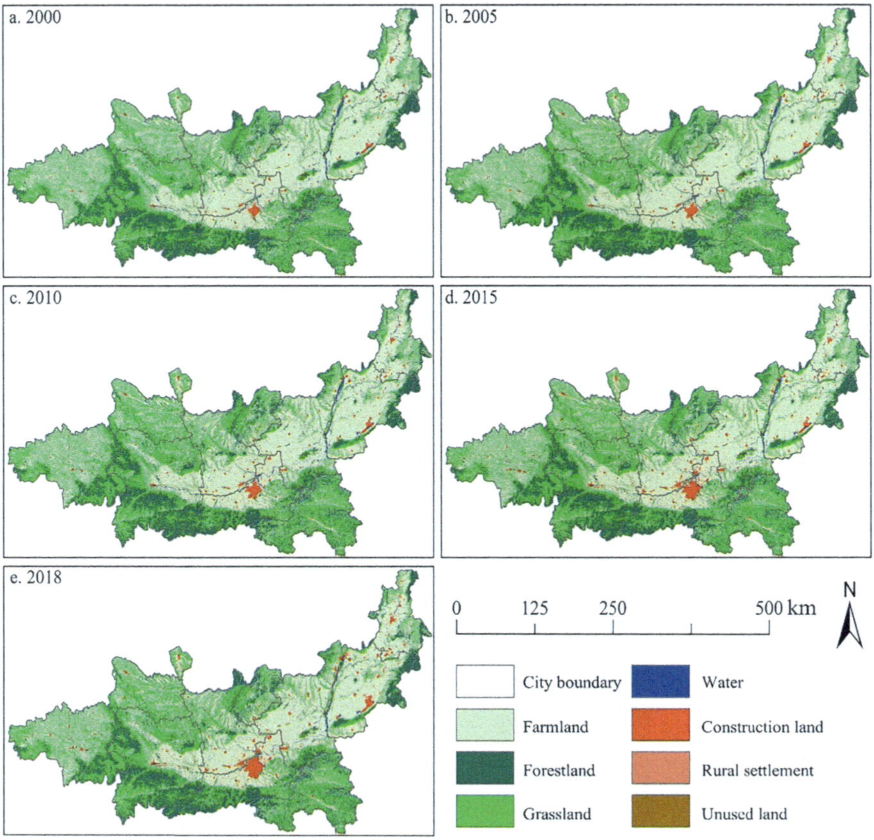

Fig. 3.2 Changes in land use type area in Guanzhong Plain urban agglomeration from 2000 to 2018 (km2)

Xianyang City, Qingyang City, and Tianshui City of the loess hilly and gully area mixed distribution with farmland. Construction land is mainly concentrated in the Xi'an–Xianyang (Xi-Xian) region, especially in the metropolitan circle composed of the two municipal districts of Xi–Xian region and its surrounding counties and cities.

It is noteworthy that in the 100 m resolution land use data used in the study, the bare land generated by the construction of the urban suburbs was not identified as unused land, and it was mainly distributed in the middle of Weinan City, as a bare land due to fallow, but this finding could not explain why this phenomenon was not seen in other farmland areas of the urban agglomeration, so the unused land distribution results in the dataset needed attention. This spatial distribution shows that the basic land use structure of urban agglomerations is comprehensively affected by topography, climate, and human economic activities, and the plain areas with moderate precipitation and ease of development are almost completely occupied by farmland and construction land, while southern mountainous areas that are difficult to develop are dominated by natural vegetation, and the northern hilly areas are mainly mixed with arable land and vegetation.

Considering the trend of change, from 2000 to 2018, only farmland has shown a decreasing trend, while forestland, grassland, and waters have shown a fluctuating trend, and only waters have shown a decreasing trend as a whole (Table 3.1), while construction land and rural settlements have been showing an increasing trend. Considering the overall amount of change, the area reduction mainly occurred in farmland, which lost 2586.97 km^2, accounting for 2.65% of the total area of the urban agglomeration, while the increase in the area mainly occurred in construction land, which rose by 1435.81 km^2, accounting for 1.34% of the total area. In terms of periods, farmland decreased the fastest between 2005 and 2010, reaching 1126.49 km^2, while the slowest decrease was 256.01 km^2 between 2015 and 2018, even taking into account the average rate of change per year. At the same time, construction land increased the fastest between 2010 and 2015, with a total increase of 655.19 km^2, slightly higher than its increase rate in 2015–2018. Among other land types, the change of grassland is also noteworthy, which increased by 271.04 km^2 and 581.21 km^2 in 2000–2005 and 2005–2010, respectively, and decreased by 39.47 km^2 and 424.21 km^2 in the next two periods, which is comparable to the above two land types.

Analysis of land use interconversion also reveals interesting patterns (Fig. 3.3). Between 2005 and 2010, when farmland saw its most significant decline, the acquired land primarily consisted of forestland and grassland, rather than the two types of construction land. In contrast, during 2010–2015, when construction land expanded rapidly, almost all newly acquired land came from farmland. These trends are more explicitly demonstrated by calculating the change transfer matrix. Moreover, the spatial shifts between farmland, forestland, and grassland have a

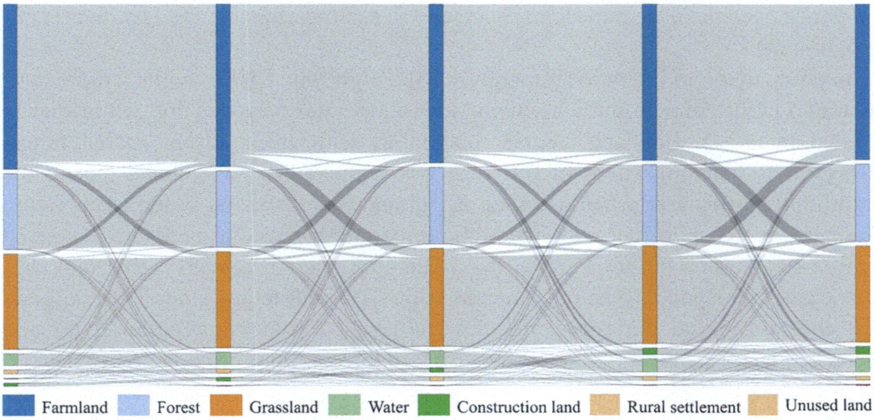

Fig. 3.3 Land use type transfer in the Guanzhong Plain urban agglomeration from 2000 to 2018

more pronounced impact than the encroachment of construction land on ecological and semi-ecological areas characterized by lower temperatures, such as farmland, forestland, grassland, and water. Notably, the continuous decrease in grassland since 2010 can be linked to the net transfer of farmland and forestland. During 2015–2018, there was a net transfer of 223.47 km^2 of grassland to farmland, with a net transfer area of 170.37 km^2. As different vegetation types feature distinct vegetation structures and soil cover types, the interconversion between them also yields specific temperature effects. Therefore, the interconversion of farmland, forestland, and grassland may play a more significant role in determining regional-scale changes in land surface temperature distribution.

3.4.2 Surface Thermal Environments

From 2000 to 2018, there were significant changes in the spatial distribution of daytime land surface temperature (LST) during summer within the urban agglomeration, influenced by alterations in landscape structure. However, the distribution of nighttime LST remained relatively consistent across different years. Notably, there were substantial differences in temperature patterns between day and night during this period (Fig. 3.4).

In 2000, regions with mid-to high-temperature daytime conditions were primarily located in the central and eastern plains of the urban agglomeration, as well as in the gullies and hills of the northwestern Loess Plateau. Conversely, areas with low temperatures were mainly found in the southwestern mountainous regions, particularly in the Qinling and Liupan mountains. The Qianshan area to the north exhibited moderate to high temperatures despite its higher vegetation cover and altitude. Over time, medium-to high-temperature areas gradually became concentrated in the densely populated plains, while high-temperature areas in the Loess Hills of Tianshui and Pingliang progressively diminished. The high-temperature areas on the plains became more prevalent, especially in regions characterized by dense construction land.

In comparison to daytime temperatures, the nighttime LST exhibits a more concentrated high-temperature zone, while low-temperature regions are still primarily located in the Qinling and Liupan mountains, with an expanded spatial range. However, the spatial extent of the very low-temperature zone shank during the nighttime, particularly when assessed using standard deviation-based temperature grading (Fig. 3.5). Furthermore, the high-temperature areas during the night consistently concentrated in the plains of the urban agglomeration, with higher temperature areas corresponding to the concentration of construction land. This suggests that the substantial anthropogenic heat emissions from construction land and increased heat retention by the urban structure have a more pronounced impact on nighttime LST.

The clustered distribution of medium-and high-temperature zones across the plains of urban agglomerations, as well as the high temperatures in the Qingyang

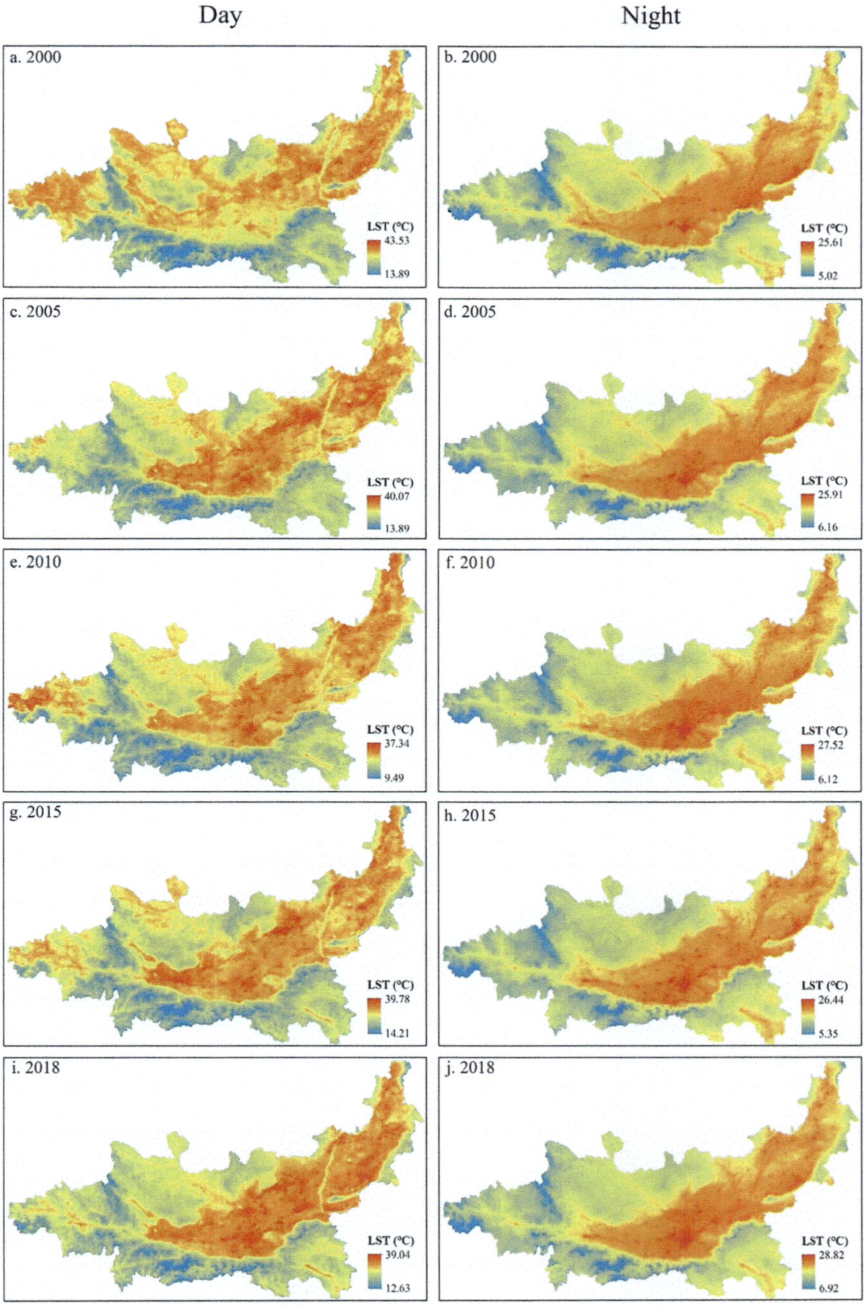

Fig. 3.4 Summer LST distribution in the Guanzhong Plain urban agglomerations from 2000 to 2018

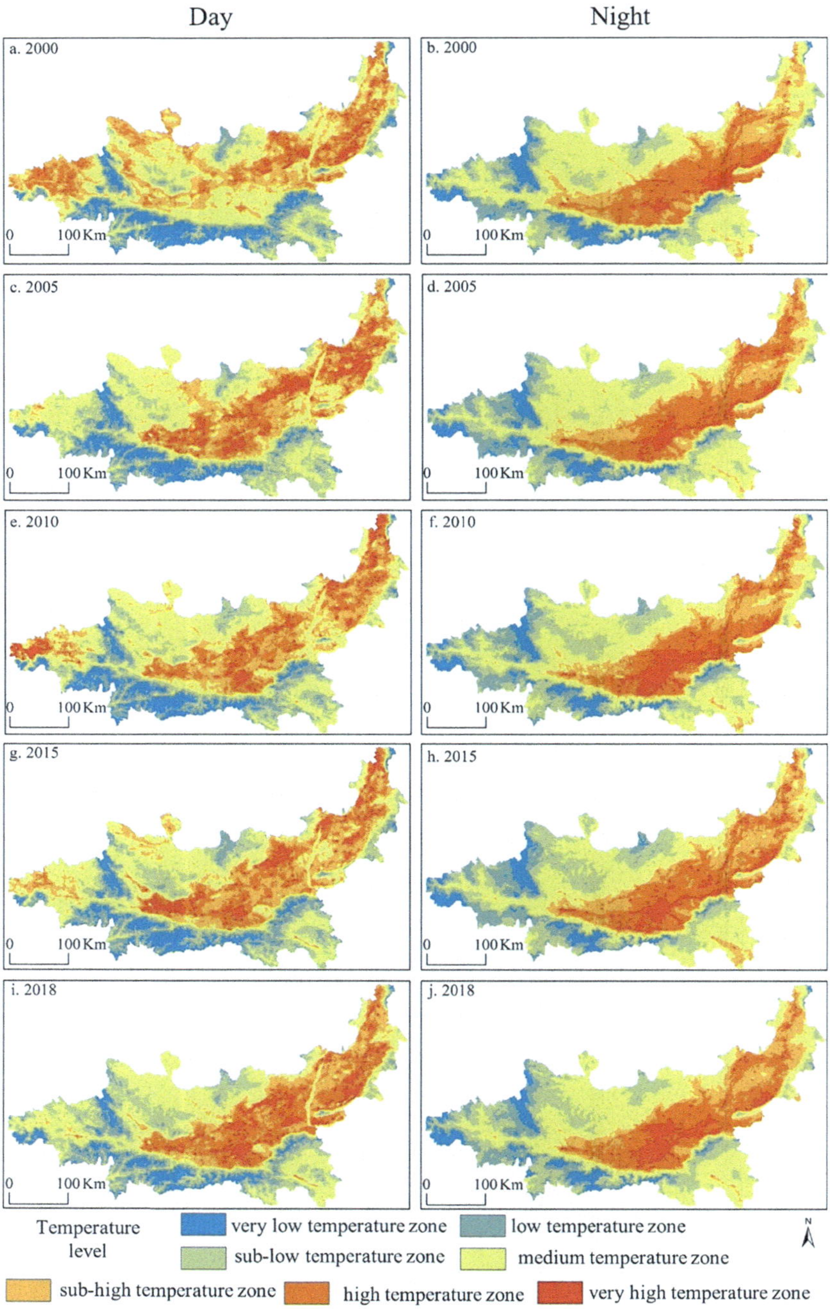

Fig. 3.5 Distribution of different levels of temperature zones in the Guanzhong Plain urban agglomeration in summer from 2000 to 2018

and Tianshui loess gully areas in 2000, indicate that the cooling effect of farmland with vegetation cover may be less significant compared to that of forestlands and grasslands. However, it is essential to note that given the mixed distribution of farmland and rural settlements in the temperature data with 1 km resolution, the temperature contribution of farmland may require a more quantitative characterization. This chapter also revealed that with the expansion of construction land, the spatial distribution pattern of daytime and nighttime temperatures is becoming increasingly similar and convergent.

In terms of the descriptive statistical characteristics of LST distribution during each year, despite the significant difference of over 10 °C between daytime and nighttime LST, the average daily and nighttime temperatures did not exhibit the expected changes over the years, as indicated in Table 3.2. For instance, the average daytime LST was 30.81 °C in 2000, but it dropped to 27.22 °C in 2010, following the rapid expansion of towns. Meanwhile, the nighttime average hovered around 18 °C in the initial years but increased to 20.02 °C in 2018.

These results suggest that even though the study employed the average LST in July and August during summer to portray the differences in the surface thermal environment of the urban agglomerations, achieving comparability of LST values between different years may be challenging. This phenomenon can be attributed to a combination of factors. First, while landscape structure influences the spatial differentiation of LST, its absolute value remains significantly affected by temporary weather systems, which might not be fully mitigated even through multi-month averages. Second, when calculating the average LST, some cloud pixels were not excluded by the cloud mask, contributing to a reduction in the average LST. Moreover, it is noteworthy that, although there are significant differences between daytime and nighttime temperatures, there is not a discernible trend in the changing standard deviation of daytime temperature. The study anticipates that as construction land expands and the number of high-temperature areas increases, differences in temperature distribution will likely grow as well.

This chapter also found that there were significant extreme values in the LST distribution in each period (Fig. 3.6), and the difference between the maximum and

Table 3.2 LST statistical values of Guanzhong Plain urban agglomeration from 2000 to 2018 (°C)

Year	Time	Mean	SD	Max	Min	98% Quantile	2% Quantile
2000	Day	30.81	4.71	43.53	13.89	37.83	20.19
	Night	18.29	2.67	25.61	5.02	22.58	12.13
2005	Day	28.47	4.48	40.07	13.89	36.36	19.18
	Night	18.22	2.83	25.91	6.16	22.79	12.03
2010	Day	27.22	3.76	37.34	9.49	33.47	19.07
	Night	18.21	2.93	27.52	6.13	23.27	12.04
2015	Day	28.84	4.21	39.78	14.21	36.09	20.16
	Night	18.21	2.95	26.45	5.35	23.26	12.09
2018	Day	28.56	4.43	39.04	12.63	36.25	19.61
	Night	20.02	3.18	28.83	6.92	25.42	13.03

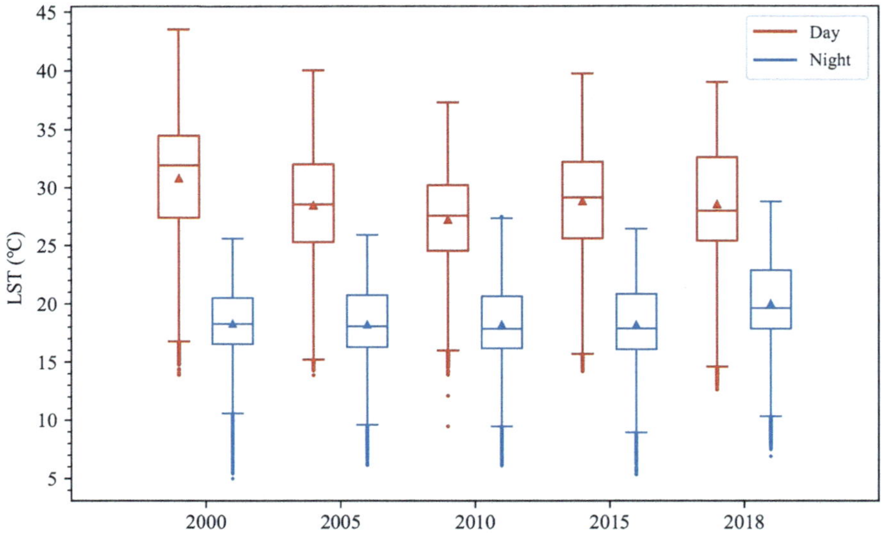

Fig. 3.6 LST distribution of Guanzhong Plain urban agglomeration during 2000–2018

minimum values of the distribution and the corresponding 98% quantile and the 2% quantile could reach an average of 7 °C. This feature is more pronounced in the box plot, and it can also be seen that at 1.5 times quartiles, the extreme values are mainly found in low temperatures, which may further amplify the cooling effect of the forestlands of the Qinling Mountains due to the elevation differentiation within the urban agglomeration.

Based on the aforementioned results, directly comparing temperatures between different years to quantitatively determine the spatial and temporal evolution of the thermal environment may not be a reasonable approach. Therefore, the study adopted a different method by categorizing temperature distribution into various classes based on mean values plus or minus 0.5 time, 1 time, and 1.5 times the standard deviations. Changes in the proportion of regions within these temperature classes and their transitions were then analyzed. On average, it was observed that the proportion of regions in different temperature classes remained fairly consistent during both daytime and nighttime. The medium-temperature region, constituting the most substantial proportion, averaged 31.47% during the day and 32.84% at night. Following this, the very high-temperature region, the high-temperature region, and the sub-low temperature region each accounted for between 10% and 20%, while the very high-temperature region had the smallest proportion, around 4% (Fig. 3.7).

During the daytime, there was an overall increase in the percentages of very high-temperature, high-temperature, and sub-low temperature zones, while at night, the proportions of low-temperature, sub-low temperature, and very high-temperature zones experienced growth. In terms of total changes, during the day, the sub-low temperature zone exhibited the most significant increase, with a total increase of

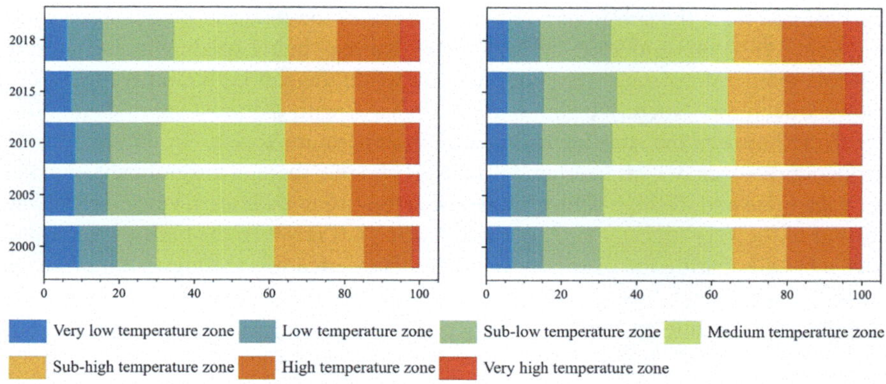

Fig. 3.7 The area proportion of different grades of temperature zones during 2000–2018

9102 km^2, equivalent to 8.50% of the total area. Conversely, the sub-high temperature zone showed the most significant decrease, down by 11,655 km^2, accounting for 10.88% of the total area, which is nearly half of its original area. During the nighttime, the sub-high temperature zone saw the greatest expansion, with a total increase of 3865 km^2 or 3.61% of the total area, while the medium-temperature zone experienced the most notable reduction, with a total decrease of 2651 km^2, amounting to 2.47% of the total area.

If the high-temperature zone is merged with the very high-temperature zone, the low-temperature zone is merged with the very low-temperature zone, and the medium-temperature zone is merged with the second temperature zone, the area of the merged high-temperature zone shows an overall increase, while the merged low- and medium-temperature zones show a decrease. Therefore, it can be said that although there is a trend of decreasing high-temperature zones and increasing low-temperature zones, the overall temperature distribution of the urban agglomeration is converging to high-temperature zones. However, it should be noted that the sub-high temperature zones show an overall decreasing trend regardless of day and night, while the sub-low temperature zones show an increasing trend, up to 10.88% of the urban agglomeration area.

During nighttime, the overall transfer volumes were more consistent across different periods, ranging from 28,000 to 29,000 km^2. This was even smaller than the minimum value recorded for overall daytime transfer volumes. The daytime transfers were primarily observed between high-temperature, sub-high temperature, medium-temperature, and sub-low temperature zones. Notably, transfers from the medium-temperature and sub-high temperature zones were significant, spanning both sides of the region. In contrast, there were fewer transfers across the very high-temperature, low-temperature, and sub-low temperature zones, owing to their smaller areas.

It is noteworthy that there is a large net daytime shift of area from the medium-temperature zone to the sub-low temperature zone, while there is also a large net shift of area from the sub-high temperature zone to the medium-temperature zone between 2000–2005 and 2005–2010, a result that may help explain why the overall sub-low temperature zone shows an increase while the sub-high temperature zone shows a decrease. Nocturnal shifts are less biased, with the most significant shifts occurring between the sub-low and medium-temperature zones. For the very high-temperature zones, which overlap with the urban area, there is a significant transfer between them and the high-temperature zones relative to their area, i.e., a significant portion of the very high-temperature zones also shift down to the high-temperature zones, but comparing the distribution images shows that this shift does not occur in the urban center and thus may not be associated with urban renewal, but more with changes on non-built-up land.

3.5 Association Between Land Use Type and Surface Thermal Environment

Although the correlation between temperature intervals and land use types can be broadly observed from the above analysis, this correlation needs to be corroborated by more quantitative evidence. Here, the study investigates the covariance between landscape structure and LST spatial and temporal patterns as well as the quantitative temperature contribution of each class to determine the association between land use type and surface thermal environment in a more refined manner.

3.5.1 Spatiotemporal Patterns of Land Surface Temperature

During 2000–2018, despite the continuous expansion of the construction land area, there were no significant alterations in its center of gravity and standard deviation ellipse. This suggests that the expansion of construction land within the urban agglomeration was relatively uniform in all directions, with a primary concentration near the center of the ellipse, particularly in the north-central section of the urban agglomeration's plain area (Fig. 3.8). The distribution of the high-temperature zone experienced notable changes during the daytime. The high-temperature zones are dispersed toward the plain region of the urban agglomeration. Simultaneously, the orientation of the ellipse gradually shifted, evolving from the direction connecting the two ends of the urban agglomeration to the direction of the urban plain. This transition resulted in a total deflection angle of 19.3°. However, during the night-time, the concentration of high-temperature areas within the urban agglomeration's plain was significant in 2000. Consequently, this concentration did not significantly affect the extent, center of gravity, or direction of the standard deviation ellipse.

By considering the covariance of spatial and temporal patterns, it is evident that construction land expansion primarily occurred in the central and northern parts of

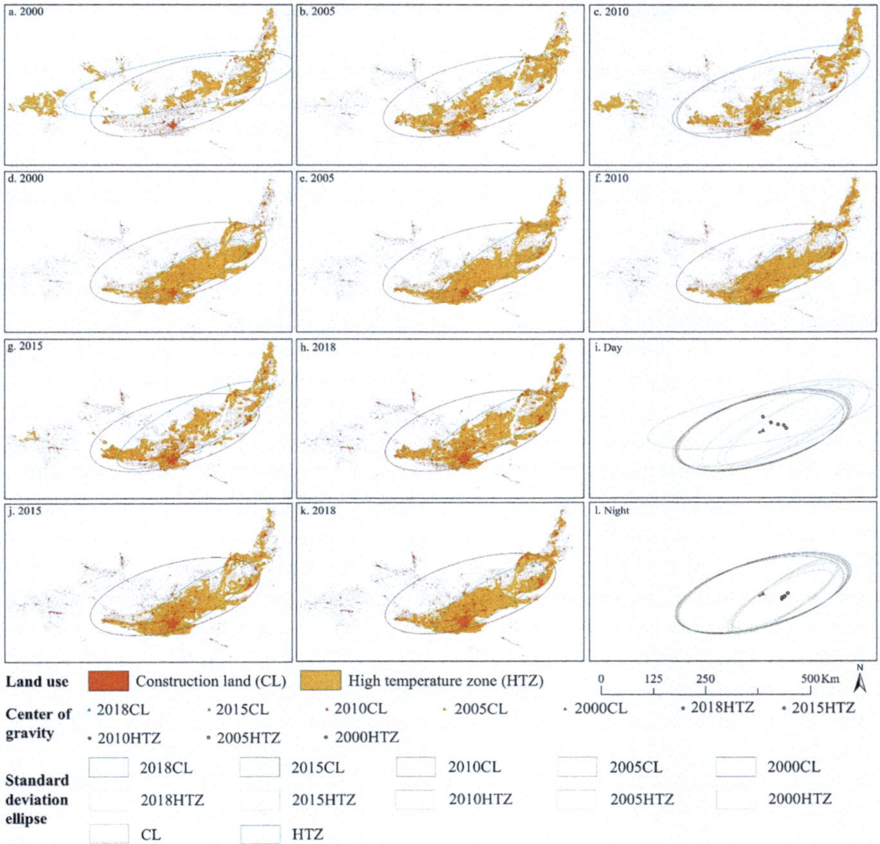

Fig. 3.8 Spatiotemporal evolution of construction land and high-temperature zones in Guanzhong Plain urban agglomeration

the urban plain. Simultaneously, high-temperature areas also clustered within the urban plain. The study also observed that the reduction of the high-temperature zone in non-construction land within the loess gullies in the northwestern part of the urban agglomeration contributed to the concentration of the high-temperature zone within the urban plain area.

3.5.2 *Temperature Contribution of Each Land Class*

The study quantified the temperature contribution of six land use types—farmland, forestland, grassland, water, construction land, and rural settlements—during 2000–2018 using a multi-task elastic net regression method. However, the temperature contribution of unused land was challenging to obtain due to its small proportion (only 0.17%) and concentrated distribution. This proportion variable was often excluded in the regression model or had coefficients that were too small. In the regression results, the explained variance fell within the range of 51–54%, indicating that the model's accuracy was maintained. Separate elastic network regressions were conducted for daytime and nighttime temperatures, with the explained variance for daytime results ranging from 55% to 60%, while the nighttime results explained approximately 40% of the variance. Consequently, the study chose to use the results of the multi-task regression. These regression results are depicted in Fig. 3.9, where each value represents the increase or decrease in temperature by increasing the proportion of the corresponding land type by 1% in a 1 km grid.

In terms of the positive or negative contribution of temperature, the study found that farmland, construction land, and rural settlements exhibited a positive temperature contribution, while forestland and grassland showed a negative temperature contribution. Water bodies, however, showed a positive temperature contribution, except in 2000 when the temperature contribution of water was almost negligible. This pattern was consistent during both daytime and nighttime. The anomaly observed in the case of waters may be due to the cooling effect of linear waters in the plains, which might be much less pronounced than the cooling effect of forestlands and grasslands located in mountainous areas, resulting in a relative warming effect at the regional scale. Additionally, the smaller area share of water makes the estimation of their temperature contribution more uncertain, showing relative warming. The anomaly in the case of farmland might be attributed to the same factors as water, with forested grassland having a more substantial cooling effect than farmland, resulting in the latter showing a relative warming effect. This difference in cooling effect can be attributed to the proximity of farmland to towns and rural settlements, making it more susceptible to the urban heat island effect, and the significantly higher soil water content in farmland due to irrigation, which can lead to higher heat storage at night.

Comparing the temperature contributions of different land types, forestland was found to have the maximum cooling contribution during both daytime and nighttime,

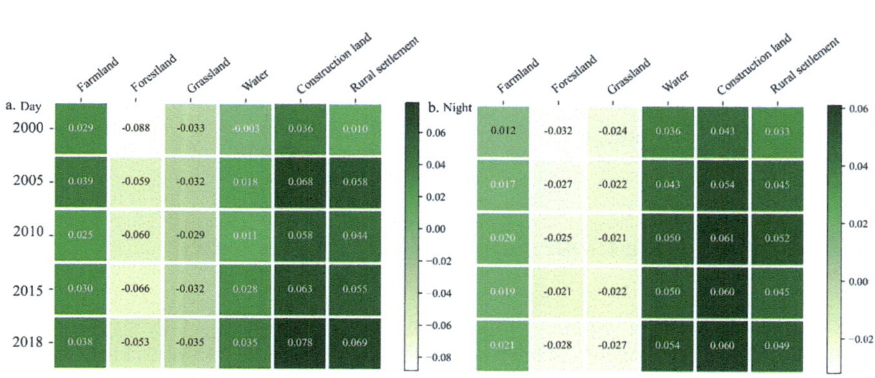

Fig. 3.9 Temperature contribution of each land use type during 2000–2018

while construction land had the maximum warming contribution. The absolute value of the cooling contribution of forestland was greater than that of the warming contribution of construction land. However, it is worth noting that the cooling contribution of forestland at night was significantly lower than during the daytime, making it roughly equivalent to that of grassland. This change may be due to the reduced intensity of the cooling effect of evapotranspiration from the forest at night compared to solar radiation during the daytime. Additionally, the lower sky openness of forestlands in comparison to grasslands at night causes them to release heat more slowly. Furthermore, it was observed that the warming contribution of waters was significantly higher at night than during the daytime, which may be attributed to the more pronounced heat storage contribution of water bodies to nighttime temperatures.

3.6 Conclusions

This chapter analyzes the land use change and surface thermal environment evolution of the Guanzhong Plain urban agglomeration from 2000 to 2018, and on this basis, the relationship between land use structure and surface thermal environment is analyzed at the urban agglomeration scale by standard deviation ellipse and elastic net regression algorithm. The results show that: (1) between 2000 and 2018, farmland as the dominant land use type decreased from 46.31% to 43.66%, while the construction land increased significantly from 0.82% to 2.16%. The mutual transfer between land use types mainly occurred between farmland, forestland, and grassland. (2) The changes in the surface thermal environment during the day are mainly manifested in the concentration of high-temperature zones to the plain area of the urban agglomeration and the expansion of high-temperature areas and the decrease in low-temperature areas. The changes at night are flatter, mainly manifested in the expansion of the high-temperature zone and the decrease of the low-temperature zone. (3) Farmland, water, construction land, and rural settlements show positive temperature contributions regardless of day and night, while forests and grasslands always have the negative temperature contributions. The temperature contribution of forests and waters changes significantly during the day and night, with the former being higher during the day and the latter higher during the night.

References

Chen, B., Xie, M., Feng, Q., Li, Z., Chu, L., & Liu, Q. (2021). Heat risk of residents in different types of communities from urban heat-exposed areas. *Science of the Total Environment, 768*, 145052. https://doi.org/10.1016/j.scitotenv.2021.145052

Cui, M., Zhou, L., Wang, W., Sun, D., Yuan, B., & Wei, W. (2024). Spatial heterogeneity and hierarchy of metropolitan area expansion and land surface temperature evolution: A Twin City perspective. *IEEE Journal of Selected Topics in Applied Earth Observations and Remote Sensing, 17*, 929–940. https://doi.org/10.1109/JSTARS.2023.3325369

Halder, B., Bandyopadhyay, J., & Banik, P. (2021). Monitoring the effect of urban development on urban heat Island based on remote sensing and geo-spatial approach in Kolkata and adjacent areas, India. *Sustainable Cities and Society, 74*, 103186. https://doi.org/10.1016/J.SCS.2021.103186

Hoh, S. T., Greenfield, D. S., Mistlberger, A., Liebmann, J. M., Ishikawa, H., & Ritch, R. (2000). Optical coherence tomography and scanning laser polarimetry in normal, ocular hypertensive, and glaucomatous eyes. *American Journal of Ophthalmology, 129*, 129–135. https://doi.org/10.1016/S0002-9394(99)00294-9

Li, J., Li, J., Yuan, Y., & Li, G. (2019). Spatiotemporal distribution characteristics and mechanism analysis of urban population density: A case of Xi'an, Shaanxi, China. *Cities, 86*, 62–70. https://doi.org/10.1016/J.CITIES.2018.12.008

Liu, H., & Weng, Q. (2009). Scaling effect on the relationship between landscape pattern and land surface temperature. *Photogrammetric Engineering and Remote Sensing, 75*, 291–304. https://doi.org/10.14358/PERS.75.3.291

Milano, V., Maisto, G., Baldantoni, D., Bellino, A., Bernard, C., Croce, A., Dubs, F., Strumia, S., & Cortet, J. (2018). The effect of urban park landscapes on soil Collembola diversity: A Mediterranean case study. *Landscape and Urban Planning, 180*, 135–147. https://doi.org/10.1016/j.landurbplan.2018.08.008

O'Malley, C., Piroozfar, P., Farr, E. R. P., & Pomponi, F. (2015). Urban Heat Island (UHI) mitigating strategies: A case-based comparative analysis. *Sustainable Cities and Society, 19*, 222–235. https://doi.org/10.1016/j.scs.2015.05.009

Peng, S., Piao, S., Ciais, P., Friedlingstein, P., Ottle, C., Bréon, F.-M., Nan, H., Zhou, L., & Myneni, R. B. (2012). Surface urban Heat Island across 419 global big cities. *Environmental Science & Technology, 46*, 696–703. https://doi.org/10.1021/es2030438

Peng, L., Zhang, L., Li, X., Wang, Z., Wang, H., & Jiao, L. (2022). Spatial expansion effects on urban ecosystem services supply-demand mismatching in Guanzhong plain urban agglomeration of China. *Journal of Geographical Sciences, 32*, 806–828. https://doi.org/10.1007/s11442-022-1973-x

Yang, Y., Guangrong, S., Chen, Z., Hao, S., Zhouyiling, Z., & Shan, Y. (2022). Quantitative analysis and prediction of urban heat Island intensity on urban-rural gradient: A case study of Shanghai. *Science of the Total Environment, 829*, 154264. https://doi.org/10.1016/j.scitotenv.2022.154264

Zhao, J., Zhao, X., Liang, S., Zhou, T., Du, X., Xu, P., & Wu, D. (2020). Assessing the thermal contributions of urban land cover types. *Landscape and Urban Planning, 204*, 103927. https://doi.org/10.1016/J.LANDURBPLAN.2020.103927

Zhou, W., Huang, G., & Cadenasso, M. L. (2011). Does spatial configuration matter? Understanding the effects of land cover pattern on land surface temperature in urban landscapes. *Landscape and Urban Planning, 102*, 54–63. https://doi.org/10.1016/J.LANDURBPLAN.2011.03.009

Chapter 4
Spatial Heterogeneity and Hierarchy of Land Surface Temperature

4.1 Introduction

Global urbanization has extensive effects on the local climate inside and around cities (Cai & Kalnay, 2004; Oke et al., 2017). The increase in the impervious surface (IS) is at the expense of loss of natural lands surrounding cities, thereby leading to deterioration of the urban thermal environment (UTE) and an increase in the temperature contrast between urban and surrounding rural areas, i.e., the urban heat island (UHI) phenomena. The deterioration of UTE has many consequences on urban sustainability, such as worsening urban environmental comfort and increasing gross energy consumption. Previous research has demonstrated that landscape transformation and land use and land cover (LULC) change due to urban expansion and urban regeneration were the main factors causing the deterioration of UTE (Deilami et al., 2018). Meanwhile, these changes had the largest contribution to the intra-city spatial variation of UTE. Therefore, studying how the local urban landscape structure quantitatively affects the variation of UTE is an important prerequisite for the improvement of UTE and the promotion of urban and regional sustainable development.

Land surface temperature (LST) is a commonly used physical parameter for characterizing the spatial variation of surface UTE. The literature has a significant amount of research on the relationship between urban landscape structure and LST with different indicators of landscape structure and modeling methods. However, there is room for hierarchical variation and the interaction effects in the relationship to obtain a more thorough understanding. This study aims to explore these characteristics in the relationship with Xi'an Metropolitan Area, which is in the arid zone of Western China and is less studied, as the case study. Thus, there are three sub-aims: (1) Derive the spatial hierarchy of the UTE from LST data, (2) explore the variation of the relationship between urban landscape structure indicators and LST across the spatial hierarchy of the UTE, and (3) examine the continuous interaction

L. Zhou et al., *Urban Surface Thermal Environments*, The Urban Book Series,
https://doi.org/10.1007/978-3-031-94499-4_4

effect indicators influencing LST. For these sub-aims, we first applied the local contour algorithm (LCT) to explore the spatial hierarchy of the UTE in the main UHI of metropolitan area. Based on the result, we examined the hierarchical relationships between landscape metrics and LST and interaction effects in the core UHI area with the most complex hierarchy of the UTE, via the extreme gradient boosting (XGBoost) model and SHapley additive exPlanations (SHAP).

4.2 Literature Review

4.2.1 Relationship Between Urban Landscape Structure and LST

At present, much previous research opened in the study of the relationship between urban landscape structure and LST (Masoudi et al., 2021; Sun et al., 2018). To characterize the urban landscape structure, previous research mainly used abundance of LULC, urban site characteristics, and landscape metrics. Moreover, a few studies even consider the landscape structure indicators in three-dimensional space (Yu et al., 2020).

Among these indicators, landscape metrics characterize the landscape pattern in terms of landscape composition and configuration. Therefore, they can measure the statistical characteristics of landscape patches and the interaction (energy exchange) between patches, which are highly relevant to the energy balance and thermal processes within and among landscape patches. In previous studies, the configurational landscape metrics at class level were mainly used, compared to the metrics at patch level and landscape level. At class level, relevant metrics could be classified into four groups, including area-related, shape-related, core-area-related, and aggregation-related metrics. The last group can be further divided into three subgroups, i.e., proximity-based, subdivision-based, and adjacency-based metrics, in terms of the definition of aggregation (McGarigal et al., 2012). Among them, the area-related, shape-related, and aggregation-related, especially the subdivision and adjacency-based metrics, are mostly used, given their relatively intuitive definition and clarified physical mechanisms for thermal processes. Conversely, the proximity-based and core-area-based landscapes were less used, given their requirement on extra parameters (e.g., searching radius). Besides, some studies also used non-patch-based indicators, e.g., local Moran's index, landscape connectivity, or Gray-level co-occurrence matrix to represent the landscape configuration.

In addition, to model the relationship, relevant research mainly used the correlation coefficients, multiple linear regression, spatial regression, Geodetector, and machine learning (ML). Among these methods, ML could capture the nonlinear responses of LST to explanatory variables and characterize them more quantitatively through ML explanation tools (Sun et al., 2020) rather than other methods.

Generally, a great number of landscape structure indicators and modeling methods have been used in relevant research; however, some limitations were also found.

4.2.2 Interaction Effects in the Relationship

Previous studies have found and explored the nonlinearity in the relationship, especially the threshold effects. However, the interaction effect or moderating effect, between influencing factors of LST, as a kind of nonlinearity, is of less concern. This term indicates that for a certain impacting factor as the moderated factor, its effects on LST depend on the valving of other factors, i.e., the moderated factors. The term is originally used in social science, but now it has been gradually explored in the drivers of UTE, although many researchers did not call the moderating effects. Considering the relationship discussed, relevant research generally divided the analytic units into groups according to the levels of a certain influencing factor (e.g., coverage or floor area ratio) to examine the inter-group difference of the relationship to explore the interaction effect (Jung et al., 2021; Lu et al., 2021). Besides, a few studies also used the Geodetector to detect the interaction effects, by testing whether the combined spatial heterogeneity of explanatory variables has greater explanatory capacity for the spatial variation of LST. Thus, this method also depends on the grouping of explanatory factors. However, further research should consider the influences of moderating factors on the thermal effect of other moderated factors in a more continuous way instead of discrete grouping. This approach may bring more refined findings about interaction effects. Moreover, it may help us find more associations between the apparent effects of landscape structure indicators on LST and the underlying physical mechanisms. Thus, in this study, the XGBoost model and SHAP method were used to examine the continuous interaction effects in the discussed relationship.

4.2.3 Spatial Hierarchy in the Relationship

In addition, to characterize the relationship more finely, some research also considered the spatial traits of the relationship including the scale dependency, location dependency (spatial heterogeneity), spatial dependency (impacts from neighboring analytic units), and spatial hierarchy dependency (nested spatial extents). To explore these traits, analytic strategies, including multiscale analysis, grouping of analytic units, local regression, spatial autoregression, and multilevel data model, are used. Among the above traits, the spatial hierarchy was less concerned. However, due to the spatial heterogeneity in the relationship, the modeled relationship (relevant parameters) may vary from the global study area to the local spatial levels with less spatial heterogeneity (Guo et al., 2020). Therefore, focusing only on the global relationship with the highest heterogeneity may be limited, and thus, the exploration of the relationship across the spatial levels is beneficial.

To the best of our knowledge, only two recent studies addressed the trait. Du et al. (2016) used the multilevel statistical model to consider the effects of influencing factors at various spatial levels on LST and the spatial heterogeneity of the effects. Guo et al. (2020) identified the spatial heterogeneity and spatial hierarchy in the relationship through a regression tree dividing the parameter space of influencing factors. In these two studies, although the spatial hierarchies of the influencing factors and relationship have been respected, it is still unclear how the UTE's own spatial hierarchy influences the relationship. Even the spatial hierarchy of UTE itself is less explored, due to the lack of appropriate analysis tools. However, these two issues are important. Identifying the spatial hierarchy of UTE could provide a new perspective for understanding the spatial structure and locating the local UHI region needing policy intervention more subtly. Moreover, exploring the variation of the relationship across the hierarchy of UTE will bring a new insight on the spatial traits of the relationship and help urban planning sections to mitigate heat stress at multiple levels.

The LCT algorithm could identify the spatially nested hierarchy in a continuous geographical field through the contour lines generated from the field. Originally, we propose to identify the geomorphic hierarchical structure. Recently, it has been also used to identify the polycentric structure of cities (Chen et al., 2017). As a generalized and data-driven method, LCT only depends on the interval between contour lines and the starting value. Based on the LST data, it could identify and delineate the spatially nested local UHI clusters which can represent the spatial hierarchy of the UTE from the perspective of local heat island. Nevertheless, only one recent study used the LCT to identify the individual high- and low-LST clusters within the UTE (Yang et al., 2021a, 2021b) and no research has been reported for the application of LCT to identify the spatial hierarchy of the UTE. Thus, in this study, the spatial hierarchy of UTEs in the main urban area of the Xi'an metropolitan area was identified using this algorithm.

4.3 Methodology

The methodology has four steps: (1) Data processing: Deriving the LST and land cover distribution from Landsat image and sentinel images through radiative transfer equation (RTE) method and random-forest-based classification, respectively; (2) Identification of spatial hierarchy: Creating contour lines from the LST data and identifying the nested local heat centers with LCT from these lines as the hierarchy of UTE; (3) Landscape metrics calculation: Calculating landscape metrics from land cover data at each analytic unit as the indicators of urban landscape structure; (4) Modeling of the multilevel relationship: Analyzing the relationship between landscape metrics and LST via the XGBoost model and SHAP method at the local heat islands with different spatial levels, respectively (Fig. 4.1).

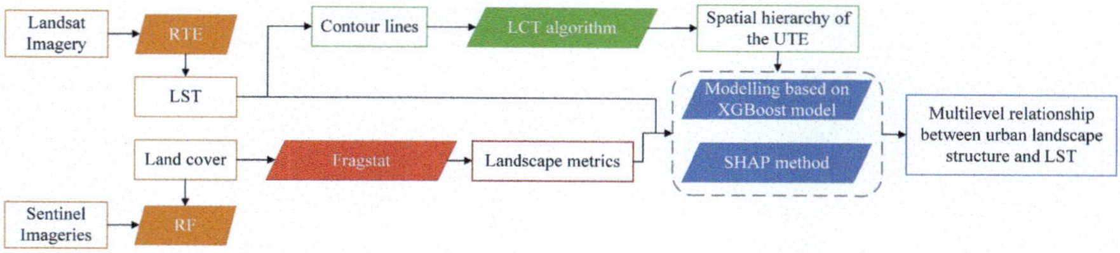

Fig. 4.1 Methodological flowchart. *RTE* Radiative transfer equation, *RF* Random forest, *LCT* Local contour tree, *LST* Land surface temperature, *UTE* Urban thermal environment, *XGBoost* Extreme gradient boosting, *SHAP* SHapley additive exPlanations

4.3.1 Study Area

Xi'an Metropolitan Area is located in the central-west Guanzhong Plain of China and is constituted by the Xi'an and Xianyang prefecture-level regions (Fig. 4.2). It is a crucial population-clustering area in the arid zone of Western China. Especially, Xi'an is a historical city of more than 2000 years and the capital of 13 dynasties. Thus, the ancient urban planning ideas, e.g., center axis symmetry and chessboard layout, have affected its urban form thoroughly, making it have significant axis characteristics. The metropolitan area has a temperate continental semi-arid climate which is controlled by the subtropical anticyclone in summer, thereby making the weather hot, calm, and arid in summer. The hottest month in Xi'an and Xianyang is July, during which the average temperature is 27.3 °C and 26.3 °C and the average precipitation is 86.0 mm and 86.1 mm, respectively. The region is undergoing a continuous process of urbanization. The built-up area and urban population increased from 327.26 km^2 and 4.13 million in 2008 to 774.57 km^2 and 7.75 million in 2018, respectively, with the increase in UHI intensity and extent. The study area includes the urban main districts of Xi'an and Xianyang and their surrounding counties and districts and excluded the mountainous areas and focuses on the main UHI areas in the Xi'an metropolitan area.

4.3.2 Data

The data used in the study mainly included Landsat8, Sentinel-2A, and Sentinel-1 GRD imageries. The Landsat8 images were from the United States Geological Survey and were used to obtain the LST. In particular, based on the Landsat 8 imagery sensed at 11:19 am, July 28th, 2019 (local time), the RTE method was used to retrieve the LST with a spatial resolution of 100 m. Among all available summertime imageries from 2016 to 2020, the thermal condition at this sensing time was most severe, given the air temperature of 34.8 °C and wind speed of 2 m/s recorded by a nearing meteorological station. The Sentinel-2A and Sentinel-1 GRD imageries obtained from the European Space Agency were used to map the land cover.

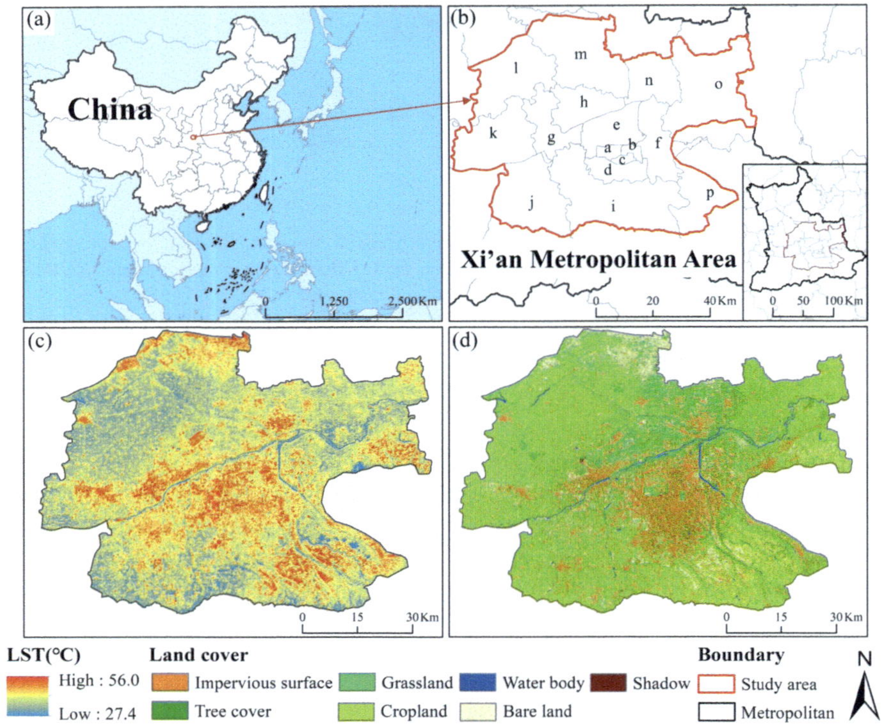

Fig. 4.2 Study area. (**a**) Xi'an Metropolitan Area location, (**b**) Xi'an Metropolitan Area range, (**c**) LST in Xi'an Metropolitan Area, and (**d**) land cover in Xi'an Metropolitan Area. Note: a: Lianhu District, b: Xincheng District, c: Beilin District, d: Yanta District, e: Weiyang District, f: Baqiao District, g: Qindu District, h: Weicheng District, i: Chang'an District, j: Huyi District, k: Xingping City, l: Liquan County, m: Jingyang County, n: Gaoling District, o: Lintong District, p: Lantian County

Specifically, based on Sentinel-2A imageries sensed on July 28th and August 12th, 2019, and Sentinel-1 GRD imageries sensed on August 10th, 2019, the random-forest-method was used to classify the land cover with a resolution of 10 m via GEE processing platform. The overall classification accuracy is 81.2% and the kappa coefficient is 0.76. Moreover, the land cover scheme includes IS, forestland, grassland, cropland, water body, bare land, and shadow. After the classification, forestland and grassland were merged into the urban green space (UGS) because in the Xi'an main urban district (core area), the grassland was minor (coverage of 7.9%) and mainly distributed in the sparse urban parks, which may create difficulties for the sequent modeling procedure if the land cover type reserves.

4.3.3 Methods

4.3.3.1 Local Contour Tree Algorithm

Local contour tree algorithm could abstract and simplify the spatial relation between contour lines derived from a continuous geographical field, to represent this field's spatially hierarchical structure (Chen et al., 2017). In principle, the algorithm first identifies the spatially inclusive relation between contour lines. Then, based on the relation, it finds the peak contours with the maximum value in their neighborhood and no other contours are inside them. The peak contours are then viewed as the leaf nodes in a contour tree, where the contours including these leaf nodes are regarded as the parent nodes and they are also the child nodes of contours with higher levels. Finally, the structure of the contour tree is constructed from bottom to top in this way. After the construction, the generated tree structure is simplified: If a parent node has a single child node, then the child node will be removed, and its children will be given to the parent node. Figure 4.3 gives an example of the algorithm. Specially, Fig. 4.3a describes 5 contour lines with inclusive relation, and Fig. 4.3b is the tree-like representation of the topological relation, where V, T, and S_2, as the sequential father–child nodes, have the same spatial level (level 1). In the simplification, the leaf node S_2 is removed first, then the node T (Fig. 4.3c). Finally, the nodes with level 1 are viewed as the elemental urban heat centers, whereas other nodes with higher spatial levels are viewed as the composite urban heat centers.

Before applying the local contour tree algorithm, we smooth the LST data via a Gaussian spatial filter with a window of 5 × 5 (the window size referred to Chen et al. (2017)) to suppress the significant local spatial heterogeneity of LST in urban areas, thereby making the hierarchical trend of LST more visible. Then, the contour lines are generated with an interval of 0.5 °C, while the interval size of 1 °C has also been tested. The result failed to represent the spatial hierarchy of UTE because the derived fundamental heat centers with the interval size of 1 °C did not match well with the real-world land covers; conversely, most of them should be considered as composite heat centers. Furthermore, the contours with a value less than the average

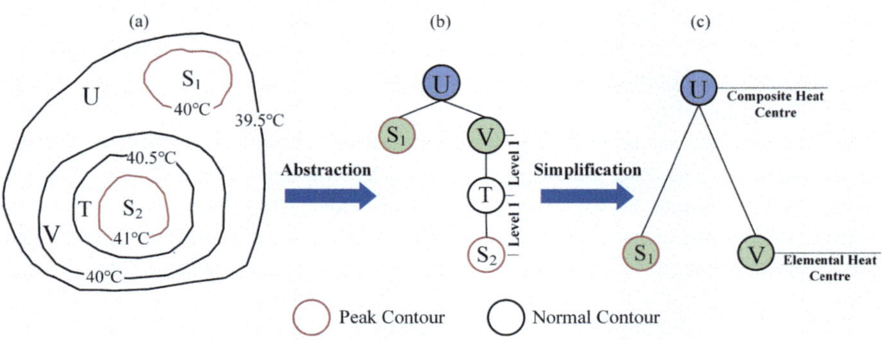

Fig. 4.3 Example for local contour tree algorithm. (**a**) LST contour lines, (**b**) original tree representation of hierarchical structure, and (**c**) simplified tree representation

temperature in built-up areas (37.5 °C) and an area less than 3 km^2, i.e., the least size of urban heat center assumed, were removed, assuming that the spatial hierarchy of UTE was only composed by local UHI clusters (heat centers), but not the cold centers. Finally, the algorithm was applied for the remaining contour lines within each contour tree to obtain the tree structures.

4.3.3.2 Landscape Metrics

Landscape metrics, as effective measurements of landscape ecological patterns, have been widely used in ecology research. This study used eight class-level landscape metrics to characterize the landscape composition and configuration of multiple land cover types (Table 4.1). Among them, only the percentage of landscape represents the landscape composition, while other metrics represent the landscape configuration in terms of area, shape, and aggregation. This study first split the land cover data into analytic units with a resolution of 300 m (square grid), then in each unit calculated the first six landscape metrics for IS, UGS, and bare land as the impacting factors. The remaining metrics (AREA_MN and ENN_AM) were only calculated in each composite heat center. Because the ENN_AM cannot be calculated within many analytic units where there was only one patch of the corresponding type. Moreover, the AREA_MN is correlated to PD and ED. The reason for using a fixed analytic unit size of 300 m is based on the balance between the variance of landscape metrics and the sample number. On the one hand, a too small unit size will make the variance of metrics small and many metrics will have a missing value in some units, considering that there may not be specific type of patches in these units due to the 10 m resolution of the land cover data. On the other hand, too large unit sizes will make the sample number not enough, thereby causing the modeling result to be unconvincing.

4.3.3.3 XGBoost Model and SHAP Explanation Method

This study used the XGBoost model and SHAP explanation method to evaluate the relative importance, quantitative effects, and the interaction effects of influencing factors on LST only at composite heat centers 15–20 because the number of analytical units in the elemental heat centers are uneven and too small (below 200) to model the relationship.

XGBoost model, as an optimized gradient boost tree model, has higher computing efficiencies than other tree structure-based ML models. Therefore, it has been widely used in studies on the drivers of UTE. Besides, the model could obtain the relative importance of impacting factors by calculating the sum of the impurity loss in each node splitting. XGBoost is performed in Python using packages "XGBoost." The dependent variable is the average LST in each analytical unit, and the independent variables are the abovementioned six landscape metrics for IS, UGS, and bare land, i.e., 18 metrics in sum. Set the parameter of XGBoost model with learning rate of 0.01, number of trees is 500, and tree complexity is 6. In addition, ten-fold cross-validation method was used to validate the modeling. We found that the mean

Table 4.1 Landscape metrics used in the study

Landscape metrics	Formula	Meaning	Landscape metrics
Percentage of landscape (PLAND, %)	$\text{PLAND}_j = \frac{\sum_{j=1}^{m} a_{ij}}{A}(100)$ a_{ij}: Area of ith patch of type i	Percentage of the areas of all patches of the corresponding type to total landscape area	Percentage of landscape (PLAND, %)
Patch density (PD, ha^{-1})	$\text{PD}_i = \frac{n_i}{A} \times 100$ n_i: Number of patches of type i	Number of patches of the corresponding type divided by total landscape area	Patch density (PD, ha^{-1})
Edge density (ED, m·ha^{-1})	$ED_i = \frac{\sum_{j=1}^{m} e_{ij}}{A_i}(10000)$ e_{ij}: Length of ith patch of type i	Total lengths of all edge segments involving the corresponding type divided by the total landscape area	Edge density (ED, m·ha^{-1})
Largest patch index (LPI, %)	$\text{LPI}_i = \frac{\max(a_{i1},\ldots,a_{im})}{A_i} \times 100$ A_i: Area of patches of type i	Percentage of area of patch with largest area of the corresponding type to total landscape area	Largest patch index (LPI, %)
Aggregation index (AI, %)	$\text{AI}_i = \left[\frac{g_{ii}}{\max \to g_{ii}}\right](100)$ g_{ii}: Number of joins between pixels of type i	Extent of aggregation of the corresponding type	Aggregation index (AI, %)
Area-weighted mean SHAPE index (SHAPE_AM)	$\text{SHAPE\_AM}_i = \frac{\sum_{j=1}^{m} \frac{0.25 p_{ij}}{\sqrt{a_{ij}}} a_{ij}}{\sum_{j=1}^{m} a_{ij}}$ p_{ij}: Perimeter of ith patch of type i	Area-weighted shape complexity of the corresponding type	Area-weighted mean SHAPE index (SHAPE_AM)
Mean patch size (AREA_MN, ha)	$\text{AREA\_MN}_i = \frac{\sum_{j=1}^{m} a_{ij}}{m}$	Mean patch size of the corresponding patch type	Mean patch size (AREA_MN, ha)
Area-weighted mean Euclidean nearest-neighbor distance (ENN_AN, m)	$\text{ENN\_MN}_i = \frac{\sum_{j=1}^{m} hi_j}{m}$ h_{ij}: Distance from patch$_{ij}$ to the nearest-neighboring patch of type i	Area-weighted distances to the nearest-neighboring patch of the same corresponding type	Area-weighted mean Euclidean nearest-neighbor distance (ENN_AN, m)

explained variances in cross-validation are 31.0%, 39.7%, 42.4%, 40.2%, 37.1%, and 49.3%, respectively, for centers 15–20. The explained variances are enough for us to interpret these models. Furthermore, the result in center 15 is relatively uncertain, given the small number of samples (201) in the center. Based on the model result, the SHAP explanation method was used to locally attribute the effect of each influential factor on each instance. The abovementioned "effect," i.e., the Shapley value, is calculated as the mean difference between the prediction functions considering the factor's value and the marginal prediction functions without considering the value, where these prediction functions are derived from the fitted ML model (Lundberg & Lee, 2017). We mainly used the SHAP dependence plot to represent the quantitative response of LST to each influencing factor and the interactions effects between factors. The moderating factors in the plot were selected based on the correlation coefficient between shapely values of the moderated factors and the candidate moderating factors.

4.4 Spatially Hierarchical Structure of UTE

4.4.1 Characteristics

The temperature contour lines (isothermal lines) could be used to evaluate the spatial pattern of regional thermal environments. In the study area, the screened isothermal lines are mainly in the built-up areas, especially in the urban main districts of Xi'an and Xianyang (Figure 4.4a). However, there are some isothermal lines

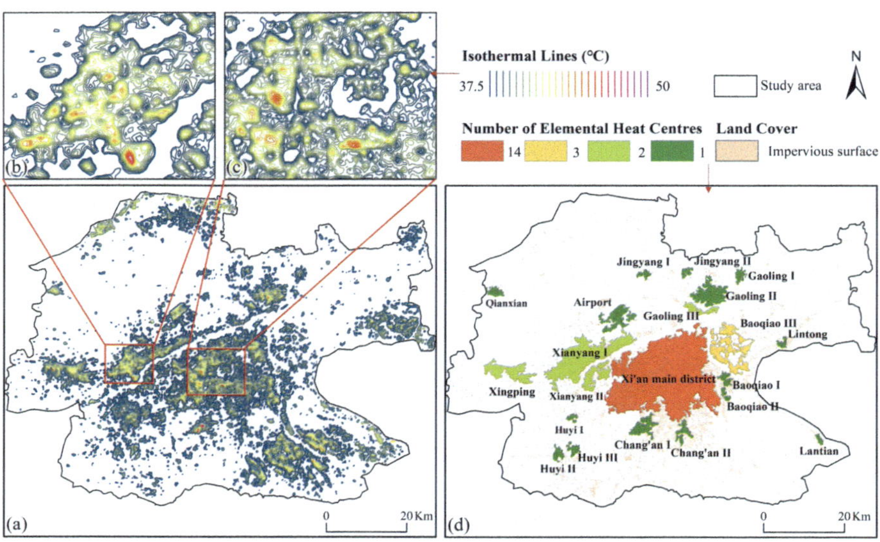

Fig. 4.4 Isothermal lines in (**a**) study area, (**b**) Xianyang urban district, (**c**) Xi'an urban district, and (**d**) derived isothermal trees with the number of elemental heat centers they include

whose spatial extents do not match that of the built-up area. For example, in Xingping, Lantian, and Huyi, the spatial extents of built-up areas are inside that of isothermal lines, whereas in the Chang'an district, there is a contrasting situation, given the high-rise and high green coverage (thus low temperature) neighborhoods in the district. More interestingly, in the north of Jingyang and Qianxian and the middle of Chang'an district, many isothermal lines are not relevant to the built-up area but to the high-temperature bare land and slopes. In addition, most of the hotspots are not in the central area of cities, but the urban edges (Figure 4.4b, c), and are mainly industrial areas and urban villages in terms of land use. Particularly, the hottest heat center with 50 °C is not in the urban main district, but in the suburban industrial area of Chang'an district, indicating that the industrial area and urban villages may have greater UHI intensity and contribution than the generally believed urban core areas.

The identification of the spatial extents of contour trees is the prerequisite for the LCT algorithm and helps us evaluate the distribution of main UHI areas in the regional thermal environment. Moreover, based on the LCT algorithm, we obtained the number of elemental heat centers in each LST contour tree (i.e., isothermal line tree), by which the hierarchical structure of each UHI area could be evaluated. A total of 21 isothermal line trees with 40 elemental urban heat centers are observed in the study area (Figure 4.4d). It is notified that the values of boundary isothermal lines for contour trees Baoqiao III and Lintong are smaller than the abovementioned threshold (37.5 °C), because their boundaries are integrated and matched to the main built-up areas. Among these trees, 15 have only a single elemental urban heat center, thereby indicating that they are UHI areas with a single-core thermal environment. In the remaining trees, 4 trees have 2 elemental urban heat centers and one tree (Baoqiao III) has 3, and these trees all have a two-level thermal environment. Meanwhile, the last main tree with spatial extent matching the Xi'an urban main district has 14 ones and 6 levels, thereby indicating its significantly polycentric structure in thermal environment. Spatially, these trees form a core-periphery structure with Xi'an urban main district as the core UHI area which contributes 52.0% (419.13 km^2) of the whole UHI area.

4.4.2 Pattern in Core Areas

Figure 4.5a depicts the distribution of urban heat centers in the core area. The elemental urban heat centers are mainly distributed on the edge of the city, especially on the northwest part of the city, and are primarily the industrial and logistics areas, the urban villages, and other low-rise commercial areas (special markets). Figure 4.5b shows the spatially nested structure of UTE in the form of tree graph, where the UTE has 6 levels, 14 elemental urban heat centers, and 6 composite urban heat centers. From level 6 to 3, there is only one composite center in each level, indicating a single-core structure in these levels, which also means that the centers 17–20 could represent the corresponding spatial levels of UTE, respectively.

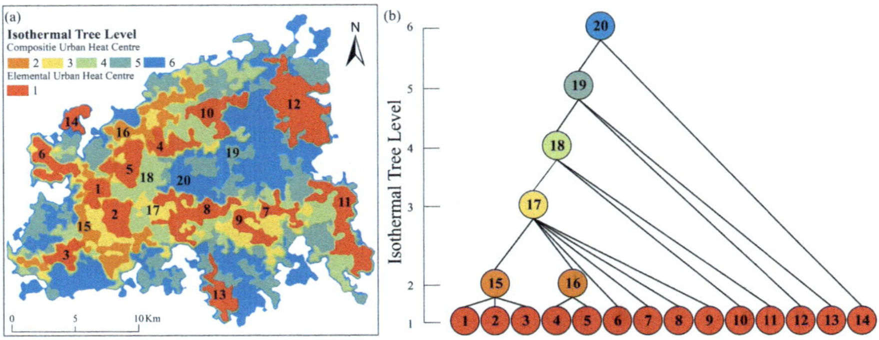

Fig. 4.5 Contour tree (isothermal tree) in the core area. (**a**) Local heat center distribution and (**b**) its tree structure representation

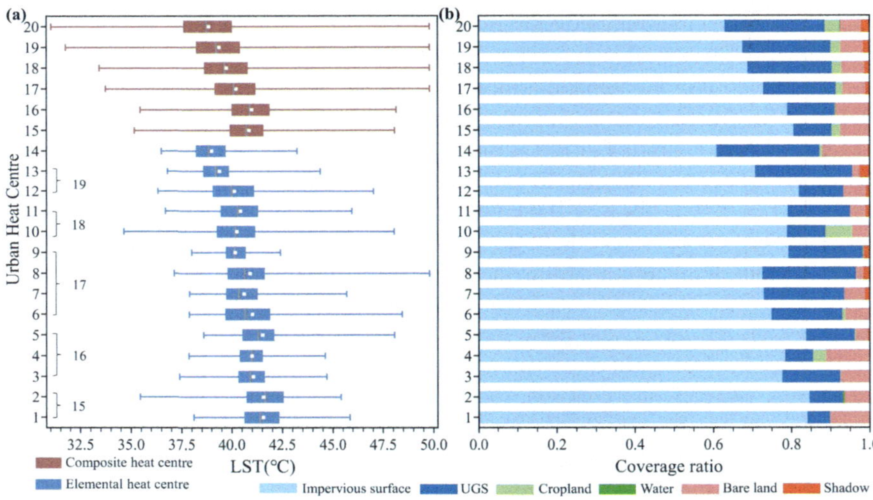

Fig. 4.6 Local heat centers' (**a**) temperature distribution and (**b**) land cover composition

Spatially, from level 6 to 3, the spatial extents of these heat centers shrink in the Lianhu, Xincheng, and Weiyang districts, while a part of two large heritage parks and the high-rise building area between them are excluded, making the center 17 (level 3) form a shape of angle toward the southwest.

With the decrease in the spatial levels of the UTE, the composite urban heat centers' LST ranges shrinks, i.e., the interval decreases from 18.8 °C to 12.7 °C, while their median and mean values increase from 38.8 °C and 38.9 °C to 40.7 °C and 40.8 °C, respectively (Figure 4.6a). Meanwhile, the percentage of IS also increases from 63.0% to 80.0% (Figure 4.6b). For the elemental urban heat centers, their mean LSTs could be grouped in terms of their parent composite centers. Moreover, the median LSTs decrease along the spatial levels of parent centers, i.e., from 41.6 °C in the group corresponding to parent center 15 (level 1) to 39.0 °C in that corresponding to center 20 (level 6). However, these elemental centers' LST ranges and coverage composition do not form similar hierarchical patterns. For example, the center 9 has the smallest temperature range and center 10 has the largest one. Besides, the percentages of IS in the group corresponding to center 18 are larger than that in the groups corresponding to centers 16 and 17, though the latter two have higher mean LSTs than the former. These findings suggest that although a hierarchical pattern of mean LST in the elemental urban heat center is observed, it could not be completely attributed to the variation of percentage of impervious surface, which may imply the impacts from the land use types and 3D building characteristics.

4.5 Effects of Urban Landscape Structure on LST

4.5.1 Relative Importance and Variation of Impacting Factors

The relative importance of influencing factors could help identify the main factors from a global perspective. Fig. 4.7 describes the relative importance of the indicators of urban landscape structure (i.e., class-level landscape metrics) on LST across composite heat centers 15–20 based on impurity and SHAP value.

Under the assumption of the impurity, the relative importance of different influencing factors significantly varies across the centers, i.e., the spatial levels of the UTE. Besides, no one factor is the main impacting factor in all levels simultaneously (Figure 4.7a). However, in the centers 18–20 (levels 4–6), the AI of IS could be viewed as the main factor. Considering the land cover type, in centers 15 and 16, the landscape metrics relevant to UGS make the main contribution, i.e., 53.2% and

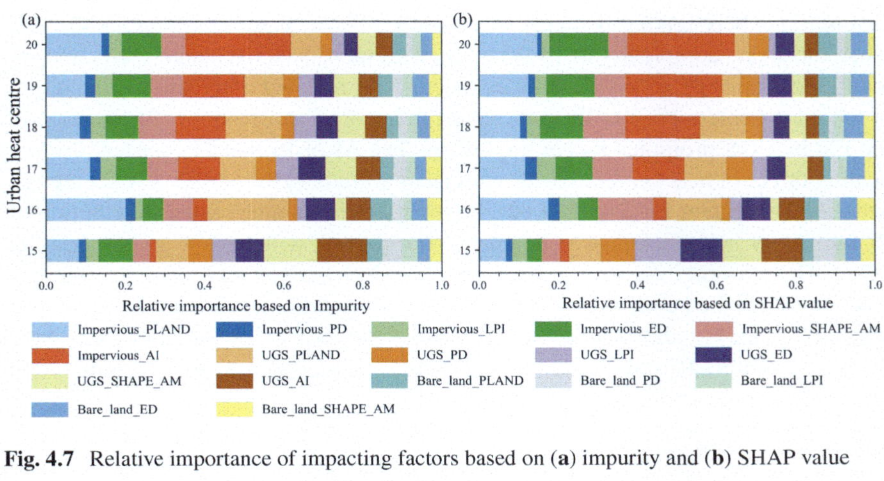

Fig. 4.7 Relative importance of impacting factors based on (**a**) impurity and (**b**) SHAP value

41.4%, respectively, and the IS follows. Conversely, in centers 17–20, landscape metrics related to IS dominated the contribution (62% at most in center 20), followed by UGS. As the spatial level of the UTE increases, the relative importance of landscape metrics related to IS increases (from 40.7% to 62.0%), whereas that related to UGS decreases significantly (from 41.4% to 25.6%), and that related to bare land decreases slightly. In addition, the relative importance of landscape metrics related to composition is always greater than the one related to configuration in all heat centers and it reaches 46.1% in heat center 20 at most. The distribution of relative importance under the assumption of SHAP value is consistent with the above result based on impurity, though the relative importance for specific indicators changes slightly, especially for the AI of IS and percentage of UGS (Figure 4.7b).

4.5.2 LST Response to Influencing Factors and Interaction Effects

Based on the SHAP dependence plot, we could analyze the responses of LST to the influencing factors and the interaction effects among them. Fig. 4.8 shows the effects of five main landscape metrics, i.e., percentage of IS, shape index of IS, AI of IS, percentage of UGS, and ED of UGS on LST across the spatial levels of the UTE in the core area.

Generally, the percentage of IS has a convex heating effect, i.e., when the percentage is larger, its increment per unit will bring more increase in the LST (Figure 4.8a). Moreover, the effect's amplitude is the largest in center 19 with 1.9 °C and the smallest in center 15 with 0.3 °C. The relevant interaction effects are observed in centers 18–20. In general, when the percentage of IS is relatively small (below 74% or 70%), the increase in the ED or percentage of UGS suppresses its heating effect. However, when the percentage of IS is relatively large, the situation will be converse. The shape index of IS has a consistent convex cooling effect with an amplitude between 0.4 °C in center 15 and 1.7 °C in center 18 (Figure 4.8b), i.e., when the shape complexity of IS is larger (with more heat exchange with environment), IS's heating effect will be reduced. The relevant interaction effects are observed in centers 18–20. When the shape index is greater than 2.6, the increase in the ED of UGS promotes its cooling effect. The AI of IS has a consistent convex heating effect in heat centers 16–20 with an amplitude between 0.2 °C in center 15 and 2.6 °C in center 20 (Figure 4.8c). In other words, an aggregated distribution of IS (with less heat exchange with other land covers) will have a greater heating effect. The relevant interaction effects are observed in centers 17–20. For both centers 17 and 18, the percentage of UGS has negative and positive moderating effects at two intervals (smaller than 80% and between 87% and 93%) of AI of IS, respectively. For centers 19 and 20, the ED of IS has a less significant positive moderating effect when the AI is smaller than 90% or 94%.

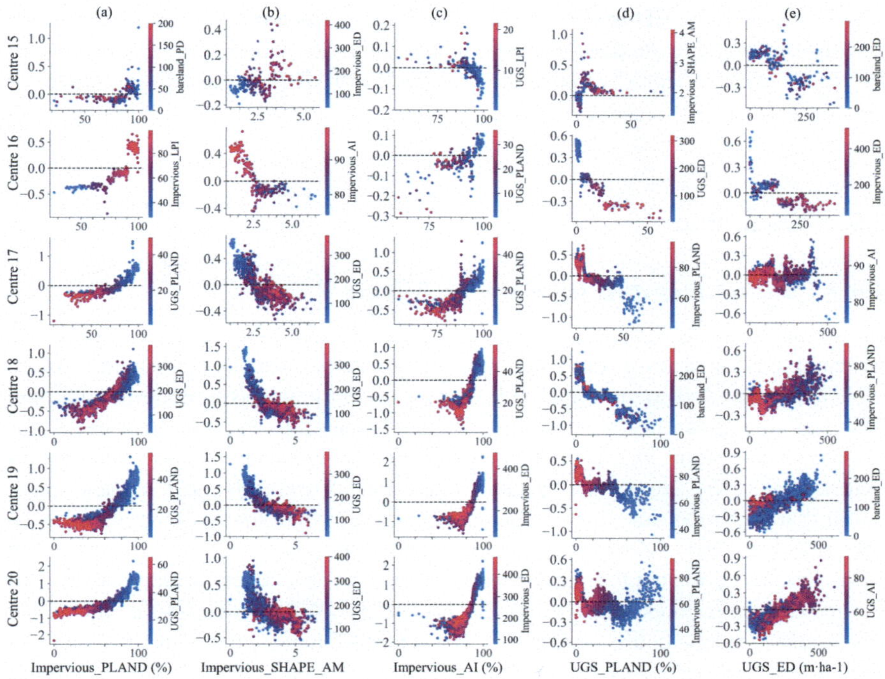

Fig. 4.8 The response of LST to (**a**) percentage of IS, (**b**) shape index of IS, (**c**) AI of IS, (**d**) percentage of UGS, and (**e**) ED of UGS. Note: Each point's value on the Y-axis stands for the effect of a certain influencing factors on LST in an analytic unit and its value on the X-axis represents the factor self. The color of points stands for the value of moderating factors with corresponding colormaps in the right of the subfigures

In general, the percentage of UGS has a cooling effect in heat centers 16–20 with an amplitude between 0.4 °C in center 15 and 1.6 °C in center 18, although the significant difference is observed among centers (Figure 4.8d). In centers 16–18, a stepwise cooling effect is observed. In centers 19 and 20, after the starting stepwise cooling effect, a heating effect is then shown when the percentage is larger than 65% or 60%. The relevant interaction effects are observed in centers 18 and 20. In center 18, when the percentage is between 27% and 47%, the increase in the ED of bare land promotes its cooling effect. Conversely, in center 20, when the percentage is between 33% and 53%, the increase in the percentage of IS will suppress its cooling effect. The ED of UGS has an inconsistent effect among heat centers (Figure 4.8e). In centers 15 and 16, it has a stepwise cooling effect with an amplitude of 0.5 °C. In center 17, the effect is not significant. In centers 18–20, it has a linear heating effect with an amplitude between 0.4 °C and 0.8 °C. The relevant interaction effect is observed in centers 17, 19, and 20. In center 17, the increase in the AI of IS has a negative and positive moderating effect when the ED is between 20 m·ha^{-1} and 180 m·ha^{-1} and between 228 m·ha^{-1} and 420 m·ha^{-1}, respectively. In center 19, when the ED is smaller than 200 m·ha^{-1}, the increase in the ED of bare land has a negative moderating effect. In center 20, when the ED is between 68 m·ha^{-1} and 180 m·ha^{-1}, the increase in the AI of UGS has a negative moderating effect.

4.6 Discussion

4.6.1 Influences and Interaction Effects

At entire UHI area level, the landscape metrics relevant to IS contribute more to LST than other land covers, and the landscape configuration is more important than landscape composition on determining LST. Considering the relatively large percentage of IS (Figure 4.7a), the former finding is consistent with previous studies (Xie et al., 2013). However, the latter finding conflicts with most studies which found that the importance of composition is greater than that of configuration on LST (Li et al., 2013; Zhou et al., 2011). Interestingly, a few studies conducted in cities with arid climate also obtained similar results for UGS or for the metrics constituted by compactness pattern index. Zhou et al. (2017) attributed the intensified effects of landscape configuration of UGS at arid climate to the increased environmental pressure in terms of air temperature and solar radiation on vegetation patches. However, in our case, the dominance of landscape configuration stands for each land cover and the total dominance mainly comes from IS but not UGS, so how to interpret the intensified effects of landscape configuration of IS in terms of the semi-arid climate and specific urban form of Xi'an warrants further studies.

With the increase in the spatial level of the UTE, the contribution from IS increased and that from UGS decreased. The finding is inconsistent with previous studies that with the increase in the percentage of IS, its importance on LST also increased (Liu et al., 2021), given that the local centers had the higher percentage of IS (Figure 4.6b). To explain it, similar to what Masoudi et al. (2021) did, we compared the general spatial patterns of IS and UGS. It is found that with the increase in the spatial levels, the IS and UGS become more fragmented and aggregated, respectively, and the connection (heat exchange) between them intensifies. Besides, the distribution of UGS becomes more spatially even (indicated by ENN). The increase in the contribution of IS mainly comes from the AI of IS, while the decrease in that of UGS is mainly attributed to the percentage of UGS. Therefore, the question is refined as: Why do the AI of IS and the percentage of UGS become more and less important, respectively, given the fragmentation of IS, the aggregation and the increasing of percentage of UGS, and the intensifying heat exchange?

A possible answer is the diminishing marginal utility effect, i.e., when the aggregation extent is relatively small, the unitary increment of the AI of IS will have a greater effect, conversely, when the percentage of UGS is relatively large, the increase in the percentage will have a smaller effect. In addition, given that the percentage of tree cover in the UGS is higher in the heat centers with lower spatial levels, the difference in the cooling capability of tree cover and grassland may explain the finding to some extent. Besides, the increase in the spatial heterogeneity of the relationship with the spatial level of UTE may affect the observed relative importance. Considering that this is an interesting finding, further investigation integrating the mechanism of thermal processes and more substantial evidence is warranted.

As for the quantitative responses of LST to single influencing factors, the effects of most factors are not linear, but with concavity, convexity, or stepwise form. Specifically, both the percentage and aggregation of IS increase the LST convexly. The finding is consistent with Xu et al. (2013) that there was an exponential relationship between IS index and LST. In addition, the effects of some factors on LST significantly vary with the spatial levels of the UTE. For example, in center 20, when the percentage is greater than 60%, the percentage of UGS has an abnormal heating effect. It may be interpreted as the difference in the composition of UGS between analytic units with high and low coverage of UGS. The analytic units with high coverage of UGS are mainly located at center 20 (entire UHI area) and the vegetations in these units are mainly sparse grassland. However, when the coverage is relatively low, the vegetation is mainly tree canopy. As mentioned above, the cooling effect of grassland is lower than that of tree cover and in high-temperature weather, the open grassland may even heat the air.

As for the interaction effects, they have significant conditionality that is not reported in previous research. For example, when the percentage of IS is relatively low (smaller than 75%), the ED of UGS will suppress its heating effect. However, when the value is relatively high, the situation will be converse, thereby indicating the possible competition in the heating/cooling effect of these two land covers (Liu et al., 2021). In addition, when the shape index of IS is relatively high (higher than 2.6), the ED of UGS will promote its cooling effect. In principle, the increases in the shape index of IS and the ED of UGS intensify the energy exchanges of these two land covers with the environment, respectively. Thus, the result indicates that these two energy exchange processes may be harmonic (mutually beneficial). Therefore, the conditionality of interaction effects observed in the study may be associated with the competitive and collaborative effects between thermal processes that exist in the urban land surface.

4.6.2 Methodological Implications

In the introduction section, assume that the spatial hierarchy in the relationship comes from the spatial heterogeneity. However, where does the heterogeneity come from and how is it addressed? (1) With the observed variables, the relationship may be highly localized and nonlinear in the parameter space, given that the measured variables are just partial representations of the terms in the surface energy balance. Thus, a simple model may not capture the global relationship well. Some research has demonstrated that with some strategies aiming to divide the parameter space, models can fit the relationship better (Guo et al., 2020). In addition, some non-parametric ML methods such as random forest and XGBoost could capture the nonlinear relationship directly. (2) Some unobserved or omitted variables existing at various spatial scales/levels control the relationship to some extent. Thus, the omission for them in modeling may cause the heterogeneity, especially for the variables related to the 3D characteristics, anthropogenic heat, and spatial dependency of LST. Considering that a site and its small neighborhood may have similar value for the unobserved variable and thus a similar relationship, the geographically weighted regression (GWR) has been used to address the issue (Su et al., 2012). In addition, some grouping strategies like the UFZ-grouping, gradient-grouping, and LCZ-grouping may address both of the issues (Jia & Zhao, 2020; Liu et al., 2021; Yang et al., 2021a, 2021b; Zhou et al., 2022), given that these groupings may make the relationship localized in the parameter space and make some unobserved variables controlled in the group level.

Generally, these strategies, except the nonlinear ML methods, all assume that the relationship at an extent (strata) with lower spatial level will be more spatially homogeneous and, thereby, could be modeled better. The assumption could be also viewed as the spatial stratified heterogeneity of relationship. Therefore, how to decompose the set of analytic units into an appropriate spatial hierarchy for modeling is a key feature of these strategies. Moreover, the explorations into the variation of the relationship across the spatial levels (spatial hierarchy), and that into the variation across the strata with the same level (spatial heterogeneity), could be viewed as two aspects of the same question, although the latter is a greater concern. In the former, the hierarchy is generally derived from the attributes (like the LCZ or gradient grouping) or the neighbor relation (like the GWR) of analytic units and is generally two-level (global versus local). Meanwhile in the latter, the hierarchy is derived from the attributes or the simple association between attributes and LST variation and is multilevel. The study used the multilevel spatial hierarchy of the UTE itself to define the hierarchy and then explored the variation of the relationship across it, which may provide a new perspective for these two types of explorations.

Practically, the methodology could be viewed as a sampling strategy that selected a series of nested local heat centers representing the spatial hierarchy of the UTE as the interesting areas for modeling, given that the centers with different spatial levels have different mean temperature (Figure 4.6a) and thus have different importance for heat-mitigation planning. In the results, the relationship varies across the spatial levels of the UTE, especially for the relative importance of IS and UGS, thereby indicating the meaning of the sampling strategy and suggesting implications for planning practice. Despite that, the modeling in the low-level heat center with small numbers of samples may bring larger uncertainty (center 15).

4.6.3 Policy Implications

The findings at the entire UHI area level imply that, at least for heat mitigation based on urban planning in Xi'an, the spatial pattern of IS, including the buildings and roads, should be prioritized over other land covers. Moreover, they imply that the optimization for the spatial configuration of IS will have a greater effect than decreasing the percentage of IS or increasing that of UGS. The comparison across the spatial levels indicates that the heat mitigation based on urban planning should use the differential optimization criteria for the different levels. In a local heat island with high level, the utility of planning resources should consider the landscape structure of the IS primarily, while in that with lower levels, the utility should favor the landscape structure of UGS. In addition, the abnormal heating effect of the percentage of UGS in the center 20 (Figure 4.8d) implies that the unplanned urban greening may not obtain the expected cooling effect. Moreover, the difference among the cooling effect and adaptability of vegetation with different species should be considered (Gál et al., 2021). The interaction effects found in the study also imply that the changes in certain landscape structure indicators not only influence the LST directly but may also promote or suppress the effects of other indicators on the LST indirectly. Thus, a deeper understanding of interaction effects will provide a larger operation space for the heat mitigation based on the optimization of landscape structure. Meanwhile, due to the significant conditionality in the interaction effects, the future practice of optimization of landscape structure may be based on the mathematical optimization of complex models and little rules of thumb could be applied.

4.6.4 Limitations and Future Work

Although the study has examined the less-explored hierarchical variation and interaction effects in the relationship between landscape structure and LST, it also has some limitations. First, only one thermal image with the worst thermal conditions was used to represent a typical summer UTE. The future research should use the thermal imageries from all seasons to examine the seasonal characteristics of the spatial hierarchy of the UTE and discuss the relationship. Second, the fallow cropland (classified as bare land in the study), bare slope, and small rural settlements around cities also have high thermal contributions to the UTE. Therefore, the landscape situation surrounding cities should be considered prudently before applying the LCT algorithm. Third, land cover data with a spatial resolution of 10 m limits the ability to explore the multiscale behavior of model relationships by varying the size of the analytical units, so further research should use very high-resolution (less than 2 m) imageries to explore it. In addition, the influence from the neighboring analytic units was omitted in the study, although the influence has been demonstrated. Because of the effects of urban landscape structure, the influence of neighborhood units could be included in modeling as an important covariate in future studies. Finally, future research should explore the spatiality traits in the relationship between 3D urban landscape structure and LST and the interaction effects between 2D and 3D influencing factors.

4.7 Conclusions

Spatial urbanization has caused and exacerbated the deterioration of UTE. Understanding the spatial variation of UTE based on the urban landscape structure will help mitigate its deterioration. This study took the Xi'an metropolitan area as the case and used LST data and LCT algorithm to identify the spatial hierarchy of UTE represented by nested heat centers, through which the XGBoost model and SHAP method were then used to analyze the relationship between urban landscape structure and LST across the spatial levels of the UTE. Moreover, the interaction effects between influencing factors on LST were also explored.

This study found that (1) In the entire metropolitan area, 21 isothermal line tree (or main) UHI areas are identified. Among them, the Xi'an urban main district's UTE has 14 elemental heat centers and 6 levels, while others have up to 3 centers and 2 levels. Spatially, these UHI areas present a core-periphery pattern with Xi'an urban main district as the core area, which contributes 52.0% of the main UHI area. Inside the core area, the elemental heat centers are mainly industrial and logistics areas and urban villages, and they are mostly distributed on the urban edge. (2) In the core area, the relationship between urban landscape structure and LST varies with the spatial levels of the UTE. Especially, with the increase in the spatial level of UTE, the importance of IS and UGS to LST increases from 40.7% to 62.0% and decreases from 41.4% to 25.6%, respectively. (3) Some landscape structure indicators' effects on LST are suppressed or promoted by the increases of other indicators, i.e., interaction effects or moderating effects are observed. Meanwhile, the suppressing/promoting impacts vary with the value of influencing factors. In other words, the interaction effects have conditionality. Especially, when the percentage of IS is relatively low, the ED of UGS will suppress its heating effect. When the value is relatively high, the situation will be converse and may be associated with the competition and synergy between underlying thermal processes.

These results demonstrate that the relationship has significant spatial and nonlinear characteristics, which should be further explored. Moreover, the study may provide a fundamental methodology for subsequent explorations. In practice, these results indicate that the heat mitigation based on landscape optimization should adopt differential optimization criteria for the different spatial levels of the UTE. Especially in locally high-temperature regions (heat centers) with a low level of UTE, the landscape structure of the IS should be considered first in planning practice to mitigate heat stress. Conversely, in regions with a higher level of UTE and thus moderate environment, the spatial pattern of green space should be emphasized. Meanwhile, the comprehensive impacts of different landscape elements should be considered together to maximize the cooling effects brought by the interaction effect.

References

Cai, M., & Kalnay, E. (2004). Impact of land-use change on climate. *Nature, 427*, 214–214. https://doi.org/10.1038/427214a

Chen, Z., Yu, B., Song, W., Liu, H., Wu, Q., Shi, K., & Wu, J. (2017). A new approach for detecting urban centers and their spatial structure with nighttime light remote sensing. *IEEE Transactions on Geoscience and Remote Sensing, 55*, 6305–6319. https://doi.org/10.1109/TGRS.2017.2725917

Deilami, K., Kamruzzaman, M., & Liu, Y. (2018). Urban heat Island effect: A systematic review of spatiotemporal factors, data, methods, and mitigation measures. *International Journal of Applied Earth Observation and Geoinformation, 67*, 30–42. https://doi.org/10.1016/j.jag.2017.12.009

Du, S., Xiong, Z., Wang, Y.-C., & Guo, L. (2016). Quantifying the multilevel effects of landscape composition and configuration on land surface temperature. *Remote Sensing of Environment, 178*, 84–92. https://doi.org/10.1016/j.rse.2016.02.063

Gál, T., Mahó, S. I., Skarbit, N., & Unger, J. (2021). Numerical modelling for analysis of the effect of different urban green spaces on urban heat load patterns in the present and in the future. *Computers, Environment and Urban Systems, 87*, 101600. https://doi.org/10.1016/j.compenvurbsys.2021.101600

Guo, G., Wu, Z., Cao, Z., Chen, Y., & Yang, Z. (2020). A multilevel statistical technique to identify the dominant landscape metrics of greenspace for determining land surface temperature. *Sustainable Cities and Society, 61*, 102263. https://doi.org/10.1016/j.scs.2020.102263

Jia, W., & Zhao, S. (2020). Trends and drivers of land surface temperature along the urban-rural gradients in the largest urban agglomeration of China. *Science of the Total Environment, 711*, 134579. https://doi.org/10.1016/j.scitotenv.2019.134579

Jung, M. C., Dyson, K., & Alberti, M. (2021). Urban landscape heterogeneity influences the relationship between tree canopy and land surface temperature. *Urban Forestry & Urban Greening, 57*, 126930. https://doi.org/10.1016/j.ufug.2020.126930

Li, X., Zhou, W., & Ouyang, Z. (2013). Relationship between land surface temperature and spatial pattern of greenspace: What are the effects of spatial resolution? *Landscape and Urban Planning, 114*, 1–8. https://doi.org/10.1016/j.landurbplan.2013.02.005

Liu, Y., Huang, X., Yang, Q., & Cao, Y. (2021). The turning point between urban vegetation and artificial surfaces for their competitive effect on land surface temperature. *Journal of Cleaner Production, 292*, 126034. https://doi.org/10.1016/j.jclepro.2021.126034

Lu, Y., Yue, W., Liu, Y., & Huang, Y. (2021). Investigating the spatiotemporal non-stationary relationships between urban spatial form and land surface temperature: A case study of Wuhan, China. *Sustainable Cities and Society, 72*, 103070. https://doi.org/10.1016/j.scs.2021.103070

Lundberg, S., Lee, S.-I. (2017). *A unified approach to interpreting model predictions*.

Masoudi, M., Tan, P. Y., & Fadaei, M. (2021). The effects of land use on spatial pattern of urban green spaces and their cooling ability. *Urban Climate, 35*, 100743. https://doi.org/10.1016/j.uclim.2020.100743

McGarigal, K., Cushman, S. A., & Ene, E. (2012). FRAGSTATS v4: Spatial pattern analysis program for categorical maps. *Computer software program produced by the authors at the University of Massachusetts, Amherst. Massachusetts, Amherst, 15*, 153–162.

Oke, T. R., Mills, G., Christen, A., & Voogt, J. A. (2017). *Urban climates*. Cambridge University Press. https://doi.org/10.1017/9781139016476

Su, Y. F., Foody, G. M., & Cheng, K. S. (2012). Spatial non-stationarity in the relationships between land cover and surface temperature in an urban heat Island and its impacts on thermally sensitive populations. *Landscape and Urban Planning, 107*, 172–180. https://doi.org/10.1016/j.landurbplan.2012.05.016

Sun, Y., Gao, C., Li, J., Li, W., & Ma, R. (2018). Examining urban thermal environment dynamics and relations to biophysical composition and configuration and socio-economic factors: A case study of the Shanghai metropolitan region. *Sustainable Cities and Society, 40*, 284–295. https://doi.org/10.1016/j.scs.2017.12.004

Sun, F., Liu, M., Wang, Y., Wang, H., & Che, Y. (2020). The effects of 3D architectural patterns on the urban surface temperature at a neighborhood scale: Relative contributions and marginal effects. *Journal of Cleaner Production, 258*, 120706. https://doi.org/10.1016/j.jclepro.2020.120706

Xie, M., Wang, Y., Chang, Q., Fu, M., & Ye, M. (2013). Assessment of landscape patterns affecting land surface temperature in different biophysical gradients in Shenzhen, China. *Urban Ecosystem, 16*, 871–886. https://doi.org/10.1007/s11252-013-0325-0

Xu, H., Lin, D., & Tang, F. (2013). The impact of impervious surface development on land surface temperature in a subtropical city: Xiamen, China. *International Journal of Climatology, 33*, 1873–1883. https://doi.org/10.1002/joc.3554

Yang, Z., Chen, Y., Guo, G., Zheng, Z., & Wu, Z. (2021a). Characteristics of land surface temperature clusters: Case study of the central urban area of Guangzhou. *Sustainable Cities and Society, 73*, 103140. https://doi.org/10.1016/J.SCS.2021.103140

Yang, J., Ren, J., Sun, D., Xiao, X., Xia (Cecilia), J., Jin, C., & Li, X. (2021b). Understanding land surface temperature impact factors based on local climate zones. *Sustainable Cities and Society, 69*, 102818. https://doi.org/10.1016/j.scs.2021.102818.

Yu, S., Chen, Z., Yu, B., Wang, L., Wu, B., Wu, J., & Zhao, F. (2020). Exploring the relationship between 2D/3D landscape pattern and land surface temperature based on explainable eXtreme gradient boosting tree: A case study of Shanghai, China. *Science of the Total Environment, 725*, 138229. https://doi.org/10.1016/j.scitotenv.2020.138229

Zhou, W., Huang, G., & Cadenasso, M. L. (2011). Does spatial configuration matter? Understanding the effects of land cover pattern on land surface temperature in urban landscapes. *Landscape and Urban Planning, 102*, 54–63. https://doi.org/10.1016/J.LANDURBPLAN.2011.03.009

Zhou, W., Wang, J., & Cadenasso, M. L. (2017). Effects of the spatial configuration of trees on urban heat mitigation: A comparative study. *Remote Sensing of Environment, 195*, 1–12. https://doi.org/10.1016/j.rse.2017.03.043

Zhou, L., Yuan, B., Hu, F., Wei, C., Dang, X., & Sun, D. (2022). Understanding the effects of 2D/3D urban morphology on land surface temperature based on local climate zones. *Building and Environment, 208*, 108578. https://doi.org/10.1016/j.buildenv.2021.108578

Chapter 5
Urban Landscape Marginal Effects on Land Surface Temperature

5.1 Introduction

Significant Impacts of UHI Rapid changes in urban form and urban landscape in the short term have exacerbated various environmental problems, such as local climate change, air pollution, ecological environment deterioration, and urban heat island (UHI) (Manoli et al., 2019; Oke, 1982). Specifically, UHI refers to the phenomenon that urban areas are warmer than surrounding non-urban areas. UHI has been intensified globally in recent years as the continuous growth and concentration of population brought tremendous pressure to natural land. In the process, natural land has been replaced by impervious built environment, which modifies the energy balance of the land surface, releases a large amount of anthropogenic heat, and removes the urban green space (UGS) which is capable of significantly alleviating UHI. Moreover, considering the shortage of horizontally available land, some cities have had to expand vertically. Three-dimensional structure and morphology of a city are important factors affecting the urban microclimate. Specifically, urban buildings and transportation facilities may exacerbate the UHI through vertical expansion, which affects local- and city-scale surface energy balance processes and airflows, thereby altering the UTE. High temperature in urban areas not only increases energy consumption (Akbari and Konopacki, 2005; Santamouris et al., 2015) and air pollution but also poses a threat to human health. In particular, the increase in the intensity, frequency, and duration of extreme heat in summer will result in more people worldwide being regularly exposed to lethal heat, which poses a challenge to the thermal comfort for urban residents and urban sustainability. Hence, how to mitigate the heat island and improve the UTE by controlling and optimizing the horizontal and vertical expansion of cities has received increasing attention from researchers and administrators.

L. Zhou et al., *Urban Surface Thermal Environments*, The Urban Book Series,
https://doi.org/10.1007/978-3-031-94499-4_5

UGS: A Cooling Engine in Cities Buildings have a significant warming effect, while UGS, as one of the most important landscape elements that regulate the local climate within the city, has an important cooling effect. In particular, it has been widely confirmed that urban trees provide effective cooling effect by the shading and evapotranspiration. Generally, researchers use various landscape metrics to analyze the relationship between UGS and LST, such as patch density, edge density, and aggregation index. Ziter et al. (2019) found that the cooling effect is the greatest when the vegetation fraction is greater than 40%. Xie et al. (2013) found that the aggregation index and the largest patch index of vegetation were negatively correlated with LST. Li et al. (2013) concluded that edge density and patch density were significantly positively correlated with LST by controlling for the percentage of UGS coverage. However, the effect on LST of some indicators remains in disagreement with previous studies. For example, Feyisa et al. (2014) demonstrated that compact and regular urban parks can provide a better cooling effect. However, some results show that shape index explains approximately 20% of cooling effect alone ($R^2 = 0.187$) (Kong et al., 2014), which suggests that the relationship between UGS structure and LST is likely related to UGS size.

Limitations in Current Studies However, previous LST research has some limitations. First, due to the lack of 3D building data and the inadequacy of metrics systems, the influence of 3D building features on LST still needs a lot of research to corroborate the existing findings. Second, most studies use buffer and grid as analytic units to explore the driving factors of LST. These units may segment UGS patches into isolated ones, and these structures cannot represent the functional boundaries of the city well. Therefore, this study adopts the blocks obtained from road segmentation as the analysis units, which can ensure the integrity of the building footprint and UGS patches and then uses boosted regression tree (BRT) model to investigate the effect of 3D building metrics and UGS metrics on LST. This study can help formulate effective solutions from a quantitative view to mitigate UHI, optimize the thermal environment based on the smallest urban management units (blocks), and improve the thermal comfort for urban residents.

5.2 Datasets

There are four main types of data used in the study (Fig. 5.1), which are building footprint, road networks, Landsat 8 image, and GF-2 satellite image. The information for the total number of floors of building footprint (2018) is derived from Baidu map (https://map.baidu.com/). Assumed that the height of each floor is constant at 3 m, so the total floor number of each building is multiplied by 3 m as the building's height. The building footprints are validated by comparing with GF-2 image, which meets the research needs. The road networks of 2019 are derived from Open Street Map, which is used to divide the study area into road-based blocks. The road classes used in this study include trunk, primary road, secondary road, and tertiary road. We

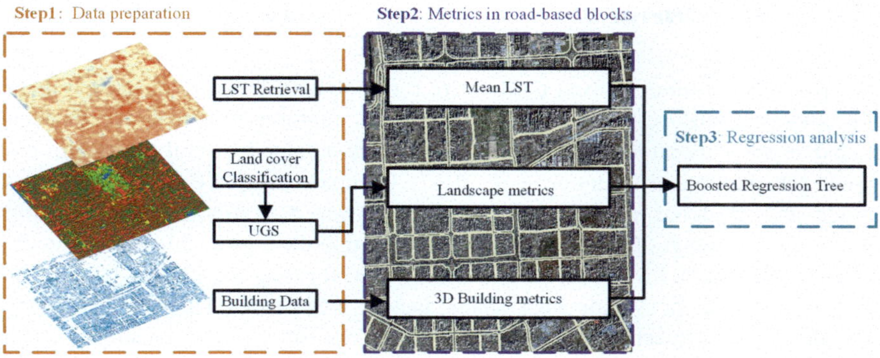

Fig. 5.1 Flowchart of the analyzing methods

finally obtained 285 road-based blocks. Landsat 8 TIRS image (https://earthexplorer.usgs.gov/) was used to retrieve LST. Due to the cloud cover and weather, only one image in summer was selected, which was obtained in Beijing time at 11:19 am, July 25, 2018. The weather on the day and before the image was obtained was sunny, and the cloud cover within the study area in the image was less than 1%. GF-2 image is derived from the Ministry of Housing and Urban-Rural Development of the People's Republic of China, which was used to map the land cover of the study area in 2018.

5.3 Methodology

5.3.1 Urban Green Space Mapping

The study adopts object-based classification methods to achieve land cover mapping. Due to the significant spatial heterogeneity of high-resolution images, direct sampling and classification on the original image is not enough to meet the research needs of exploring the driving factors of LST at road-based blocks. Therefore, we applied initial object segmentation of GF-2 images using eCongnition 9.0 software, and pre-classification of the images into eight classes, including bare land, roads, buildings, water, shade, urban trees, lawn, and cultivated land. The preliminary land cover classification results are obtained based on support vector machine (SVM) method. Then, eight classes were merged into six land cover types according to their physical properties and effects on LST: Bare land, impervious surface, water body, urban trees, lawn, and cultivated land. A random sample of 415 subjects was selected to verify the accuracy of the classification. The final overall accuracy is 84.1%, with Kappa of 0.80. In this study, urban trees and lawns are considered as UGS.

5.3.2 LST Retrieval

The two most commonly used parameters for analyzing UHI are air temperature (AT) and land surface temperature (LST). AT is captured and recorded through ground weather monitoring stations, while LST is obtained from thermal infrared remote sensing images (Estoque et al., 2017). Compared with AT monitored by the limited and scattered weather stations, LST retrieved from medium resolution remote sensing images has the advantage of reflecting the temperature of a continuous large-scale area. LST, therefore, is widely used for long-term heat island intensity analysis and local-scale heat island factors attribution (Yang et al., 2020). Though the surface temperature of building façade may create uncertainty, the advantage of LST as the proxy of local UHI has been demonstrated in previous studies and it is assumed that the mentioned uncertainty is unimportant at road-based block. Therefore, LST was acquired using Landsat 8 TIRS band 10 with a spatial resolution of 100 meters, which represents the general thermal condition of the analytical unit and can be calculated by the radiative transfer equation (Weng et al., 2004).

5.3.3 Influential Metrics

The 13 indicators in this study all refer to the relevant conclusions in existing research (Chen et al., 2014; Zhou et al., 2011). The study focuses on how the 3D buildings metrics and UGS metrics separately and together influence LST. Therefore, the influential metrics are two groups: (1) The seven 3D building metrics, including building coverage ratio (BCR), mean building height (MBH), floor area ratio (FAR), sky view factor (SVF), low-rise building area ratio (LRAR, building height < 10 m), mid-rise building area ratio (MRAR, 10 m < building height < 24 m), and high-rise building area ratio (HRAR, building height > 24 m). The categories of last three metrics are obtained from building height according to the "Chinese Code for Design of Civil Buildings" (Liu et al., 2017). (2) Six landscape metrics of UGS, including green coverage ratio (GCR), patch density (PD), largest patch index (LPI), edge density (ED), aggregation index (AI), and mean patch size (MPS), are used to describe the UGS composition and configuration. Three-dimensional building metrics are calculated in ArcGIS 10.5. SVF is obtained by using the horizon package in R software according to building height.

5.3.4 Boost Regression Tree

Methods of analyzing the driving factors of LST are mainly correlation coefficient and linear regression models in existing research. However, linear regression models cannot capture the complex nonlinear relationship between LST and its driving factors. Therefore, this study adopts boosted regression tree (BRT) model to analyze the relative contribution and marginal effects of driving factors on LST.

BRT model is fundamentally different from the traditional linear regression model (Clarke & Johnston, 1999; Sun et al., 2019). BRT can explain the influence of each predictor variable on the response variable and has been widely used in related research on urban expansion and thermal environment (Alavipanah et al., 2018). In this study, BRT is performed in R software by using packages "gbm" and "dismo." The dependent variable is the average LST in each road-based block, and the independent variable is the 13 metrics. According to the recommendations by Elith et al. (2008), the parameters in BRT are set with a learning rate of 0.005, bagging fraction of 0.5 and tree complexity of 5.

In addition, BRT uses 50% of the training data and ten-fold cross-validation to select the optimal model. The final correlation for training data of building metrics and UGS metrics is 0.728 and 0.657, respectively. The final correlation for cross-validation of building metrics and UGS metrics is 0.641 and 0.505, respectively. The correlation for training data of combined metrics is 0.784, and the correlation for cross-validation of combined metrics is 0.666.

5.4 LST Characteristics

LST in the main urban areas of Xi'an is strongly related to buildings and urban trees during the high-temperature period as shown in Fig. 5.2. Overall, the average LST in the study area is 41.2 °C, and LST shows a characteristic of three circle structures with "High–Low–High" from the city center to the fringe area, which is highly related to the urban morphology with "Compact–Open–Compact." Besides, it is noted that LST of the west side of city is relatively higher than other areas, which is mainly because these high-temperature areas include considerable color-steel houses and bare land. Even though these materials present a relatively high surface albedo, low thermal admittance makes its temperature rise fast. In addition, built-up

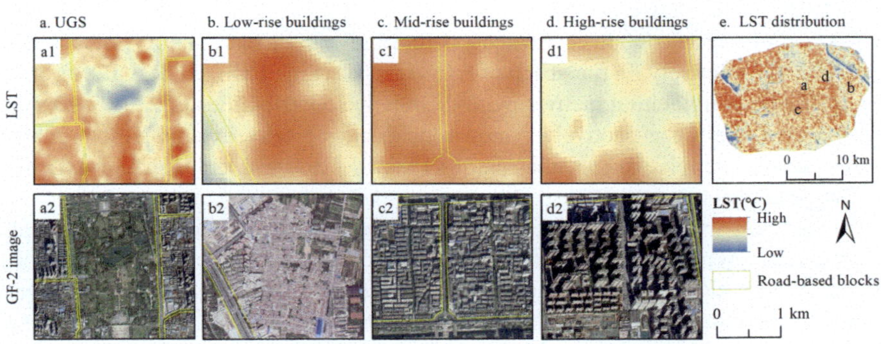

Fig. 5.2 The distribution of LST in different local zones

areas always show high surface temperatures, especially in high-density building areas, while water bodies, high-rise building areas, and large green spaces show low surface temperatures. Specifically, a high-temperature zone appears on the north–south axis in the Damming Palace National Park site (Fig. 5.2a). The land cover of the high-temperature zone is impervious trails and squares, which have a low vegetation coverage on two sides, thus exhibiting extremely high surface temperature than in the surrounding areas, while the LST of the artificial lake in the park is obviously lower than the other areas, and the lake and trees form a distinct cold island. The compact structure and less vegetation coverage in low-rise building areas (Fig. 5.2b) and mid-rise buildings (Fig. 5.2c) are often not conducive to temperature diffusion, creating local heat islands. Besides, Fig. 5.2c is a high-resolution image of the internal buildings of the Xi'an Circumvallation, and the surface temperature of the entire area is relatively high, with an average of 43.1 °C. The high-rise building area (Fig. 5.2d), however, exhibits a low surface temperature with an average of 40.4 °C in this block, which is 0.8 °C lower than the average surface temperature of the entire study area. High-rise buildings could cast more shadows to directly reduce LST in the surrounding area. Moreover, high-rise buildings, especially the isolated or open ones, could increase the surface roughness to deflect high-velocity cooler air mass down into pedestrian level, thereby enhances the heat exchanges at the vertical and horizontal dimensions.

5.5 Relative Contributions of Factors to LST

Buildings and UGS are the two most dominant components of the urban landscape, and analyzing their composition and configuration is essential to alleviate LST. The relative contributions of the 13 metrics obtained by BRT to LST are shown in Fig. 5.3. In the group of building metrics (Fig. 5.3a), the dominant factor is BCR, which contributes 44.6%, far exceeding the other six metrics. In other words, BCR explains 44.6% of the LST change in the study area, which shows that BCR is significantly correlated with LST. HRAR, LRAR, and MBH are much less important than BCR, with relative contributions of 15.8%, 10.4%, and 10.1%, respectively. MRAR, SVF, and FAR are the lowest, with relative contributions of 7.4%, 6.7%, and 5.0%, respectively. In the group of UGS metrics (Fig. 5.3b), GCR, DE, PD, and AI are the most influential metrics, with the relative contributions of 21.0%, 20.9%, 20.4%, and 18.9%, respectively. These four metrics totally explain 81.2% of the LST change. The relative contribution of LPI is 12.1%, and MPS is the lowest with a relative contribution of 6.6%.

Figure 5.3c shows the relative contributions of nine metrics to LST. Four weak metrics were removed (the contributions of metrics to LST are less than 10%, including MRAR, SVF, FAR, and MPS) regarding the contributions of dominant metrics. The results show that the most influential metric is BCR with a relative contribution of 28.3%, followed by HRAR, LRAR, ED, and MBH, with relative contributions of 12.8%, 11.0%, 10.0%, and 9.1%, respectively. The remaining four

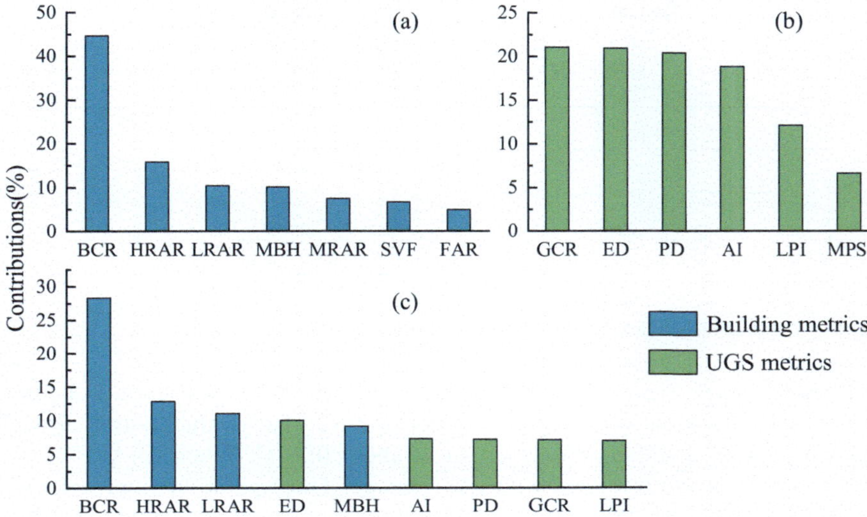

Fig. 5.3 Contribution of metrics to LST. (**a**) Contributions of each metric in the group of building metrics. (**b**) Contributions of each metric in the group of landscape metrics of UGS. (**c**) Contributions of each metric in the combined group of building metrics and landscape metrics

metrics: AI, PD, GCR, and LPI have a relative contribution of 7.2% to LST, approximately. In the combined group of nine metrics, the total relative contribution of four building metrics reaches 61.2%, which indicates that considering the cooling effect of UGS, the building metrics can still explain 61.2% of the LST change. Thence, it can be said that buildings are the determining factor that changes the ambient temperatures. Besides, although the five UGS metrics explain a total of 38.8% of the LST change, their relative contribution is significantly reduced compared to the separate impact on LST, which illustrates that the cooling effect of UGS is restricted by the proportion of buildings in local regions.

5.6 Marginal Effects of Factors

In order to further analyze the relative influence of different metrics on LST, the study draws the marginal effects curves, which could show the potential impacts of each independent variable with changes in its magnitudes. When the value of influence is greater than 0, the variable has a positive effect on LST, while a value less than 0 indicates a negative effect. A value of 0 means that the effect of the variable on LST is not significant (Sun et al., 2019).

As presented in Fig. 5.4, the marginal effects show that BCR is positively correlated with LST change. When BCR < 44%, for every 10% increase in BCR, LST increases by about 0.8 °C. HRAR is negatively correlated with LST, and

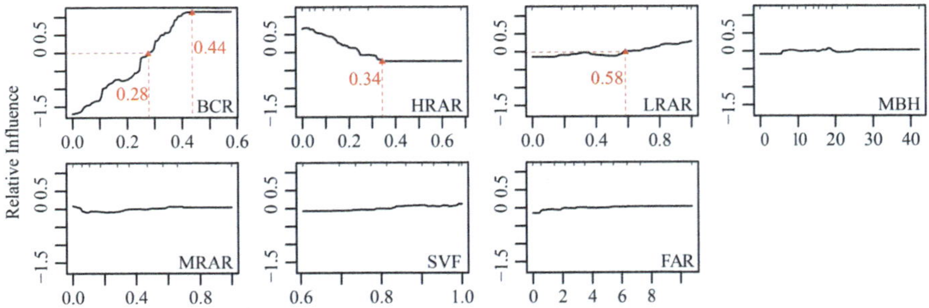

Fig. 5.4 The relationship between 3D building metrics and LST

LST significantly decreases with HRAR increases. When HRAR >34%, the increase of HRAR shows a negative effect on LST, that is, the cooling effect, and can provide a stable cooling effect of about 0.3 °C. When LRAR >58%, LRAR is positively correlated with LST change. The regulation amplitudes of the BCR, HRAR, and LRAR for the block average temperature are 2.8 °C, 0.8 °C, and 0.4 °C, respectively. In addition, the other variables have little relative influence, with all remaining in [−0.2, 0.2], but show a positive correlation with LST.

Among UGS metrics, GCR, ED, PD, and AI show a significant influence on LST (Fig. 5.5). Specifically, GCR and AI are negatively correlated with LST change. The regulation amplitudes of the GCR and AI for the block average temperature are 1.3 °C and 0.5 °C, respectively. When GCR > 64% and AI > 64, this can provide a stable cooling effect of about 0.8 °C and 0.3 °C, respectively. ED and PD are positively correlated with LST change. The regulation amplitudes of the ED and PD for the block average temperature are 1.1 °C and 0.8 °C, respectively. Specifically, when ED > 175 $m{\cdot}ha^{-1}$ and PD > 27 m^{-2}, this could provide a clear warming effect. Moreover, LPI > 60 can provide a stable cooling effect of about 0.2 °C. MPS shows almost no effect on LST.

Figure 5.6 shows the marginal effects of nine combined metrics on LST. Building metrics still maintain the strong effect on LST, while the influence of UGS metrics changes markedly. Specifically, BCR is the most influential metric. Comparing Fig. 5.3, the influence range of BCR on LST decreases from [−1.7, 1.1] to [−1.5, 1.0], the cooling effect decreases 0.3 °C, but BCR remains significantly positively correlated with LST. Besides, the influence of HRAR and LRAR on LST has hardly changed compared to Fig. 5.3, and MBH shows little effect. However, the influence of ED, PD, GCR, and AI on LST decreases dramatically. The regulation amplitudes of them for the block average temperature are 0.7 °C, 0.5 °C, 0.7 °C, and 0.1 °C, respectively.

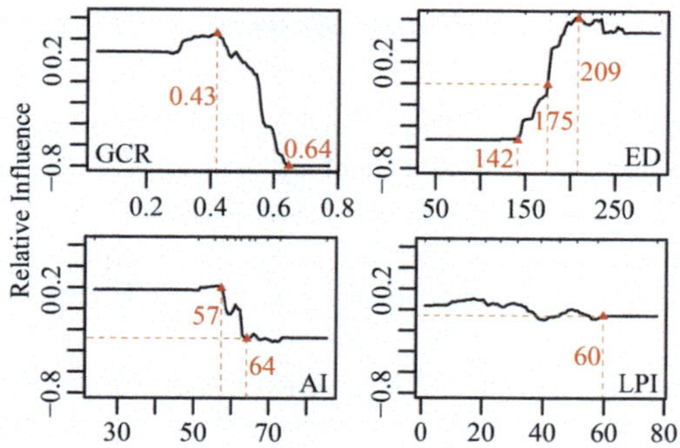

Fig. 5.5 The relationship between UGS metrics and LST

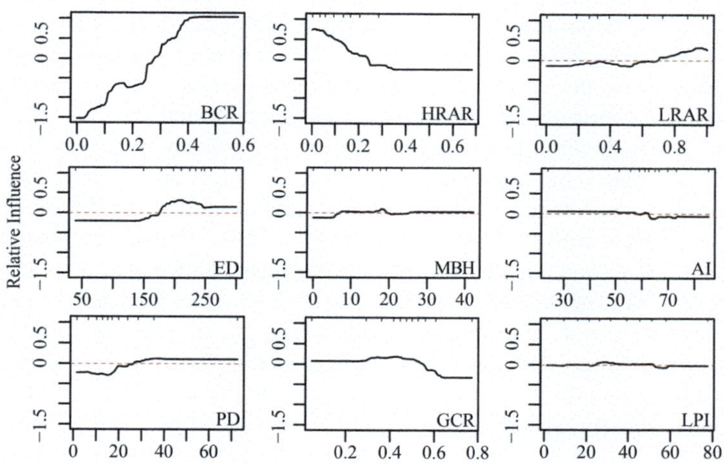

Fig. 5.6 The relationship between combined metrics and LST

5.7 Discussion

Heat balance mechanisms within cities are complex, and the built environment is often considered a significant heat source, while UGS can effectively cool the surroundings. However, the quantitative role of both spatial patterns in the warming/cooling effect on the surroundings is still difficult to determine. This study quantifies the relative influence of 3D building metrics and landscape metrics of UGS on LST through the BRT model. The marginal effect curves in BRT model are capable to reflect the quantitative effect of each metric on LST in the form of a threshold value.

5.7.1 Relationship Between 3D Building Metrics and LST

In this study, LST increases with BCR raise. The replacement of the physical properties of natural surface by the built environment results in a significant weakening of the interaction of ground and near-surface energy, and the impervious surface absorbs more solar radiation that leads to an increase in surface temperature. This study finds that when BCR > 44%, the continued increase of BCR can raise the temperature by about 1 °C in a block. Further analysis reveals that LRAR and MRAR within these blocks were 55.9% and 39.4%, respectively, which illustrates that low- and mid-rise buildings can significantly increase LST at a block scale. The secondary influential factories with HRAR increases can alleviate LST. Especially when HRAR <30%, a cooling effect of 0.7 °C is observed. Generally, high-rise buildings are usually scattered, with large spacing between buildings, which is more conducive to heat dissipation, thereby exhibiting a lower temperature than the surrounding low-rise buildings. Besides, the relatively large coverage of UGS and long buildings' shadows provides effective cooling effect. Compared with the scattered high-rise buildings, the study also finds that concentrated and compact low-rise buildings can increase the temperature, which is confirmed by Li et al. (2011). However, MBH exhibits little effect on LST, which agrees with the results by Sun et al. (2020) that MBH has a smaller effect on LST in spring and summer than in fall and winter, with a difference of nearly 10% in relative contribution. Moreover, SVF has little correlation with LST in summer, which is different from the results of Hu et al. (2020). They concluded that when building-based SVF > 0.8, there is a negative effect on LST. The possible reason is that SVF was averaged at the road-based block scale in this study, causing low spatial disparity. In addition, FAR shows a small effect on LST, possibly because buildings with higher FAR generally occupy a large building area. But BCR has counteractive effects on LST, which diminishes the influence of FAR on LST.

5.7.2 Relationship Between UGS Metrics and LST

The configuration and composition of UGS can effectively alleviate the UHI. In particular, urban trees can effectively cool the ambient temperature through the shading effect and evapotranspiration. This study finds that when GCR > 43%, its cooling effect improved dramatically. That finding is generally consistent with the results by Ziter et al. (2019), which showed that vegetation has a significant negative correlation with LST when vegetation fraction is greater than 40%. Besides, when GCR > 0.64, it can keep the ambient temperature lower by 0.8 °C, indicating that UGS has an important role and great application potential in reducing urban temperature and mitigating UHI. Theoretically, ED and PD of UGS have a negative correlation with LST due to the increased opportunity to exchange energy with the external environment, but we find that ED is positively correlated with LST, consistent with the results of Li et al. (2013), but diverging from the results of

Gómez-Muñoz et al. (2010). The latter results showed that higher ED provides more shading and thus a cooling effect. Moreover, a significant increase in LST, when ED > 175 m·ha^{-1} and each 34 m·ha^{-1} increases, could raise the surrounding LST by approximately 0.6% (Fig. 5.5). One possible reason suggested is that dispersed and small UGSs within road-based blocks diminish the cooling effect. Li et al. (2013) noted that PD and ED are highly correlated. The large ED increases the cooling effect of UGS to surroundings. Further decreasing the MPS will reduce the cooling effect of UGS patch itself. Additionally, a previous study found a significant negative correlation between LPI and LST (Xie et al., 2013). However, this study finds that LPI shows little effect, most likely due to the small differences in area among landscape patches at the block scale.

5.7.3 Relationship Between Combined Metrics and LST

Combined analysis of UGS metrics and buildings metrics can explore their interaction of cooling or warming effect. Comprehensive analysis of nine metrics shows that although the contribution of BCR to LST decreases, BCR is still the dominant factor. In contrast, the contribution of GCR to LST decreases by about 17.8% and ranks eighth, which indicates that the reduction of buildings is a determining factor cooling the surroundings. As presented in Fig. 5.6, the cooling effect of UGS is greatly influenced by architectural factors. The cooling effect of PD and AI largely disappears, and the probable reason for this phenomenon is that the warming effect of buildings on the surroundings is significantly greater than the cooling effect of UGS. Although limited UGS can cool the surroundings, the warming effect of buildings on LST far exceeds the maximum cooling effect that could be provided by UGS. LRAR can heat the surroundings, which is contrary to the research conducted by Yu et al. (2020). They found that low-rise buildings tend to alleviate LST. However, the cooling effect of LRAR is only 0.1 °C in this study. In their study, low-rise buildings are mostly cottages in suburban areas where the background temperature is lower than that of the central city. They also considered the 3D vegetation structure, which is confirmed by Gage and Cooper (2017), which can provide an effective cooling effect in structurally similar areas, such as low-rise building clusters.

5.7.4 Implications for Urban Planning and Management

Based on the findings of this study, the landscape composition and configuration should be targeted based on units like blocks to propose planning and management solutions. It seems less realistic to regulate and optimize the landscape from a grid perspective. Therefore, feasible recommendations to mitigate the effects of UHI are presented from two perspectives. First, the existing built environment needs to be optimized. Generally, the coverage of mid- and low-rise buildings in areas where BCR exceeds 44% often far exceeds that of high-rise buildings, and these areas should appropriately increase the high-rise buildings coverage to decrease the regional surface temperature through the shadows of high-rise buildings and vegetation together. Hu et al. (2020) also proposed that more high-rise buildings (more than 68 m) should be built in areas with low-density vegetation (NDVI < 0.5) to reduce the heat absorption by impervious surface through their shadows. For high-density vegetation areas, the number of low- and high-rise buildings should be appropriately increased to coordinate the demand for housing from population growth and to ensure land intensification. Second, there is a need to rationalize the planning of development zones. It is an unavoidable reality that land supply restrictions lead to the growth of building height. Therefore, for historical cities like Xi'an, on the basis of protecting heritage sites, the building coverage within the development zone should be controlled to be less than 44%, with vegetation coverage higher than 43%. Mid-rise buildings should be built to divert the population from the central city to the development zone and ensure a comfortable thermal environment. In addition, based on the result that high-rise buildings alleviate LST, the micro layout of high-rise buildings needs to be designed to form effective ventilation corridors.

5.7.5 Limitations and Future Works

Several limitations should be considered in future research. First, the study used LST data imaged at 11:19 a.m., but the highest daily temperature occurs from 14:00 to 16:00 p.m. Therefore, the warming/cooling effect of urban landscapes in that duration needs more attention. Besides, all available Landsat images should be used to obtain average LST to improve the representativeness of the results. Moreover, the study uses roads to generate the analysis unit, but it ignores the warming effect of main roads. Roads can easily absorb solar radiation to make their temperature abnormally high. Whether the hotter air mass on this local high-temperature belt will be transferred into the road-based block and thereby have an effect on the temperature inside the block needs to be further explored. Finally, as urban trees and high-rise buildings are both capable of producing varying degrees of shading, what is the degree of cooling effect of both? Whether the shading cast by buildings on urban trees canopies reduces their cooling effect requires continued research.

5.8 Conclusions

This study analyzes and compares the separate and combined effects of 3D building metrics and UGS metrics on LST using BRT model. (1) BCR is the dominant factor of increasing surface temperature at the road-based block, and the increase in BCR can significantly increase LST. When BCR is less than 44%, each 10% increase in

BCR raises LST by about 0.8 °C. HRAR is negatively correlated with LST, especially when HRAR is less than 34%, and the increase in HRAR can play a role in alleviating LST. (2) GCR and AI can provide effective cooling effects when ignoring the building metrics. However, when the ED is greater than 175 m·ha^{-1} and PD > 27 m^{-2}, it will heat the surroundings. Therefore, the composition and configuration of UGS should consider the cooling threshold of these metrics, thereby maximally alleviating LST. (3) The cooling effect provided by UGS is limited, and the key to alleviating UHI is to optimize the composition and configuration of the 3D built environment. (4) Using a road-based block as the analysis unit can ensure consistent building shapes, which may provide a useful insight for urban planning and managers to optimize and regulate the UTE from the block perspective.

References

Akbari, H., & Konopacki, S. (2005). Calculating energy-saving potentials of heat-island reduction strategies. *Energy Policy, 33*, 721–756. https://doi.org/10.1016/j.enpol.2003.10.001

Alavipanah, S., Schreyer, J., Haase, D., Lakes, T., & Qureshi, S. (2018). The effect of multi-dimensional indicators on urban thermal conditions. *Journal of Cleaner Production, 177*, 115–123. https://doi.org/10.1016/j.jclepro.2017.12.187

Chen, A., Yao, L., Sun, R., & Chen, L. (2014). How many metrics are required to identify the effects of the landscape pattern on land surface temperature? *Ecological Indicators, 45*, 424–433. https://doi.org/10.1016/j.ecolind.2014.05.002

Clarke, A., & Johnston, N. M. (1999). Scaling of metabolic rate with body mass and temperature in teleost fish. *The Journal of Animal Ecology, 68*, 893–905. https://doi.org/10.1046/j.1365-2656.1999.00337.x

Elith, J., Leathwick, J. R., & Hastie, T. (2008). A working guide to boosted regression trees. *The Journal of Animal Ecology, 77*, 802–813. https://doi.org/10.1111/j.1365-2656.2008.01390.x

Estoque, R. C., Murayama, Y., & Myint, S. W. (2017). Effects of landscape composition and pattern on land surface temperature: An urban heat island study in the megacities of Southeast Asia. *Science of the Total Environment, 577*, 349–359. https://doi.org/10.1016/j.scitotenv.2016.10.195

Feyisa, G. L., Dons, K., & Meilby, H. (2014). Efficiency of parks in mitigating urban heat island effect: An example from Addis Ababa. *Landscape and Urban Planning, 123*, 87–95. https://doi.org/10.1016/j.landurbplan.2013.12.008

Gage, E. A., & Cooper, D. J. (2017). Relationships between landscape pattern metrics, vertical structure and surface urban heat island formation in a Colorado suburb. *Urban Ecosystem, 20*, 1229–1238. https://doi.org/10.1007/s11252-017-0675-0

Gómez-Muñoz, V. M., Porta-Gándara, M. A., & Fernández, J. L. (2010). Effect of tree shades in urban planning in hot-arid climatic regions. *Landscape and Urban Planning, 94*, 149–157. https://doi.org/10.1016/j.landurbplan.2009.09.002

Hu, Y., Dai, Z., & Guldmann, J.-M. (2020). Modeling the impact of 2D/3D urban indicators on the urban heat island over different seasons: A boosted regression tree approach. *Journal of Environmental Management, 266*, 110424. https://doi.org/10.1016/j.jenvman.2020.110424

Kong, F., Yin, H., James, P., Hutyra, L. R., & He, H. S. (2014). Effects of spatial pattern of greenspace on urban cooling in a large metropolitan area of eastern China. *Landscape and Urban Planning, 128*, 35–47. https://doi.org/10.1016/j.landurbplan.2014.04.018

Li, J., Song, C., Cao, L., Zhu, F., Meng, X., & Wu, J. (2011). Impacts of landscape structure on surface urban heat islands: A case study of Shanghai, China. *Remote Sensing of Environment, 115*, 3249–3263. https://doi.org/10.1016/j.rse.2011.07.008

Li, X., Zhou, W., & Ouyang, Z. (2013). Relationship between land surface temperature and spatial pattern of greenspace: What are the effects of spatial resolution? *Landscape and Urban Planning, 114*, 1–8. https://doi.org/10.1016/j.landurbplan.2013.02.005

Liu, M., Hu, Y.-M., & Li, C.-L. (2017). Landscape metrics for three-dimensional urban building pattern recognition. *Applied Geography, 87*, 66–72. https://doi.org/10.1016/j.apgeog.2017.07.011

Manoli, G., Fatichi, S., Schläpfer, M., Yu, K., Crowther, T. W., Meili, N., Burlando, P., Katul, G. G., & Bou-Zeid, E. (2019). Magnitude of urban heat islands largely explained by climate and population. *Nature, 573*, 55–60. https://doi.org/10.1038/s41586-019-1512-9

Oke, T. R. (1982). The energetic basis of the urban heat island. *Quarterly Journal of the Royal Meteorological Society, 108*, 1–24. https://doi.org/10.1002/qj.49710845502

Santamouris, M., Cartalis, C., Synnefa, A., & Kolokotsa, D. (2015). On the impact of urban heat island and global warming on the power demand and electricity consumption of buildings—A review. *Energy and Buildings, 98*, 119–124. https://doi.org/10.1016/j.enbuild.2014.09.052

Sun, F., Mejia, A., & Che, Y. (2019). Disentangling the contributions of climate and basin characteristics to water yield across spatial and temporal scales in the Yangtze River basin: A combined hydrological model and boosted regression approach. *Water Resources Management, 33*, 3449–3468. https://doi.org/10.1007/s11269-019-02310-y

Sun, F., Liu, M., Wang, Y., Wang, H., & Che, Y. (2020). The effects of 3D architectural patterns on the urban surface temperature at a neighborhood scale: Relative contributions and marginal effects. *Journal of Cleaner Production, 258*, 120706. https://doi.org/10.1016/j.jclepro.2020.120706

Weng, Q., Lu, D., & Schubring, J. (2004). Estimation of land surface temperature–vegetation abundance relationship for urban heat island studies. *Remote Sensing of Environment, 89*, 467–483. https://doi.org/10.1016/j.rse.2003.11.005

Xie, M., Wang, Y., Chang, Q., Fu, M., & Ye, M. (2013). Assessment of landscape patterns affecting land surface temperature in different biophysical gradients in Shenzhen, China. *Urban Ecosystems, 16*, 871–886. https://doi.org/10.1007/s11252-013-0325-0

Yang, J., Wang, Y., Xiu, C., Xiao, X., Xia, J., & Jin, C. (2020). Optimizing local climate zones to mitigate urban heat island effect in human settlements. *Journal of Cleaner Production, 275*, 123767. https://doi.org/10.1016/j.jclepro.2020.123767

Yu, S., Chen, Z., Yu, B., Wang, L., Wu, B., Wu, J., & Zhao, F. (2020). Exploring the relationship between 2D/3D landscape pattern and land surface temperature based on explainable eXtreme gradient boosting tree: A case study of Shanghai, China. *Science of the Total Environment, 725*, 138229. https://doi.org/10.1016/j.scitotenv.2020.138229

Zhou, W., Huang, G., & Cadenasso, M. L. (2011). Does spatial configuration matter? Understanding the effects of land cover pattern on land surface temperature in urban landscapes. *Landscape and Urban Planning, 102*, 54–63. https://doi.org/10.1016/j.landurbplan.2011.03.009

Ziter, C. D., Pedersen, E. J., Kucharik, C. J., & Turner, M. G. (2019). Scale-dependent interactions between tree canopy cover and impervious surfaces reduce daytime urban heat during summer. *Proceedings of the National Academy of Sciences, 116*, 7575–7580. https://doi.org/10.1073/pnas.1817561116

Chapter 6
Dynamics of Urban Thermal Environments Under Dry and Wet Conditions

6.1 Introduction

Importance of Seasonal LST Variations The seasonal LST variation is a significant aspect of climate change and surface processes research. It reflects the seasonal differences in surface energy balance and the response of land cover types to climate change. LST is influenced by various factors, including seasonal changes in solar radiation, atmospheric conditions, land cover type, soil moisture, and human activities (Hardwick Jones et al., 2010; Rhee et al., 2014; Shafizadeh-Moghadam et al., 2020). Investigating the seasonal variation of LST is crucial for understanding the surface energy balance under global warming, improving climate model predictions, urban planning, optimizing resource management, and environmental protection.

Characteristics of Seasonal LST Variations Relevant studies show that the seasonal variation of LST is more pronounced in high-latitude regions than in low-latitude regions. LST in spring and autumn is similar, while it significantly decreases in winter. The distribution of high-temperature areas in the four seasons has distinct characteristics, mainly concentrated in urban centers, whereas low-temperature areas are concentrated in urban green spaces, parks, and other vegetated areas, which are less affected by seasonal changes (Chen et al., 2023). Although there have been studies on the seasonal variation of LST, few have analyzed the impact of relative humidity on LST. Seasonal variation studies typically focus on how solar radiation affects LST, while studies on air humidity concentrate on how atmospheric moisture regulates energy exchange between the surface and the atmosphere (Zhao & Fong, 2017). Research on seasons, relative humidity, and their combined effects is crucial for understanding the driving forces behind LST variations. However, there is still a lack of studies exploring the impact of urban dry and wet conditions on the seasonal variation of LST in conjunction with relative humidity.

L. Zhou et al., *Urban Surface Thermal Environments*, The Urban Book Series,
https://doi.org/10.1007/978-3-031-94499-4_6

Comprehensive research on both factors is essential for fully understanding urban microclimates and developing effective urban planning and management strategies to adapt to and mitigate climate change.

Study Objectives This chapter further explores the characteristics of LST in various local climate zones (LCZs) during different seasons and analyzes the regulatory role of relative humidity on LST in different seasons. By using the least squares method, the study delves into the contributions of different LCZ types to LST under varying seasonal changes and humidity conditions, providing new insights into mitigating urban heat environments and regulating microclimates.

6.2 Datasets and Methods

6.2.1 Data Processes

This section utilizes Landsat remote sensing imagery from 2022, with cloud cover below 15% (Table 6.1). The single-window algorithm in ENVI was used for LST retrieval, following the steps below: (1) Radiometric calibration of the TIR band 10 to obtain brightness temperature. (2) Calculation of the normalized difference vegetation index (NDVI). (3) Extraction of vegetation coverage based on NDVI. (4) Calculation of surface emissivity using atmospheric profile parameters. (5) Calculation of LST. The nearest neighbor method was used to resample the LST data to 100 m to match the spatial resolution of the LCZ data.

6.2.2 Urban Boundary Extraction

This chapter employs a combination of multiple data sources to extract urban boundaries with high accuracy (Ding & Chen, 2022). The primary datasets used include population data, nighttime light data, and land use data, each providing different perspectives on urban development and human activities. Population data reflects residential density, nighttime light data reveals activity intensity, and land use data indicates land development patterns. Detailed descriptions of these datasets are provided in Table 6.2. To ensure consistency and comparability, all data were

Table 6.1 Details of Landsat images used in this study

Season	Date	Local time	Satellite	Cloud cover (%)
Spring	2022.04.07	11:19	Landsat-9	0.03
Summer	2022.08.05	11:20	Landsat-8	8.09
Autumn	2022.11.01	11:20	Landsat-9	3.18
Winter	2022.01.17	11:20	Landsat-9	0.23

Table 6.2 Data sources

Data	Source	Resolution
Administration boundary	https://www.resdc.cn	–
Landsat-8, 9	https://earthexplorer.usgs.gov	30 m
Population	https://hub.worldpop.org	100 m
Nighttime light	https://doi.org/10.7910/DVN/YGIVCD	500 m
Land use	https://www.resdc.cn	100 m

processed using ArcGIS 10.6. The spatial resolution of all datasets was resampled to 500 m, which ensures precise data processing and effectively captures the subtle differences between urban and rural areas. Then, a min–max normalization method was applied to eliminate differences arising from various data types. The normalized data were then used for K-means clustering analysis. K-means clustering is a widely used unsupervised learning algorithm that discovers inherent structures and patterns by partitioning data points into clusters. In this study, K-means clustering effectively distinguishes urban and rural areas based on indicators, such as population density, nighttime light intensity, and land use types. The advantage of this method lies in its simplicity and intuitive nature, allowing automatic grouping of data points into corresponding clusters, thereby delineating urban boundaries. Finally, the urban and rural boundaries were vectorized to obtain the precise urban boundaries.

6.2.3 Calculation of Urban Heat (Cold) Islands and Wet (Dry) Islands

To thoroughly investigate the differences in thermal and humidity environments between urban and rural areas in the study area, we calculated the spatial distribution patterns of heat islands, cold islands, wet islands, and dry islands (Eqs. 6.1 and 6.2). The urban heat island phenomenon refers to higher temperatures in urban centers compared to surrounding rural areas, typically caused by the heat absorption and retention of buildings and roads in cities. Additionally, urban areas often exhibit lower relative humidity due to reduced vegetation coverage, leading to decreased natural evaporation. Conversely, rural areas, with more green vegetation, can more effectively maintain ground moisture, resulting in higher relative humidity than urban areas. These phenomena significantly impact urban climate, environmental quality, and residents' quality of life. To quantify the environmental differences, this chapter uses heat island intensity and wet island intensity as the primary metrics. Heat island intensity is measured by comparing the average temperature difference between urban and surrounding rural areas, while wet island intensity is defined by comparing the differences in average relative humidity (Balke et al., 2017). The ground relative humidity data is derived from the "Spatially Interpolated Datasets of Meteorological Elements in China" released in the Resource and Environment

Science Data Platform by the Institute of Geographic Science and Natural Resource, CAS (https://www.resdc.cn/DOI/doi.aspx?DOIid=104). Based on observations from 2421 meteorological stations, this grid-form dataset with spatial resolution of 1 km is produced via thin plate spline interpolation with elevation as the covariable. Additionally, to facilitate the comparison between LST and relative humidity, this humidity data is resampled into 100 m. These indicators not only help in understanding the temperature and humidity disparities between urban and rural areas but also reveal the underlying environmental, social, and economic factors contributing to the differences. By analyzing the spatial distribution patterns of heat islands and wet islands, we can better understand the impact of urbanization on regional climate, providing scientific evidence for urban planning and sustainable development.

$$SUHII_i = LST_i - LST_{\mathrm{rural}} \tag{6.1}$$

$SUHII_i$ is the surface urban heat island intensity for grid i, LST_i is the LST of grid i, and LST_{rural} is the average LST of the rural areas.

$$UDII_i = RH_i - RH_{\mathrm{rural}} \tag{6.2}$$

$UDII_i$ is the urban dry island intensity for grid i, RH_i is the RH of grid i, and RH_{rural} is the average RH of the rural areas.

6.2.4 Least Squares Method

The least squares method is a mathematical optimization technique (Björck, 1990). It seeks the best fit for data by minimizing the sum of the squares of the errors. Geometrically, the least squares method represents the projection of a vector in high-dimensional space onto a lower-dimensional subspace. It simplifies finding unknown data such that the sum of the squared differences between the observed and computed values is minimized. This method is a standard approach for obtaining approximate solutions to linear systems of equations, especially when the number of equations exceeds the number of unknowns, as in regression analysis. In the overall solution, the least squares method minimizes the total sum of the squared residuals (the differences between observed and model-provided fitted values). Its most significant application is in curve fitting, where it achieves the best fit by minimizing the sum of the squared residuals. The method can directly compute model parameters analytically without complex iterative processes, making it computationally efficient. It is not only applicable to simple linear regression but also to multiple linear regression, polynomial regression, and various other statistical models (Wang et al., 2020). Therefore, this chapter employs the least squares method to explore the contribution of each LCZ to LST. To establish the quantitative relationship between LCZ and LST, the study area is resampled into 1 km × 1 km grids. The proportion of each LCZ within the grid is then calculated, along with the grid's average LST. Finally, the least squares method is used to determine the impact of LCZ on LST.

6.3 Spatial Autocorrelation of Thermal and Humidity Environments

To better understand the spatial association and dependency characteristics between LST and relative humidity, this study employs the bivariate Local Moran's I index (Lee & Li, 2017). This method effectively reveals the spatial correlation between two variables. Figure 6.1 shows the significant spatial distribution differences by analyzing the seasonal variations of LST and relative humidity. In spring, there is no obvious consistency or discrepancy in the spatial distribution of LST and relative humidity within the ring road. Most of the northwestern region of the study area exhibits dry-hot characteristics, while a few areas show a dry-cold environment. During summer, the area within the ring road predominantly displays a dry-hot environment, likely due to high urbanization levels, with fewer green spaces and water bodies leading to high temperatures and low humidity. Conversely, the southern area outside the highway shows a dry-cold environment. In autumn, the central-northern area becomes a wet-hot environment, while most of the eastern region is a dry-hot area. In winter, the city center is dominated by wet-hot and wet-cold environments, possibly due to the high building density and heat emissions in these areas.

The northwestern part shows dry-hot characteristics, and the western region is mainly dry-cold. Throughout spring, summer, and autumn, the southern part of the study area predominantly exhibits a wet-cold environment, likely influenced by the extensive vegetation cover. The interspersed wet-hot areas are mainly occupied by LCZs characterized by building types. However, this pattern changes in winter when the vegetated areas in the south primarily exhibit a wet-hot environment. This phenomenon might be related to the geographical location and climate type of Xi'an. Xi'an is located in the warm temperate deciduous broadleaf forest zone; amount of bare ground appears in the non-growing season. This exposed ground can more effectively absorb solar radiation, leading to an increase in LST.

6.4 Seasonal LST Differences Across Various LCZs

To explore the LST differences across different LCZs during various seasons, box plots were used to display the LST characteristics of each LCZ under different climatic conditions (Fig. 6.2). In spring, summer, and autumn, the LST of built-up type LCZs is higher than that of land cover type LCZs. However, in winter, the average LST of land cover type LCZs exceeds that of built-up type LCZs by 0.30 °C. Compact building zones (LCZ 1–3) have higher LSTs compared to open building zones (LCZ 4–6), and high-rise buildings (LCZ 1, 4) have lower LSTs than mid- and low-rise

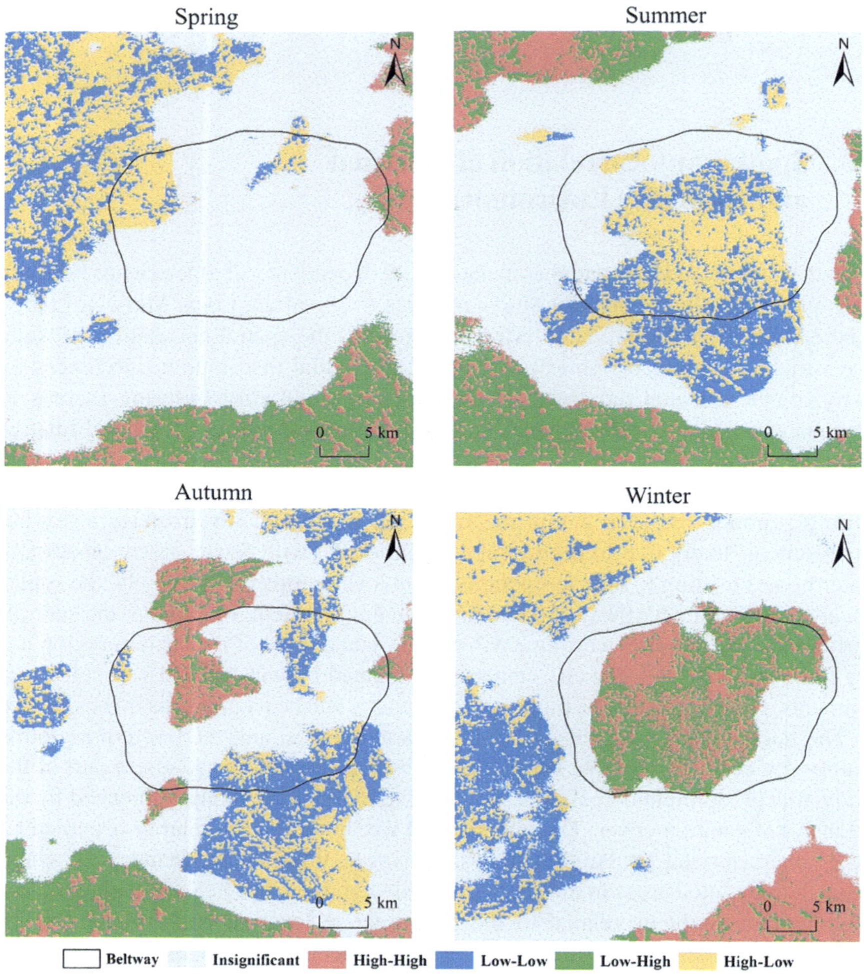

Fig. 6.1 Bivariate spatial autocorrelation of LST and relative humidity

buildings. In spring, summer, and autumn, LCZ 8 (large low-rise) exhibits the highest LST, while LCZ G (water bodies) shows the lowest LST. The average LST of LCZ 8 is 1.57 °C higher than the average LST of built-up type LCZs, and the average LST of LCZ G is 2.33 °C lower than the average LST of land cover type LCZs. Among the land cover type LCZs, LCZ E (paved surfaces) stands out with an average LST of 25.46 °C, second only to LCZ 8. Overall, except in winter, the LSTs of built-up type LCZs are generally higher than those of land cover type LCZs. LST increases with the reduction in building height and the increase in building density, while an increase in vegetation density leads to a decrease in LST.

Compared to LST, most LCZs exhibit a larger interquartile range (IQR) for relative humidity, indicating a more dispersed distribution of relative humidity data within the same type of LCZ and greater differences in relative humidity across

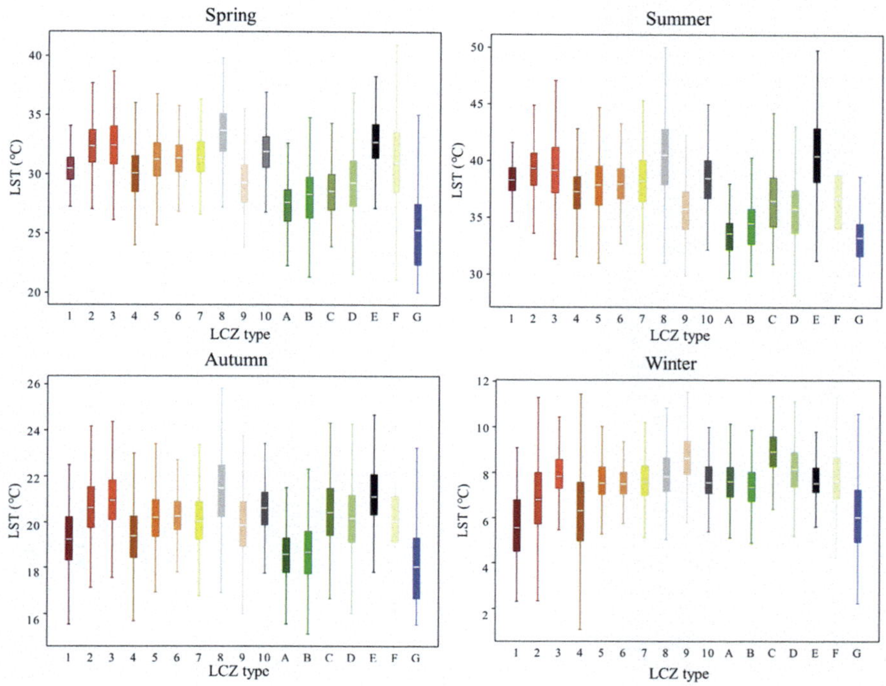

Fig. 6.2 The seasonal variations of different LCZ types

different spatial locations (Fig. 6.3). In spring, summer, and autumn, the relative humidity of land cover type LCZs is higher than that of built-up type LCZs, while it is reverse in winter. Vegetation-type LCZs have the highest relative humidity in spring, summer, and autumn. In contrast, compact mid- and high-rise buildings (LCZ 1, 2) exhibit higher relative humidity in winter. In addition, built-up type LCZs tend to form wet-cold environments in winter, whereas they predominantly experience dry-hot environments in other seasons. Among built-up type LCZs, LCZ 9 (sparse low-rise) has the highest average relative humidity at 60.99%, likely due to its sparse building distribution and dense vegetation, which creates a relatively humid environment. In contrast, LCZ F (bare soil) within land cover type LCZs has the lowest relative humidity at 58.10%. Therefore, increasing vegetation cover in drier urban areas can not only mitigate urban heat but also enhance humidity.

6.5 LST Characteristics Under Dry and Wet Environment

This chapter conducted a detailed statistical analysis of LST across four seasons, revealing the patterns of LST variations in different warm and humid conditions (Fig. 6.4). The results unveiled an interesting phenomenon: Under different climatic conditions, LST in arid areas and humid areas shows significant differences. In

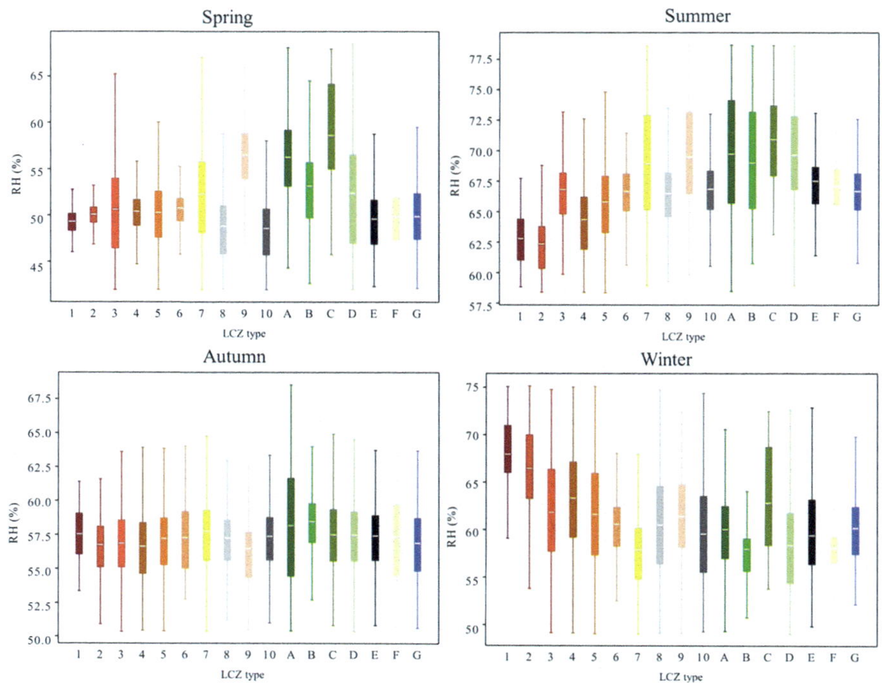

Fig. 6.3 The seasonal variations of humidity across various LCZ types

colder environments, it was observed that the LST in arid areas is generally higher than in humid areas. This may be because the lower moisture content in arid areas reduces the surface cooling effect. In contrast, humid areas, with better vegetation cover, can effectively lower LST through vegetation transpiration. However, the phenomenon is reversed in warmer conditions. The LST in humid areas is typically higher than that in arid areas, especially in seasons other than spring. This might be due to the increased heat retention of surface moisture in high humidity conditions, leading to higher LST in humid areas. Finally, the research identified a consistent trend by calculating the average LST of arid and humid areas across the four seasons in the entire study area. LST in arid areas is higher than in humid areas in all seasons. This trend shows significant variations across different seasons, particularly in spring, where the LST in arid areas is 1.35 °C higher than in humid areas. These results indicate that seasonal changes have a significant impact on LST.

6.6 Relationship Between LST and LCZ Proportion

The relationship between LCZs and urban wet (dry) islands and heat (cold) island effect was analyzed to deeply understand the variations of LST and relative humidity. For each of the four seasons, patches of dry-hot, dry-cold, wet-hot, and wet-cold

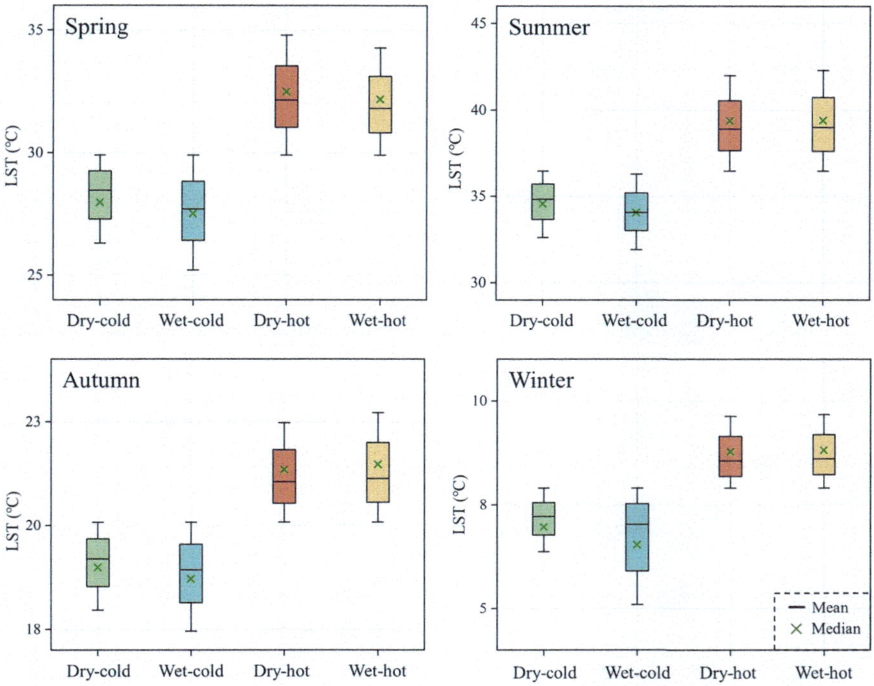

Fig. 6.4 LST distribution under different warm and humid conditions

conditions were identified. The proportion of LCZ types within these patches and the correlation between the average LST and relative humidity of these patches were then analyzed using statistical methods (Fig. 6.5). Since different patches contain varying LCZ types, some LCZ types might be absent in certain patches. To ensure the accuracy and reliability of the analysis, patches with no same LCZ types were excluded from the analysis. The results indicate that for LST, there is a negative correlation between land cover type LCZs and LST in dry-wet and cold conditions during spring and summer. Additionally, built-up type LCZs (such as commercial and residential areas) generally show a positive correlation with LST across all four seasons. Notably, LCZ 8 (large low-rise buildings) exhibits a significant positive correlation with LST in wet-hot conditions during spring (0.71) and in dry-cold conditions during autumn (0.90). This highlights the crucial role of large low-rise building areas in regulating urban microclimates, particularly under specific seasonal and climatic conditions.

The impact of different LCZ types on relative humidity shows significant variation across seasons. Specifically, in spring, summer, and autumn, most built-up type LCZs, such as commercial areas, residential areas, and industrial zones, exhibit a negative correlation with relative humidity. This may be due to human activities and cooling systems, which tend to reduce the surrounding air humidity. In contrast, land cover type LCZs, such as parks, forests, and farmland, generally show a positive

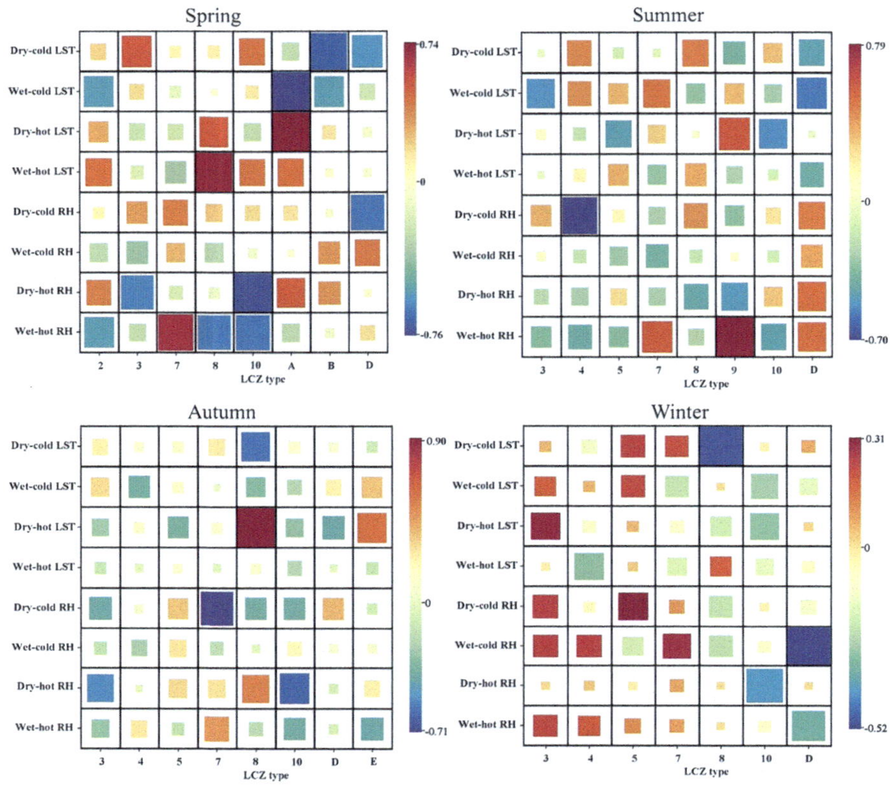

Fig. 6.5 The correlations between LCZ and LST across different seasons

correlation with relative humidity. These areas often maintain higher humidity due to the vegetation transpiration and the lower percent of buildings. In summer, the positive correlation is highest, approximately around 0.5, indicating the critical role of vegetation in maintaining urban humidity. However, the pattern changes in winter. Most built-up type LCZs show a positive correlation with relative humidity in winter. Meanwhile, land cover type LCZs exhibit an opposite trend in winter. Notably, LCZ D (low plants) in wet-cold conditions shows a significant negative correlation with relative humidity, as high as 0.52. This may be related to the fallow period of most farmlands during winter, resulting in lower relative humidity.

6.7 Seasonal Variations in LCZ Contributions to LST

Seasonal variations in the relationship between LCZ and LST were explored using the least squares regression method (Table 6.3). In spring, most LCZs show a negative contribution to LST in dry-cold and wet-hot conditions, indicating that an increase in LCZ types may lead to a decrease in LST. However, LCZs have a

Table 6.3 The coefficients of the least squares regression

	LCZ	2	3	4	5	6	7	8	9	10	A	B	D	E
Spring	Dry-cold	–	3.43	−7.87	−7.66	–	−10.02	−8.97	−5.62	−15.62	−9.45	−11.55	−7.97	–
	Wet-cold	–	15.32	8.27	12.30	–	16.76	7.70	8.41	–	6.68	−5.67	11.36	–
	Dry-hot	−2.88	1.09	1.69	−0.88	–	0.60	3.86	−6.15	2.77	85.36	–	0.23	–
	Wet-hot	−46.80	−16.22	−11.63	−17.04	–	−16.34	10.28	−16.32	55.13	−15.41	–	−15.85	–
Summer	Dry-cold	20.73	1.27	8.25	8.17	–	4.99	10.12	–	15.64	–	–	7.97	–
	Wet-cold	–	−15.08	−1.25	−1.39	–	−0.61	−2.88	–	–	–	−7.45	−2.42	–
	Dry-hot	0.92	−16.08	−18.42	−18.15	–	−15.36	−15.37	−13.24	−19.58	–	–	−16.29	–
	Wet-hot	−46.61	11.40	15.92	20.87	–	14.74	19.42	–	6.09	–	–	14.45	13.78
Autumn	Dry-cold	40.31	21.51	12.66	16.64	15.58	14.26	–	–	23.14	−9.00	−14.70	17.29	19.08
	Wet-cold	–	13.65	6.24	7.77	–	6.39	4.12	–	4.77	–	–	7.02	13.80
	Dry-hot	–	−0.71	−0.98	−1.52	–	0.35	1.74	–	2.33	–	–	−1.06	−1.20
	Wet-hot	−6.43	10.95	2.57	3.58	–	4.24	3.50	–	−24.78	−0.08	–	3.13	21.04
Winter	Dry-cold	–	3.52	0.87	2.65	–	–	0.10	–	0.63	–	–	1.78	3.66
	Wet-cold	–	5.81	5.71	6.51	–	4.67	–	2.52	–	6.78	3.79	4.18	–
	Dry-hot	–	1.04	0.56	0.25	–	0.31	0.08	–	−1.20	–	–	0.51	−0.36
	Wet-hot	–	0.00	−1.12	−0.26	3.86	–	−0.17	–	−0.42	–	–	−0.34	−0.15

positive contribution to LST in wet-cold and dry-hot conditions, indicating that some LCZ types promote an increase in LST. Additionally, some LCZ types with higher vegetation coverage play a crucial role in cooling effect. For example, LCZ B (scattered trees) in wet-cold conditions and LCZ 9 (sparsely built) in dry-hot conditions significantly contribute to lowering LST. In summer, all LCZs exhibit a positive contribution to LST in dry-cold and wet-hot conditions, promoting an increase in LST. Conversely, in wet-cold and dry-hot conditions, except for compact mid-rise buildings (LCZ 2), other LCZ types contribute to a decrease in LST. These findings highlight the complex and dynamic interactions between LCZ types and LST across different seasons and environmental conditions, providing valuable insights for urban planning and climate adaptation strategies.

In autumn, built-up type LCZs have a positive effect on increasing LST in dry-cold and wet-cold conditions. Specifically, an increase in LCZ 2 (compact mid-rise) is significantly associated with higher LST in dry-cold conditions. For industrial zones (LCZ 8 and LCZ 10), an increase in proportion by 10% is correlated with an approximate 0.2 °C increase in LST. Conversely, in wet-hot conditions, a 10% increase in proportion of LCZ 10 can result in a decrease of about 2 °C in LST. In winter, all LCZs in dry-cold, wet-cold, and dry-hot conditions show a positive contribution to LST, although the contribution is relatively low. However, in wet-hot conditions, most LCZs contribute to lowering LST, showing an opposite trend. This indicates that the impact of LCZs on LST is not only regulated by climatic and seasonal factors but also closely related to relative humidity.

6.8 LST Strategies for Improving Thermal Environment

The variation in LST appears to be regulated by multiple factors, including relative humidity, the spatial distribution of LCZs, and climatic conditions. To address the impact of these factors on the thermal environment, a series of seasonal mitigation measures are proposed.

Spring: Increasing humidity can effectively reduce LST in heat island areas. This can be achieved by expanding urban blue-green spaces, such as developing more parks (LCZ A), sidewalk trees (LCZ B), and gardens (LCZ C), as well as improving wetlands and water bodies (LCZ G). These measures not only provide the necessary cooling effect but also increase urban biodiversity and recreational spaces for residents.

Summer: Due to generally high LST and overall high humidity in the study area, increasing humidity alone might not significantly regulate LST. Implementing ventilation corridors for heat dissipation is crucial. However, increasing humidity remains an effective solution in compact building areas (LCZ 1–3), such as residential and commercial zones. For instance, green roofs or rooftop gardens on buildings can help absorb solar radiation and cooling surroundings through vegetation evaporation.

Autumn: Ventilation corridors in open mid- and high-rise building areas (LCZ 4, 5) will help regulate temperature. In vegetated areas, it is advisable to use water resources moderately and maintain plant health through irrigation systems to enhance evaporation efficiency.

Winter: While the urban heat island effect is often viewed negatively due to its contribution to increased temperatures in urban areas, it can have some benefits during the winter season, such as reduced heating costs for residents and businesses. Overall, increasing blue-green spaces in urban areas is an effective strategy. This not only reduces solar energy absorption but also dissipates surface heat through evaporation, creating a more comfortable and healthy urban living environment.

6.9 Conclusions

This chapter delves into the spatial distribution characteristics of LST and relative humidity in urban areas across different seasons, revealing the regulatory role of dry and wet conditions on LST and their seasonal variations. In spring, summer, and autumn, due to the impact of vegetation cover, the southern part of the study area primarily exhibits a wet and cold environment. Except in winter, the LST of building type LCZ is generally higher than that of land cover type LCZ. Statistical analysis of LST across the four seasons reveals significant differences between dry and wet areas. In cold island environments, the LST in dry areas is generally 1 °C higher than in wet areas. In heat island environments, the LST in wet areas is usually 0.5 °C higher than in dry areas. The contribution of LCZ to LST varies across different seasons and humidity conditions. In dry and hot summer environments, an increase of 0.1 in open high-rise areas can reduce LST by 1.82 °C, whereas it is opposite in winter. Based on the identified seasonal characteristics, this chapter proposes a series of targeted thermal environment mitigation measures. Specially, increasing urban blue-green spaces, such as parks and wetlands, can effectively reduce LST and provide a more comfortable urban living environment. Additionally, rational urban design and planning, such as enhancing building energy efficiency and optimizing urban layout, are crucial for mitigating urban heat island effects and improving urban environmental quality.

References

Balke, T., Lõhmus, K., Hillebrand, H., Zielinski, O., Haynert, K., Meier, D., Hodapp, D., Minden, V., & Kleyer, M. (2017). Experimental salt marsh islands: A model system for novel metacommunity experiments. *Estuarine, Coastal and Shelf Science, 198*, 288–298. https://doi.org/10.1016/j.ecss.2017.09.021

Björck, Å.B.T.-H. of N.A. (1990). *Least squares methods* (pp. 465–652). Elsevier. https://doi.org/10.1016/S1570-8659(05)80036-5

Chen, Y., Yang, J., Yu, W., Ren, J., Xiao, X., & Xia, J. C. (2023). Relationship between urban spatial form and seasonal land surface temperature under different grid scales. *Sustainable Cities and Society, 89*, 104374. https://doi.org/10.1016/j.scs.2022.104374

Ding, W., & Chen, H. (2022). Urban-rural fringe identification and spatial form transformation during rapid urbanization: A case study in Wuhan, China. *Building and Environment, 226*, 109697. https://doi.org/10.1016/j.buildenv.2022.109697

Hardwick Jones, R., Westra, S., & Sharma, A. (2010). Observed relationships between extreme sub-daily precipitation, surface temperature, and relative humidity. *Geophysical Research Letters, 37*, 1–5. https://doi.org/10.1029/2010GL045081

Lee, J., & Li, S. (2017). Extending Moran's index for measuring spatiotemporal clustering of geographic events. *Geographical Analysis, 49*, 36–57. https://doi.org/10.1111/gean.12106

Rhee, J., Park, S., & Lu, Z. (2014). Relationship between land cover patterns and surface temperature in urban areas. *GIScience & Remote Sensing, 51*, 521–536. https://doi.org/10.1080/15481603.2014.964455

Shafizadeh-Moghadam, H., Weng, Q., Liu, H., & Valavi, R. (2020). Modeling the spatial variation of urban land surface temperature in relation to environmental and anthropogenic factors: A case study of Tehran, Iran. *GIScience & Remote Sensing, 57*, 483–496. https://doi.org/10.1080/15481603.2020.1736857

Wang, L., Hou, H., & Weng, J. (2020). Ordinary least squares modelling of urban heat island intensity based on landscape composition and configuration: A comparative study among three megacities along the Yangtze River. *Sustainable Cities and Society, 62*, 102381. https://doi.org/10.1016/j.scs.2020.102381

Zhao, T. F., & Fong, K. F. (2017). Characterization of different heat mitigation strategies in landscape to fight against heat island and improve thermal comfort in hot–humid climate (Part I): Measurement and modelling. *Sustainable Cities and Society, 32*, 523–531. https://doi.org/10.1016/j.scs.2017.03.025

Chapter 7
Land Surface Temperature Variability during Heatwaves and Non-heatwaves

7.1 Introduction

Heat Waves in Urban Areas Heat wave is a period of abnormally hot weather generally lasting more than 2 days. The deterioration of UTE causes numerous adverse consequences for urban living conditions and sustainability. Increased temperatures can lead to heat-related illnesses, such as heat exhaustion, heatstroke, and dehydration (Ellena et al., 2020), and vulnerable populations like the elderly, children, and individuals with certain medical conditions are at higher risk during heat waves (Mora et al., 2017). The energy demand for indoor cooling in cities also significantly increases during heat waves, thereby increasing overall urban energy consumption and carbon emissions (Li et al., 2019). This situation places urban ecological spaces under greater environmental stress (Müller et al., 2014), affecting their ecosystem services and raising maintenance costs. The adverse effects resulting from the deteriorating UTEs create substantial impediments to the achievement of sustainable development goals (SDG), particularly Goal 11.3 (enhancing inclusive and sustainable urbanization), Goal 11.6 (reducing per capita negative environmental impacts in cities), and Goal 13.1 (strengthening resilience and adaptive capacity to climate-related disasters). These challenges are expected to intensify as UTEs deteriorate over time.

7.2 Methodology

7.2.1 Three-Dimensional Landscape Structure Index System

The study considered nine landscape structure indices, of which the first five were landscape indices and the last four were building indices. The last three building indices reflected certain 3D characteristics (Table 7.1).

L. Zhou et al., *Urban Surface Thermal Environments*, The Urban Book Series,
https://doi.org/10.1007/978-3-031-94499-4_7

Table 7.1 Three-dimensional landscape structure index system

Landscape structure indicators	Abbreviations	Formula	Meaning
Percent of landscape	PLAND	$\text{PLAND}_i = \sum_{j=1}^{n} a_{ij} / A(100)$	Patch type landscape ratio
Patch density	PD	$\text{PD} = \frac{n_i}{A} \times 100$	The degree of landscape fragmentation
Edge density	ED	$\text{ED} = \sum_{k=1}^{m} e_{ik} / A(10{,}000)$	The total length of a certain type of landscape
Mean patch area	AREA_MN	$\text{AREA\_MN} = \sum_{j=1}^{m} a_{ij} / m$	The average size of a certain type of landscape patch
Aggregation index	AI	$\text{AI} = \left[\frac{g_{ij}}{\max \quad g_{ij}} \right](100)$	The degree of aggregation of a certain type of landscape path
Area-weighted mean shape index	SHAPE_AM	$\text{SHAPE\_AM} = \sum_{j=1}^{m} \frac{0.25 p_{ij}}{\sqrt{a_{ij}}} a_{ij} / \sum_{j=1}^{m} a_{ij}$	Certain landscape path shape complexity
Mean building area	MBA	$\text{MBA} = \sum_{i}^{m} A_i / m$	The average size of buildings
Floor area ratio	FAR	$\text{FAR} = \sum_{i}^{m} A_i h_i / S$	Plot building volume to area ratio
Mean building height	MBH	$\text{MBH} = \sum_{i}^{m} h_i / m$	The average height of buildings
Architecture height standard deviation	AHSD	$\text{AHSD} = \sqrt{\sum_{i}^{m} (h_i - \text{MBH})^2 / m}$	Building height standard deviation

To limit the number of explanatory variables in modeling, the study calculated the landscape indices solely for impervious surfaces and green spaces, which represent the prominent features of the urban landscape. The research partitioned land cover data in the study area into 400 m × 400 m grids and used Fragstats software to calculate landscape indices for impervious surfaces and green spaces, grid by grid. Likewise, building footprint data was calculated in the same grids, and building indices were determined for each grid.

7.2.2 SHAP Interaction Values

The SHAP interaction value offers a deep understanding of how two factors interact. Unlike the interaction definition presented in the previous chapter, where one factor's effect is altered by the other factor's value, the SHAP-valued interaction depicts the interaction between two factors as the portion of their joint effect that cannot be clarified by the sum of their individual effects alone. Similar to other models, this is an efficient method for documenting and presenting interactions (Friedman & Popescu, 2008; Hu & Li, 2020). In principle, for the interaction between the *i* and *j* influences, this study calculated the mean of $\delta_{ij}(S)$ based on Eq. (7.1) by displacing the set *S* of influence variables outside *i* and *j*. Based on Eq. (7.2), $\delta_{ij}(S)$ can be interpreted as the difference between the joint effect of factors *i* and *j* minus the sum of their individual effects for a given set of external variables *S*.

$$\phi_{i,j} = \sum_{S \subseteq \backslash \{i,j\}} \frac{|S|!(M-|S|-2)!}{2(M-1)!} \delta_{ij}(S) \tag{7.1}$$

$$\begin{aligned} \delta_{ij}(S) &= \hat{f}_x(S \cup \{i,j\}) - \hat{f}_x(S \cup \{i\}) - \hat{f}_x(S \cup \{j\}) + \hat{f}_x(S) \\ &= \left(\hat{f}_x(S \cup \{i,j\}) - \hat{f}_x(S)\right) - \left(\hat{f}_x(S \cup \{i\}) - \hat{f}_x(S)\right) - \left(\hat{f}_x(S \cup \{i\}) - \hat{f}_x(S)\right) \end{aligned} \tag{7.2}$$

7.3 Surface Temperature Variation

Despite variations in the range and fluctuations of LST, the distribution and proportion of low- and high-temperature zones exhibited striking similarities during heat wave and non-heat wave periods. The LST range extended from 30.67 to 57.00 °C and 27.37 to 57.90 °C, with corresponding mean values of 41.41 °C and 37.78 °C and standard deviations of 2.43 °C and 2.58 °C during two periods, respectively. Although the temperature range and variability were comparatively narrower during heat wave periods when evaluated in absolute terms, LST was generally higher during the heat wave period. The geographical distribution of low- and high-temperature zones remained consistent during both heat wave and non-heat wave periods in spatial terms (Fig. 7.1).

7.3.1 High- and Low-Temperature Distribution

High-temperature zones are primarily concentrated in the northwest and southwest parts of the study area, along with industrial logistics areas and urban villages at the periphery. Notably, the contribution of the older urban areas in the city center to

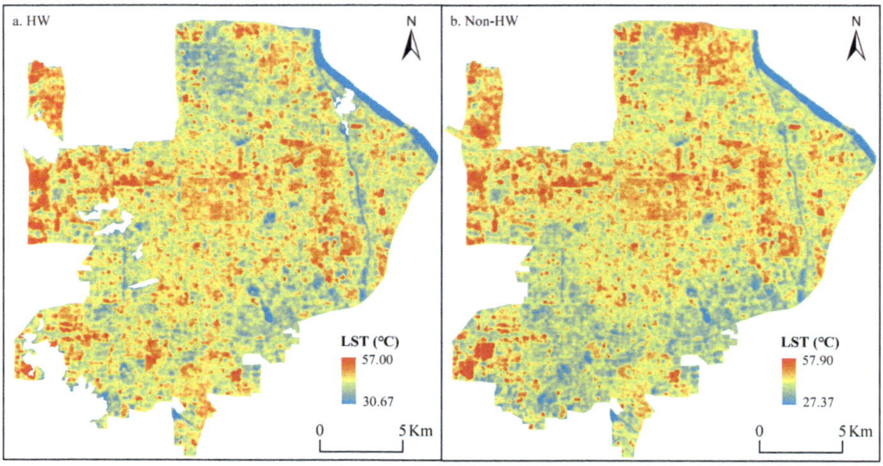

Fig. 7.1 LST distribution during heat wave and non-heat wave

high temperatures remained relatively low, potentially owing to factors, such as vegetation coverage, building shading, and the low thermal inertia of roofing materials. The presence of high anthropogenic heat emissions and materials with low thermal inertia in these regions might account for this phenomenon. Conversely, low-temperature zones tended to be situated in proximity to large bodies of water and green areas, particularly in the case of the Bahe River in the northeast corner of the study area. This can be attributed to the high thermal inertia and the cooling effect resulting from evapotranspiration in these environments. When temperatures were ranked according to standard deviation, this trend became even more pronounced (Fig. 7.2).

7.3.2 Vegetation Effects

The study identified that areas with low-temperature and sub-low-temperature characteristics, except for those encompassing water bodies and green spaces, were predominantly situated in regions boasting abundant vegetation to support medium- to high-rise structures, notably in the north-central part of the study area. The research analyzed the proportions of various temperature zone categories and found that these proportions remained consistent during heat wave and non-heat wave periods. Specifically, during the heat wave, the proportions of low-temperature, sub-low-temperature, medium-temperature, sub-high-temperature, and high-temperature zones were 13.1%, 15.4%, 42.6%, 15.4%, and 13.5%, respectively, while during non-heat wave periods, they were 13.5%, 16.2%, 42.2%, 15.3%, and 12.9%, respectively. All these values adhered to a normal distribution and remained within a 1% margin. Additionally, the study uncovered that, despite having extensive vegetation

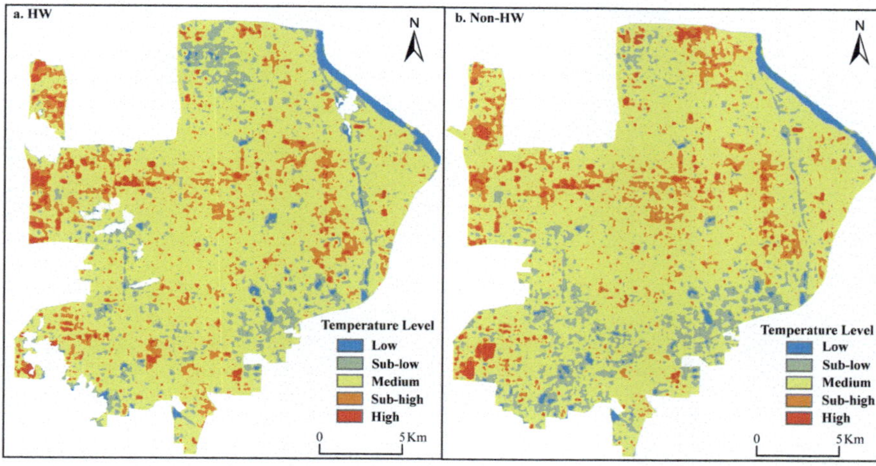

Fig. 7.2 Distribution of temperature zones during heat waves and non-heat waves

cover, Daminggong heritage park did not effectively mitigate the urban heat island effect within its central impervious surface. This could be attributed to the predominant presence of grass as the primary vegetation type in the park.

The distribution of LST exhibits notable variations between heat wave and non-heat wave periods, particularly in specific regions. During heat waves, the relative urban heat island effect diminishes in the northern part of the study area, while the opposite trend emerges in the western and southwestern regions (Fig. 7.2). This pattern becomes particularly pronounced in examining normalized LST differences (Fig. 7.3). These differences highlight that regions experiencing warming, with differences greater than 0, account for 60.2% of the total, indicating a considerably higher intensity of extreme warming compared to cooling. Upon closer examination of actual characteristics, it becomes evident that areas displaying more substantial warming intensity, defined as a change exceeding 0.1, are primarily locations with ongoing land construction or development projects. Nevertheless, a significant number of areas also exhibit substantial changes in LST despite the absence of concurrent land cover alterations.

7.3.3 Variable Impacts of Landscape Structure on Cooling

Conversely, in certain cooling regions, there were areas where changes in the landscape structure did not correspond to significant temperature variations. Additionally, it was observed that areas with substantial changes were more prevalent in cooling regions, while areas with negligible cooling exhibited fewer changes. Furthermore, the cooling regions in the northern areas displayed significant and contiguous temperature reductions compared to the central zone, which corresponds to the old part

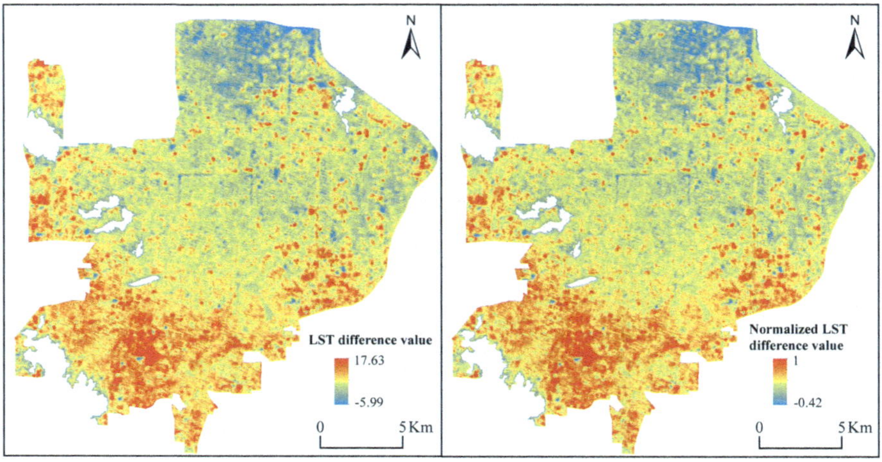

Fig. 7.3 Normalized LST difference distribution during heat wave and non-heat wave. The left figure is the original LST difference during heat wave and non-heat wave. The right figure is the normalized LST difference during heat wave and non-heat wave

of the city. This pattern remains unclear at present. To comprehend the temperature variation disparity between the northern and southern parts during both heat wave and non-heat wave periods, the study conducted an analysis of the variance using landscape structure indicators as explanatory factors. However, the model was able to account for only 10% of the differences, suggesting that the north–south divergence pattern cannot be wholly attributed to landscape patterns. It is worth noting that even though cirrus cloud contamination was corrected in the thermal images from the heat wave period, remnants of cloud-contaminated elements may still exist, potentially introducing some uncertainties in the findings.

7.4 Effect of Landscape Structure on Surface Temperature

7.4.1 Relative Importance of Influencing Factors

The study employed the XGBoost model to examine the influence of 2D and 3D landscape structure factors on LST during heat waves and non-heat waves. The explained variance of the modeled data during heat waves was 54.5%, significantly lower than the 69.7% during non-heat waves. SHAP interpretation revealed that the relative importance of urban landscape's influence on LST during heat waves and non-heat waves remains relatively similar (Fig. 7.5).

Impervious Surfaces and Building Factors Dominate The impact of the impervious surface ratio and the MBH was notably more significant than that of other factors during heat wave and non-heat wave periods, combining to contribute almost half, specifically 47.0% and 51.6%, respectively. After accounting for the various types of influencing factors, the study concluded that the contribution ratio of these factors was approximately 4:1:5. Building-related factors comprised roughly half of the total, followed by proximity to impervious surfaces, with green spaces contributing approximately 10%. It is worth noting that during the heat wave period, the contribution of green spaces was slightly higher (10.9%) than that of during the non-heat wave period (8.3%). The study observed that when dividing influencing factors into two and three dimensions, the influence of the 3D factor was marginally less significant than that of the 2D factor, as the influence of the 3D factor accounted for 42% in two periods.

Upon examining the relative importance ranking of individual factors, the analysis revealed that while the ranking of pivotal variables, such as the impervious surface proportion and MBH, remained consistent between the two periods, the contributions and rankings of other factors displayed some variation between the periods. For instance, during the non-heat wave period, the relevance of AHSD and the impervious surface patch shape index notably decreased in comparison to the heat wave period. In contrast, the significance of MBA and the impervious surface patch shape index increased, while the importance of green spatial landscape patterns declined. Furthermore, the research also found that during heat waves, the significance of the specific landscape configuration of green spaces exceeded that of the green space proportion itself.

Robust Effects of Impervious Surfaces on LST The study not only computed SHAP values but also determined the relative importance of each factor based on the loss of impurity during modeling (Fig. 7.4). The investigation revealed that the importance of impervious surface remained nearly constant, accounting for 37.7% and 37.6% during the heat wave and non-heat wave periods, respectively. However, during the heat wave, the importance of green space increased by approximately 5% compared to the non-heat wave period, resulting in a 5% reduction in the relative importance of the building factors.

In contrast, the patch shape index saw a significant decrease during the non-heat wave period when compared to the heat wave period. Conversely, the importance of MBA and the impervious surface shape index increased, while the significance of green spatial landscape pattern features decreased. Furthermore, the investigation unveiled that during a heat wave, the significance of individual green space landscape configuration features surpassed that of the green space proportion itself. Alongside the SHAP values, the study calculated the relative importance of each factor based on the loss of impurity during modeling (Fig. 7.4). The investigation found that the impervious surface ratio remained nearly constant at 37.7% and 37.6% during the heat wave and non-heat wave periods, respectively. During the heat wave, the importance of green space increased by nearly 5% compared to the non-heat wave period, causing a corresponding 5% decrease in the relative importance of the building index.

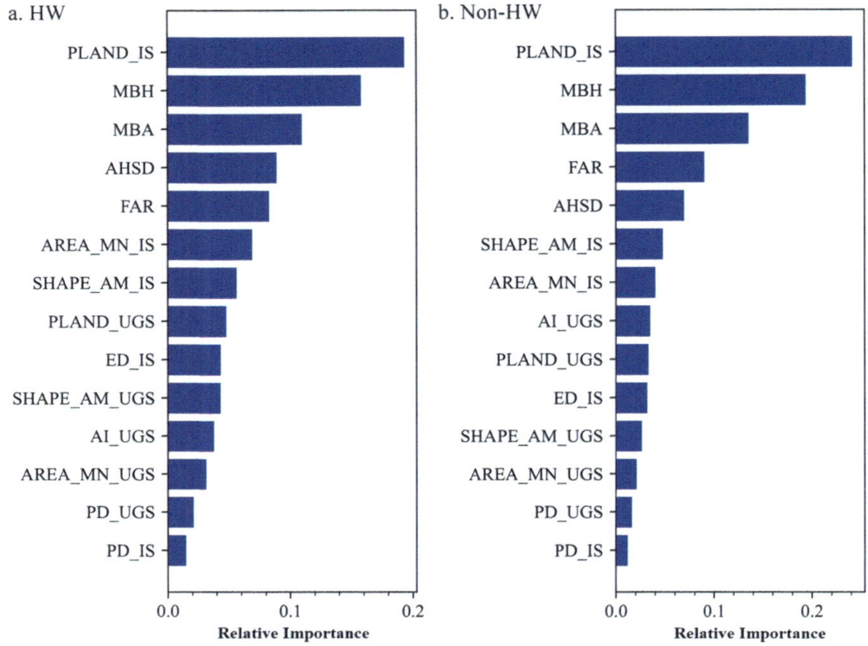

Fig. 7.4 Relative importance of influencing factors during HW (**a**) and Non-HW (**b**) based on impurity loss

Regarding the influence of individual factors, the impact of all factors during the heat wave and non-heat wave was similar (Fig. 7.5). The direction of all factors mostly aligned with the theoretical expectation of the study, except for the impervious surface patch shape index, impervious surface edge density, and average green space patch area. Specifically, for the first seven influencing factors, it was discovered that the impervious surface proportion, FAR, average impervious surface patch, and building average patch size had a warming effect, while MBH, AHSD, and complexity of impervious surface shape had a cooling effect.

7.4.2 Quantitative Role of Influencing Factors on Surface Temperature

The study examined the influence of the previously mentioned factors on LST and their potential interactions using SHAP dependence plots (Fig. 7.6). The investigation revealed that almost all the first nine factors had nonlinear effects on LST, with specific thresholds. Furthermore, most of these factors exhibited similar trends and thresholds of influence in both time periods, although the magnitude of their effects varied somewhat between the two periods. To be specific, during the non-heat wave

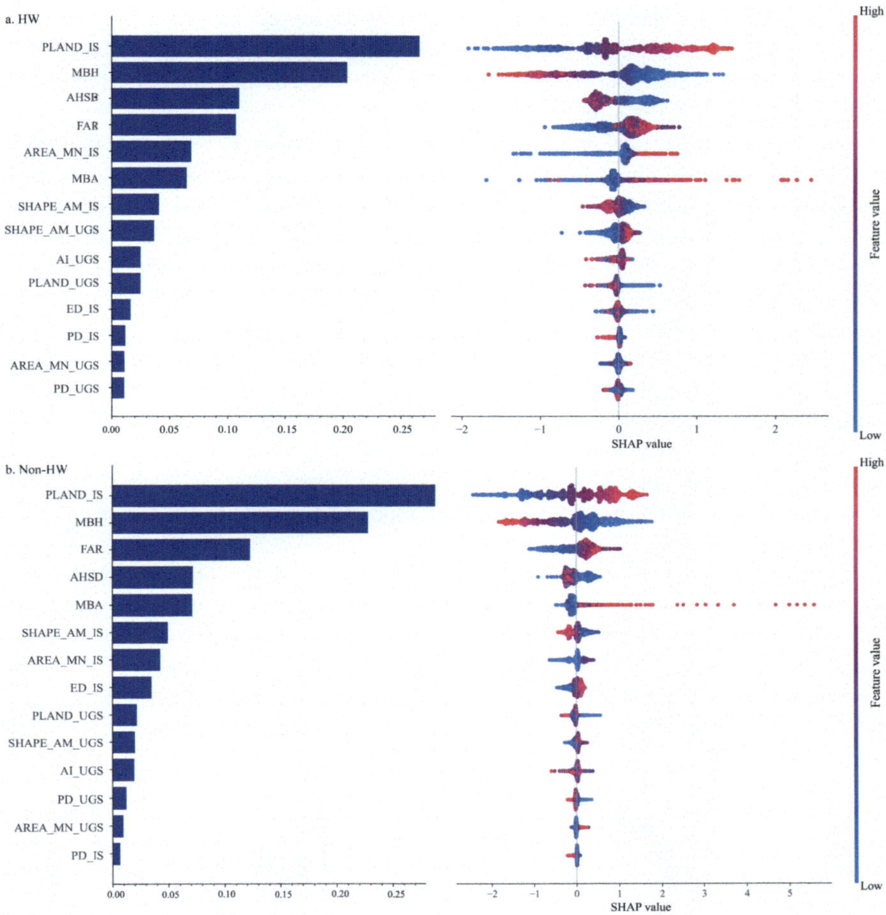

Fig. 7.5 Relative importance of influencing factors during HW (**a**) and Non-HW (**b**) based on SHAP values

period, several significant influencing factors, such as FAR, MBH, and the impervious surface proportion, had more substantial effects on LST. In contrast, during the heat wave period, minor factors, such as the proportion of urban green space, the average impervious surface patch size, and the green space patch shape index, had larger effects on LST.

The investigation revealed that for MBA, the SHAP value exhibited a significant increase when it exceeded 4000 m^2; however, it is worth noting that only 26 analysis units in the study had MBA greater than 4000 m^2, leading to a high degree of uncertainty when we extrapolated the temperature effect of MBA exceeding 4000 m^2. On the contrary, MBA had a warming effect when it was less than 4000 m^2, with a notable threshold effect at 750 m^2 that caused the marginal warming effect to increase considerably. Regarding FAR, it predominantly had a warming effect.

Nevertheless, after reaching 1.5, the marginal warming effect diminished significantly. Similarly, MBH exhibited a convex cooling effect, with the boundary cooling effect significantly decreasing as height increased. The marginal cooling effect became insignificant once the height exceeded 50.

The impervious surface ratio has a linear warming effect. However, its marginal effect is not significant for values falling below 20% or exceeding 85%. As for UGS, marginal cooling is particularly pronounced for values below 20%, but for values greater than 20%, the marginal cooling effect is nearly negligible. The threshold effect alters not only the magnitude but also the direction of the marginal impact. The study reveals that landscape structure indicators, including AHSD, the impervious surface patch shape index, and the green space patch shape index, have distinct directions of influence depending on the range of values. For example, both the 4 and 20 m thresholds of AHSD resulted in a change in the direction of their influence. The threshold effect of the impervious surface patch shape index varies between heat wave and non-heat wave periods. During heat waves, marginal temperature effects become zero after the threshold point, whereas during non-heat wave periods, the direction of the marginal temperature effect changes from positive to negative.

Figure 7.6 reveals potential interaction effects, particularly for the FAR, AHSD, and impermeable surface patch shape index. The temperature effect of the FAR shows two similar patterns based on whether the AHSD exceeds 20 m; in both cases, there is a threshold with a warming effect. However, the impact is less pronounced when the AHSD is below 20 m. Additionally, increasing the AHSD reduces the FAR's warming effect when its value is above 0.6 and vice versa when the FAR

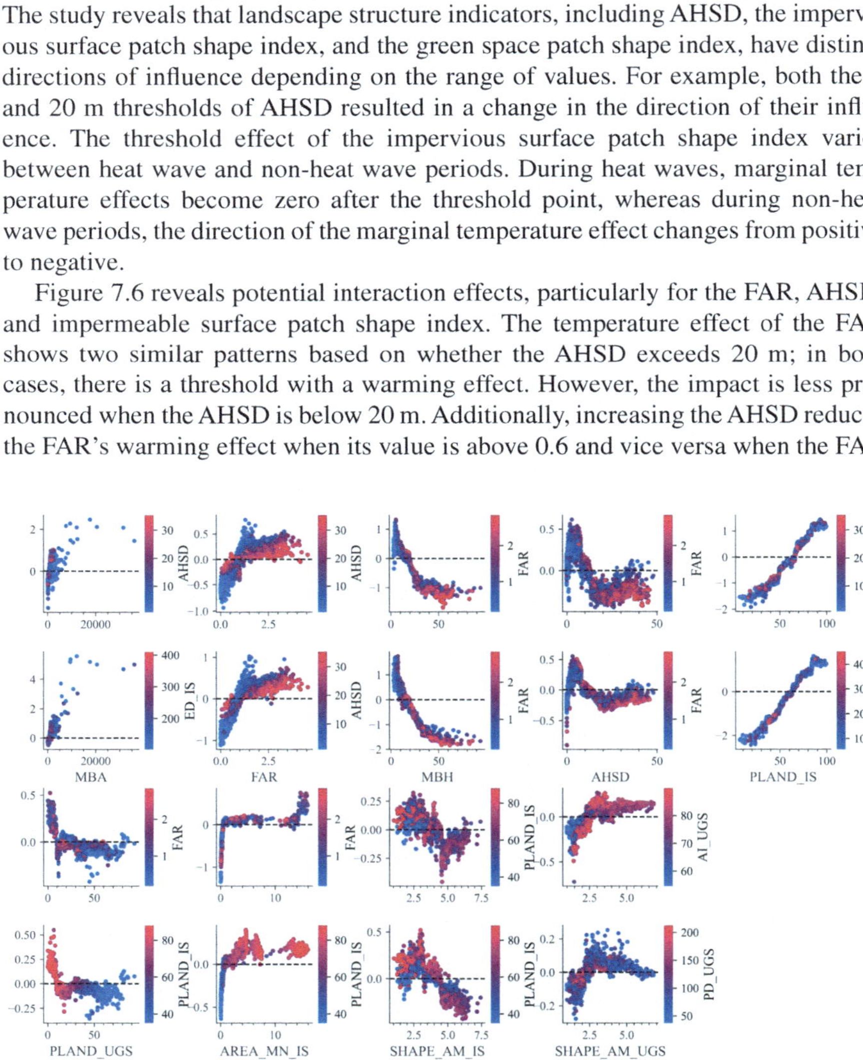

Fig. 7.6 Quantification of the main influences on LST

is below 0.6, identified based on the *Y*-axis divergence. Similarly, when the FAR is set to 1.5, the temperature effect of the AHSD follows two patterns. When the FAR exceeds 1.5, the temperature effect forms a V-shape with a threshold at 20 m. When the FAR falls below 1.5, another threshold point emerges. The Y-axis divergence indicates that the threshold value is at an AHSD of 17 m, hence increasing the FAR beyond the threshold value reduces the cooling effect. Although it is challenging to discern patterns for the impervious surface patch shape index based on moderating variables, the *Y*-axis divergence reveals that higher impervious surface patch densities correspond to higher SHAP values. This relationship occurs for both heat wave and non-heat wave periods, provided that the impervious surface patch shape index is less than 4.5, with a cut-off value of 2.7. This effect suppresses the warming effect and promotes the cooling effect. For the non-heat wave period, increasing PD_IS promotes cooling when the impermeable surface patch shape index is greater than 4.5.

Other Promotion and Inhibition Effects The green spatial patch shape index had an interaction with *Y*-axis divergence when below 2.2, with increasing AI_UGS inhibiting its cooling effect. Nonetheless, some less significant interactions can be difficult to detect from graphs due to other factors influencing moderating variables. To address this limitation, the study employed SHAP interaction plots, which revealed significant interactions among FAR vs. MBH, impervious surface ratio vs. MBH, FAR vs. AHSD, MBA vs. AHSD, MBA vs. MBH, and FAR vs. MBA (Fig. 7.7). These interactions were analyzed further using interaction plots (Fig. 7.8).

When considering the interaction between MBH and FAR, with FAR being treated as the primary controlling factor, two distinct patterns emerge when FAR = 1.5 is used as the threshold. If FAR exceeds 1.5, an increase in MBH results in a decrease in the interaction value, shifting from 0.1 to −0.3. This implies a more pronounced cooling effect with the elevation of MBH, aligning with the primary impact of MBH. However, for FAR values less than 1.5, the influence of MBH on the interaction remains insignificant until MBH exceeds 11 m. In this scenario, the divergence on the Y-axis demonstrates that an increase in FAR can transition the interaction from negative to positive. Beyond 11 m, the warming effect of increased MBH contradicts the primary effect, suggesting that the heightened structure of an isolated high-rise building may lead to a warming effect. These changes are more notable during non-heat wave periods compared to heat waves.

Interactions between FAR and MBH When we consider FAR as the controlling factor, two distinct patterns emerge based on MBH with a threshold at 28 m. For MBH values less than 28 m, an increase in FAR shifts the interaction from negative to positive, aligning with the primary effect. However, if FAR exceeds 28 m, the interaction transitions from positive to negative, exhibiting bifurcation with one branch moving away from 0 and the other converging toward it. These effects are more prominent during heat wave periods.

Furthermore, when we analyze the interaction between the impervious surface ratio and MBH with the impervious surface ratio as the controlling factor, two

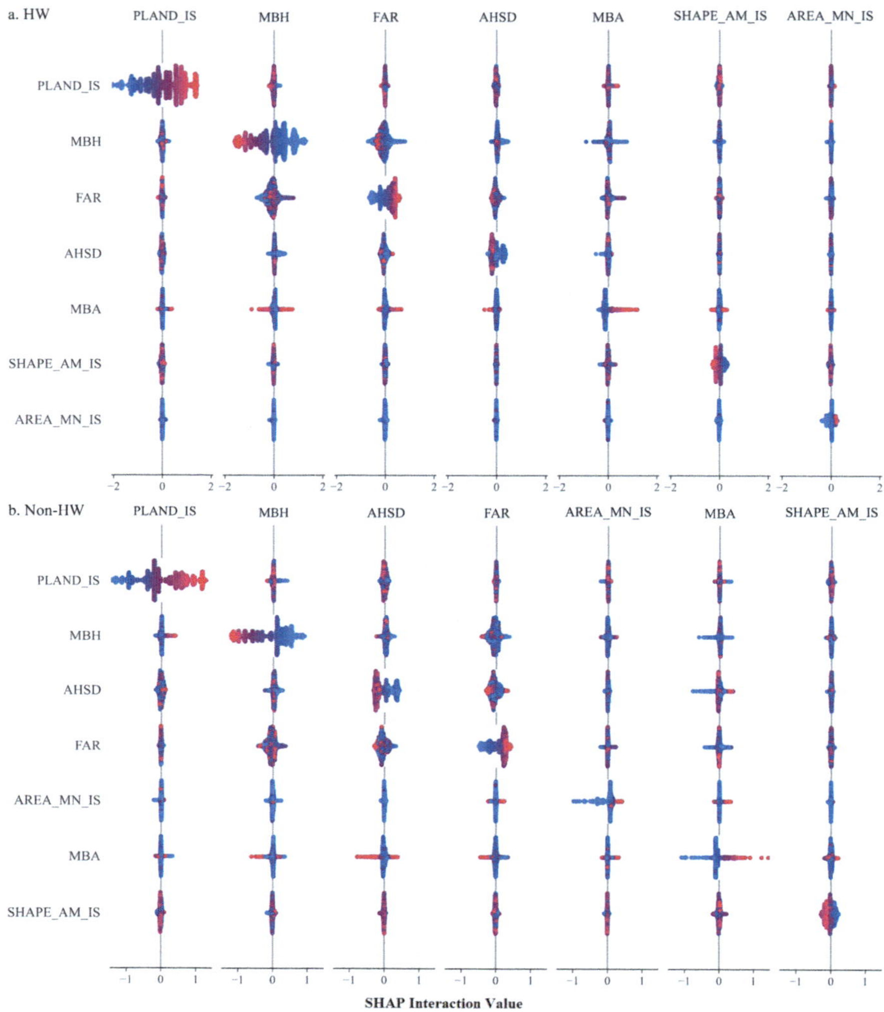

Fig. 7.7 Summary of the interaction of LST influences

patterns emerge during the heat wave period. MBH can be divided into two distinct types at 27 m: when MBH exceeds 27 m, there is a positive interaction value indicating a warming effect, although this interaction rapidly decreases as the impervious surface ratio approaches 0. This trend in the interaction differs from the main effect. In contrast, for MBH values less than 27 m, the interaction ranges between −0.1 and 0.1, making it less distinguishable. During the non-heat wave period, a similar separation can be made based on a MBH threshold of 20 m.

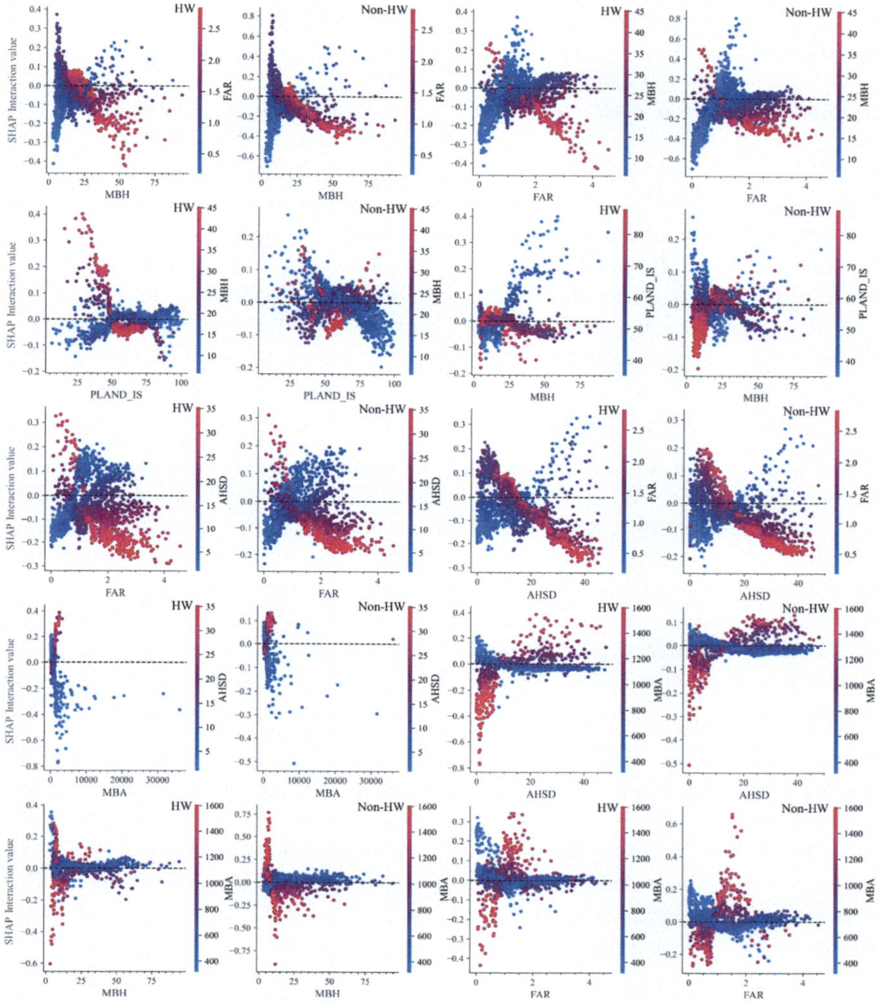

Fig. 7.8 Significant interactions between factors influencing LST

Interactions between MBH and Impervious surface Ratio Similarly, when MBH is used as the controlling factor, two distinct categories emerge based on the impervious surface ratio. During heat waves, when the ratio exceeds 50%, an increase in MBH leads to a minor shift in the interaction value from positive to negative. In contrast, when the ratio is less than 50%, an increase in MBH results in an elevation of the interaction value from −0.05 to 0.35. Conversely, if MBH is less than 20 m, the interaction value transitions from positive to negative with an increase in the impervious surface proportion, although neither of these effects is substantial. However, when MBH exceeds 20 m, the interaction exhibits a "V" shape with the percentage of impervious surface, initially moving from positive to negative and then back to positive.

Interactions between FAR and AHSD The interaction pair of FAR and AHSD exhibits two patterns based on AHSD when FAR is used as the controlling factor. When AHSD is less than 20 m, an increase in FAR changes the interaction value from negative to positive. Conversely, when AHSD exceeds 20 m, an increase in FAR results in an opposite change trend. These patterns are more pronounced during non-heat waves. If the FAR is below 1, an increase in AHSD causes the interaction value to shift from negative to positive. When the FAR ranges between 1 and 1.5, the interaction value varies in a "V" shape with AHSD, initially transitioning from positive to negative and then back to positive. However, when the FAR exceeds 1.5, an increase in AHSD changes the interaction value from positive to negative.

For the interaction between MBA and AHSD, if MBA controls the factor, it has almost no effect on the interaction value. Conversely, AHSD significantly affects the Y-axis divergence. When AHSD is under 10 m, the interaction value primarily lies in the negative range, while an AHSD greater than 10 m causes the interaction value to become positive. In the scenario where AHSD serves as the controlling factor, the changes in interaction values are divided into two categories based on MBA. When MBA exceeds 700 m^2, the interaction values shift from negative to positive with an increase in AHSD, with the turning point at 10 m. Conversely, when MBA is less than 700 m^2, the interaction values are concentrated around zero, although there is a transition from positive to negative with increasing AHSD.

Interactions Between MBH and MBA This study primarily focuses on the interaction between MBH and MBA, as the interactional trends are unclear when controlling for MBA. Therefore, this study exclusively analyzes the results when MBH is the controlled variable. During heat wave conditions, when MBH is less than 15 m, the interaction value exhibits a significant divergence on the Y-axis. However, this divergence is not adequately explained by changes in MBA. Conversely, when MBH exceeds 15 m, the interaction values are centered and distributed around zero.

In non-heat wave conditions, the interaction values can be categorized based on MBA into two distinct trends. When MBA exceeds 900 m^2, as MBH increases, interaction values transition from positive to negative, with a zero-point shift occurring at 10 m. MBA values less than 900 m^2, however, result in interaction values concentrated around zero, with minimal reaction to changes in MBH. The interaction between FAR and MBA was also analyzed, with FAR considered as the controlled variable. The study observed two MBA categories with respect to interaction trends. For MBA exceeding 800 m^2, an increase in FAR leads to interaction values shifting from negative to positive. When MBA is below 800 m^2, an increase in MBA results in interaction values declining to around zero.

7.5 Conclusions

This chapter deeply investigated the impact of 2D and 3D landscape structure on summer LST in the urban core of Xi'an during heat wave and non-heat wave periods, utilizing a landscape structure index system. The analysis relies on the XGBoost model and the SHAP value interpretation method to investigate the interactions among various landscape structure indices. The key findings are as follows:

The distribution of summer LST in the urban core of Xi'an during heat waves and non-heat wave periods is quite similar. Hotspots are primarily concentrated in industrial logistics areas and urban villages, while low-temperature areas are situated around extensive water bodies and green spaces. Moreover, the proportions of temperature zones remain relatively consistent across all categories.

Nonetheless, during heat waves, the UHI effect becomes more pronounced in the southern part of the study area while diminishing in the northern regions. This pattern is complex to understand solely through landscape indicators. Despite the limitations of these indicators, an examination of the 3D landscape structure's influence reveals that the relative significance of each influencing factor on LST remains relatively consistent during both heat waves and non-heat wave periods. The relative importance ratio of impervious surface, urban green space, and the building index is approximately 4:1:5. However, considering the quantitative effects of individual factors on LST, most of the influencing factors display nonlinear effects, characterized by concavities and thresholds. These factors exhibit similar trends and thresholds in both heat wave and non-heat wave periods, although their magnitudes differ to some extent. Building indicators demonstrate significant interaction effects, but a notable interaction between the impervious surface percentage and average building height is also observed, with the interaction's strength being influenced by the values of both the moderating and moderated variables. Moreover, with respect to the previous findings, the enhancement of the interaction between heat wave and non-heat wave periods is substantial.

References

Ellena, M., Breil, M., & Soriani, S. (2020). The heat-health nexus in the urban context: A systematic literature review exploring the socio-economic vulnerabilities and built environment characteristics. *Urban Climate, 34*, 100676. https://doi.org/10.1016/j.uclim.2020.100676

Friedman, J. H., & Popescu, B. E. (2008). Predictive learning via rule ensembles. *The Annals of Applied Statistics, 2*, 916–954. https://doi.org/10.1214/07-AOAS148

Hu, L., & Li, Q. (2020). Greenspace, bluespace, and their interactive influence on urban thermal environments. *Environmental Research Letters, 15*. https://doi.org/10.1088/1748-9326/AB6C30

Li, X., Zhou, Y., Yu, S., Jia, G., Li, H., & Li, W. (2019). Urban heat Island impacts on building energy consumption: A review of approaches and findings. *Energy, 174*, 407–419. https://doi.org/10.1016/j.energy.2019.02.183

Mora, C., Dousset, B., Caldwell, I. R., Powell, F. E., Geronimo, R. C., Bielecki, C. R., Counsell, C. W. W., Dietrich, B. S., Johnston, E. T., Louis, L. V., Lucas, M. P., McKenzie, M. M., Shea, A. G., Tseng, H., Giambelluca, T. W., Leon, L. R., Hawkins, E., & Trauernicht, C. (2017). Global risk of deadly heat. *Nature Climate Change, 7*, 501–506. https://doi.org/10.1038/nclimate3322

Müller, N., Kuttler, W., & Barlag, A. (2014). Analysis of the subsurface urban heat Island in Oberhausen, Germany. *Climate Research, 58*, 247–256. https://doi.org/10.3354/cr01195

Chapter 8
Urban Heat Risk Assessment Incorporating Local Climate Zones

8.1 Introduction

Significant impacts of SUHI: Compared to the period of 1861–1880, global temperatures are projected to increase by approximately 2.0–4.9 °C by 2100 (IPCC, 2021), with the rate of warming in global cities being twice that of the global average. The urbanization process transforms natural and semi-natural surfaces into impervious urban structures, disrupting the Earth's surface radiation and energy balance. Changes in land cover, building materials, and colors, along with anthropogenic heat emissions during urbanization, lead to the surface urban heat island (SUHI) effect, where urban areas exhibit higher temperatures than their surrounding rural areas. The SUHI effect not only exacerbates air pollution (Zhou et al., 2021) and increases energy consumption but also raises the frequency and intensity of extreme heat wave events, leading to higher heat-related morbidity and mortality rates among populations (Ye et al., 2012). This seriously threatens the health and well-being of urban residents and poses a significant challenge to the sustainable development of global cities.

Heat risk assessment framework: A heat risk map is crucial for preventing and addressing the adverse impacts of heat hazards. Since high temperatures can increase heat-related morbidity and mortality, early studies on heat risk primarily focused on the spatial and temporal distribution characteristics of heat wave events. However, these studies relied solely on temperature as a variable, which does not accurately reflect human thermal comfort. Some studies have used indices such as physiological equivalent temperature to measure the energy balance of the human body (Koopmans et al., 2020). Nevertheless, heat risk assessment is a complex process that should also consider other environmental variables (Aubrecht & Özceylan, 2013). The Intergovernmental Panel on Climate Change (IPCC) introduced the vulnerability assessment framework in its Third Assessment Report (AR3) and Fourth Assessment Report (AR4) to quantify the vulnerability of coupled

L. Zhou et al., *Urban Surface Thermal Environments*, The Urban Book Series,
https://doi.org/10.1007/978-3-031-94499-4_8

human–environment systems. Vulnerability is defined as a function of exposure, sensitivity, and adaptive capacity, significantly advancing research on health vulnerability related to high temperatures (Grigorescu et al., 2021). Further, based on Crichton's Risk Triangle framework, the IPCC's Fifth Assessment Report (AR5) shifted its focus to a risk-centered assessment framework, identifying hazard, exposure, and vulnerability as the three fundamental elements of risk assessment. Compared to traditional single-indicator and vulnerability assessment frameworks, this risk assessment framework is more systematic and comprehensive, and it is widely used in the field of risk assessment.

Spatial and temporal scales of heat risk assessment: As an emerging research field, heat risk assessment in rapidly urbanizing small-and medium-scale regions using multi-source data and the disaster-exposure-vulnerability framework is still relatively limited. The statistical accuracy of data sources constrains detailed heat risk assessment research. For example, census data typically provide statistics at the district or county level, leading to heat risk assessments that focus mainly on national or city levels and lack finer spatial scale analysis (Elizabeth Loughnan et al., 2014). The local climate zone (LCZ) concept proposed by Stewart and Oke offers a solution for studying the spatial heterogeneity of urban thermal environments. LCZs are defined as areas with uniform surface cover, structure, materials, and human activity, with horizontal extents ranging from hundreds to thousands of meters. They consist of ten types of building zones (LCZ 1–10) and seven types of land cover zones (LCZ A–G). LCZs provide a standard framework for describing urban morphology and are widely used in studies of urban surface temperatures or surface urban heat islands (SUHI). Currently, LCZs are also being applied in heat risk identification studies, but there is a lack of research combining them with the disaster-exposure-vulnerability framework for risk assessment. Moreover, most existing heat risk studies focus on single-time analysis, neglecting the temporal dynamic evolution of heat risk. However, long-term studies are crucial for a comprehensive understanding of the impact of rapid urbanization on heat risk and its mechanisms.

Study objectives. To address the aforementioned issues, this chapter selects the Xi'an metropolitan area as a case study. It combines the LCZ concept with the IPCC risk assessment framework to explore the changes in heat risk in the Xi'an metropolitan area from 2000 to 2020 at a 100 m grid scale. This chapter aims to answer the following questions: (1) What are the spatial and temporal changes of LCZs in the Xi'an metropolitan area during the urbanization? (2) What are the spatial and temporal distributions of heat risk in the Xi'an metropolitan area? (3) Are there significant differences in heat risk among different LCZ types? The research findings will provide theoretical references and practical guidance for mitigating disaster risks and enhancing urban resilience.

8.2 Methods

The framework for assessing the heat risk includes three main steps (Fig. 8.1). First, SAGA-GIS is used to create LCZ distribution maps for the Xi'an metropolitan area for the years 2000, 2010, and 2020. Second, the heat storage capacity index is derived based on certain attribute parameters of the LCZ and used as the vulnerability index within the risk assessment framework. The SUHI intensity (SUHII) is calculated using the retrieved LST and LCZ as the hazard index, and the population from WorldPop, adjusted by census data, is used as the exposure index. These three indices are then combined to produce a heat risk distribution map. Finally, a spatio-temporal pattern analysis is conducted based on the LCZ maps and heat risk maps.

8.2.1 Local Climate Zone Classification

The local climate zone (LCZ) is a climate-based classification framework proposed by Stewart and Oke in 2012, which has become a standard method for analyzing urban thermal environments. To standardize the landscape classification of the study area, this research utilizes the WUDAPT L0 method for LCZ classification at a 100 m grid scale. The process comprises four main steps: (1) preprocessing Landsat images; (2) creating LCZ training samples; (3) classifying LCZs using a random forest classifier in SAGA-GIS; and (4) evaluating the accuracy of the classification results. Steps (2) and (3) are iterative processes conducted until satisfactory accuracy is achieved (Bechtel et al., 2015). In this study, the classification accuracies for

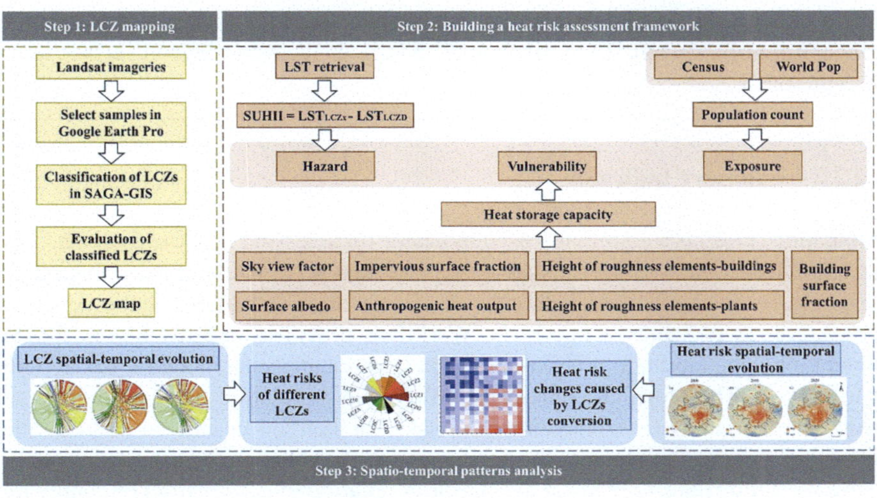

Fig. 8.1 The framework for assessing the heat risk of population

LCZs in 2000, 2010, and 2020 were 72.5%, 77.1%, and 84.6%, respectively, meeting the mapping requirements for heat risk assessment.

8.2.2 Heat Risk Assessment Framework

This study uses hazard, vulnerability, and exposure as the three main dimensions for understanding and assessing heat risk.

Hazard: Reflects the extent to which populations are affected or disturbed by high temperatures, quantified using the surface urban heat island intensity (SUHII).

Exposure: Indicates the number of citizens potentially threatened by high temperatures, represented by the population count.

Vulnerability: Represents the capacity of the urban area to cope with and adapt to high-temperature impacts, quantified by heat storage capacity.

8.2.2.1 Hazard Indicator

The urban heat island effect exacerbates heat waves, negatively impacting human well-being and health (Estrada et al., 2017). Therefore, SUHII is selected as the indicator for hazard. Stewart and Oke proposed describing SUHII through the temperature difference in LCZs, which is the difference between the LST value of any LCZ pixel and the average LST value of LCZ D (Stewart & Oke, 2012), as shown in Eq. 8.1.

$$\mathrm{SUHII}_{\mathrm{LCZx}} = \mathrm{LST}_{\mathrm{LCZx}} - \mathrm{LST}_{\mathrm{LCZD}} \tag{8.1}$$

where SUHII is the surface urban heat island intensity, $\mathrm{LST}_{\mathrm{LCZx}}$ is the land surface temperature of a specific LCZ pixel, and $\mathrm{LST}_{\mathrm{LCZD}}$ is the average land surface temperature of LCZ D.

8.2.2.2 Exposure Indicator

Population (POP) is a commonly used indicator for assessing heat exposure and has been applied in many heat risk studies. Census data is relatively accurate but relies on administrative units, which do not reflect population spatial distribution characteristics at small scales. Conversely, WorldPop provides high-resolution 100 m population data, facilitating visualization and spatial analysis, though it has certain inaccuracies. Therefore, the unadjusted WorldPop population data was corrected using census data before analyzing heat risk. The correction process includes the following steps: (1) Calculate the total population within each district or county using WorldPop data. (2) Divide the census population of each district or county by the corresponding WorldPop population to obtain correction factors. (3) Adjust the

WorldPop population raster distribution using these correction factors to make it more accurate and closer to real data. Additionally, this study applies a logarithmic transformation to the population data for nonlinear scaling, which helps avoid the influence of extreme values on the results analysis (Hu et al., 2019).

8.2.2.3 Vulnerability Indicator

In general, urban warming trends are related to the heat absorption and release properties of the underlying surfaces (Lemoine-Rodríguez et al., 2022). Compared to blue and green spaces, the heat storage capacity of artificial surfaces creates more high-temperature zones (Oke, 1982). Analyzing the heat storage capacity of underlying surfaces can enhance the understanding of regional energy and heat balance and the related temperature conditions. Heat storage capacity typically describes a material's ability to store and release energy and includes three types of storage: sensible heat storage, latent heat storage, and thermochemical reaction storage. For artificial materials in urban areas, the primary influence on the local climate is their ability to absorb (or release) heat when temperatures rise (or fall), which is known as sensible heat storage capacity. In this study, considering the interaction between LCZs and their thermal environments, each LCZ unit is abstracted as a medium for heat storage and release. The more heat released by the underlying surface within a unit, the more intense the human activities, and the more compact the spatial layout making it harder for heat to dissipate, the stronger the heat storage capacity of the unit is. Consequently, the heat risk faced by the population within the unit is higher. Therefore, this study extends the concept of LCZs and their interaction with the thermal environment in three aspects: **Physical properties of building materials**: Different types of building surfaces have varying specific heat capacities and thermal conductivity rates, leading to differences in heat storage capacity within the unit. **Spatial configuration of building clusters**: The arrangement of buildings can cause heat accumulation or ventilation and cooling (Guo et al., 2021), resulting in variations in local heat storage capacity. **Heat release from human activities**: The extent of heat released due to human activities within the unit.

In this study, relevant LCZ parameters were selected, and the maximum, minimum, and median values of these parameters were used to describe the maximum (maxHSC), minimum (minHSC), and average heat storage capacity (avgHSC) for each LCZ type. Based on previous research, it is considered that building density, impervious surface (IS) ratio, roughness element height (buildings), surface albedo, and anthropogenic heat emissions have positive contributions to heat storage capacity. Conversely, sky view factor and roughness element height (vegetation) have negative contributions. A higher heat storage capacity value indicates poorer heat dissipation ability within the spatial unit and, consequently, higher heat risk. Each parameter was normalized and then equally weighted and summed to calculate the maximum, minimum, and average heat storage capacities, as shown in Table 8.1. The sum of the equally weighted maximum, minimum, and average heat storage

Table 8.1 The maxHSC, minHSC, and avgHSC values for each LCZ type

LCZ types	maxHSC	minHSC	avgHSC
LCZ 1	1.00	1.00	1.00
LCZ 2	0.71	0.66	0.69
LCZ 3	0.65	0.55	0.61
LCZ 4	0.70	0.73	0.72
LCZ 5	0.58	0.51	0.56
LCZ 6	0.47	0.35	0.43
LCZ 7	0.72	0.69	0.71
LCZ 8	0.49	0.46	0.48
LCZ 9	0.27	0.21	0.24
LCZ 10	0.62	0.65	0.64
LCZ A	0.10	0.12	0.11
LCZ B	0.01	0.15	0.13
LCZ C	0.25	0.22	0.23
LCZ D	0.15	0.19	0.16
LCZ E	0.44	0.46	0.45
LCZ F	0.24	0.27	0.25
LCZ G	0.00	0.00	0.00

capacities represents the overall heat storage capacity (HSC), which characterizes vulnerability (Eq. 8.2).

$$\mathrm{HSC} = \left(\mathrm{maxHSC} + \mathrm{minHSC} + \mathrm{avgHSC}\right) / 3 \tag{8.2}$$

8.2.2.4 Risk Indicator

According to the risk concept framework proposed by IPCC AR5, this study uses equal weighting to avoid making value judgments on the relative importance of the risk components (Magnan et al., 2021).

$$\mathrm{Risk} = \left(\mathrm{SUHII} + \mathrm{POP} + \mathrm{HSC}\right) / 3 \tag{8.3}$$

To eliminate the impact of different dimensions among the indicators, the min–max normalization method is used to process the raw data for all 3 years before calculating each index.

8.3 Spatiotemporal Characteristics of LCZ

The distribution and conversion of LCZs in 2000, 2010, and 2020 are shown in Fig. 8.2. In 2000, built-up type LCZs (LCZ 1–10) accounted for 45.1% of the area, mainly distributed in the urban core areas of Xi'an and Xianyang. There were no

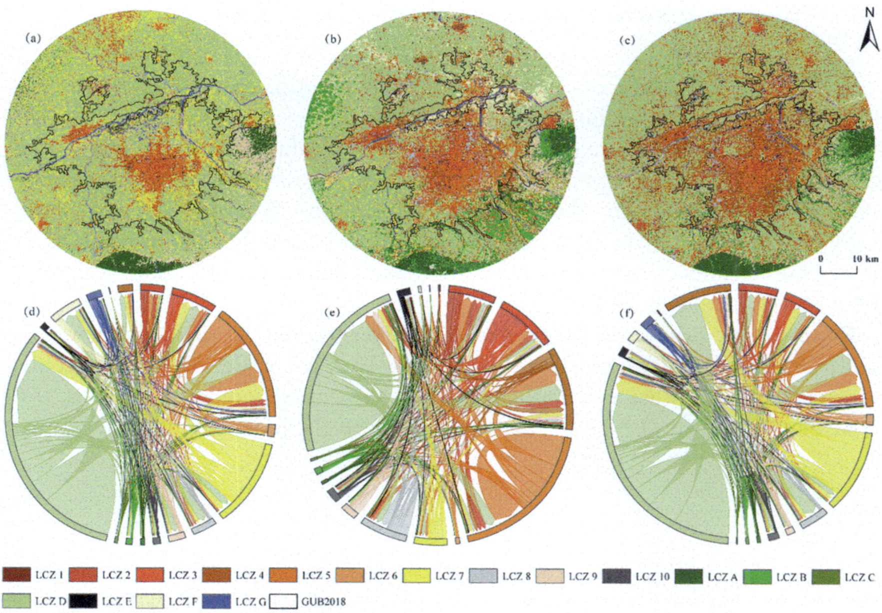

Fig. 8.2 The spatial distribution of LCZs in the Xi'an metropolitan area for the years 2000, 2010, and 2020 (**a**–**c**). The LCZ conversions in 2000–2010 (**d**), 2010–2020 (**e**), and 2000–2020 (**f**)

high-rise buildings (LCZ 1, 4) in the study area. The dominant built-up type was LCZ 7 (lightweight low-rise), primarily located at the interface between buildings and vegetation, accounting for 21.0% of all LCZ types. The second most prevalent built-up type was LCZ 5 (open mid-rise), spread across urban construction areas, comprising 11.7%. LCZ D (low plants) was the most prevalent natural type, mainly found on fringe areas of urban construction areas. By 2010, the construction area had rapidly expanded outward. The most common built-up type became LCZ 5 (23.8%). High-rise buildings only accounted for 4.4%, primarily located on the southwest and north sides of Xi'an's built-up area, which was mainly translated from the conversion of LCZ D and LCZ 7. By 2020, 59.9% of LCZ D had been expanded into built-up types, with built-up type LCZs increasing to 69.7%. The built-up areas of Xi'an and Xianyang had merged, achieving an integrated built-up area. LCZ 4 (open high-rise) became the most common built-up type, accounting for 19.9%. Compact buildings (LCZ 1–3) increased from 6.3% in 2000 to 16.3% in 2020. Overall, the construction area in the study region expanded outward from the central areas of Xi'an and Xianyang in a concentric pattern and gradually merged. The most prevalent built-up type translated from LCZ 7 to LCZ 5 and finally to LCZ 4. The study area experienced significant vertical growth, with an increasing trend in compact buildings, indicating a densification of the study area.

8.4 Spatiotemporal Characteristics of Heat Risk

Figure 8.3 illustrates the spatial patterns of hazard, exposure, vulnerability, and heat risk in the Xi'an metropolitan area from 2000 to 2020.

Hazard: Over the 20 years, the average hazard value increased from 0.54 to 0.57, indicating a slight numerical change but significant spatial distribution changes. In 2000, hazards were concentrated only in urban areas. From 2010 to 2020, heat hazards spread across the entire study area, with hotspots shifting to the edges of urban areas. This phenomenon may be related to the relocation of factories (LCZ 8) to these areas.

Exposure: Exposure levels were most prominent in the Lianhu, Xincheng, and Beilin districts. In 2000, exposure was relatively concentrated but gradually dispersed to the south, west, and north after 2000. The average exposure levels in 2000, 2010, and 2020 were 0.29, 0.32, and 0.37, respectively, indicating not only a rapid

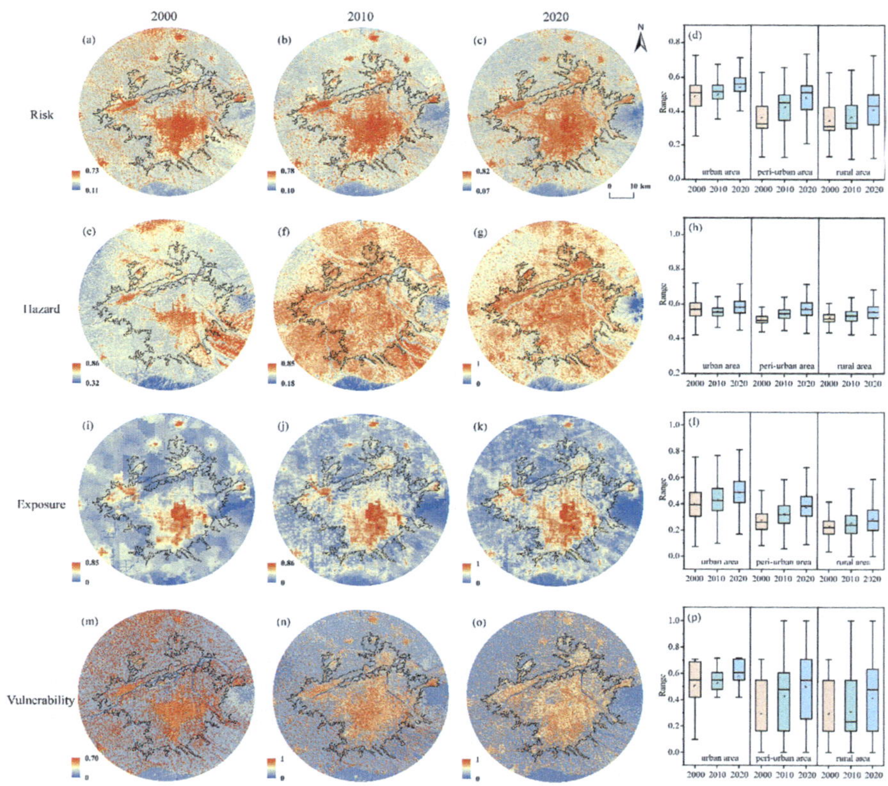

Fig. 8.3 The spatial distribution of risk, hazard, exposure, and vulnerability for the years 2000, 2010, and 2020, along with box plots showing the urban–rural gradient trends for these indicators

increase in population but also a trend toward more dispersed population distribution.

Vulnerability: Different LCZ types exhibited varying vulnerability characteristics. Generally, land cover types with higher heat capacity had lower vulnerability values because they retained less heat. In contrast, building types with higher and denser structures had higher vulnerability values. LCZ 1 (compact high-rise) had the highest vulnerability value at 1, followed by LCZ 4 (open high-rise) with a value of 0.72. LCZ G (water bodies) had the lowest vulnerability.

Risk: The risk areas closely matched the spatial distribution of LCZ building types, showing an expansion trend outward from the urban centers of Xi'an and Xianyang. The Bahe and Weihe rivers were identified as low-risk areas. The average risk values in 2000, 2010, and 2020 were 0.40, 0.43, and 0.48, respectively. The areas with risk values above the average increased from 44.2% in 2000 to 52.9% in 2010 and 58.2% in 2020, indicating an intensification of heat risk and a rapid increase in areas under high heat risk.

Figure 8.3 shows the urban–rural gradient trends of risk, hazard, exposure, and vulnerability in the Xi'an metropolitan area from 2000 to 2020. Spatially, there are significant urban–rural gradient effects in the risk, exposure, and vulnerability indicators. The high concentration of population and dense distribution of buildings in urban areas lead to high-value clusters of risk. The mean values of risk, exposure, and vulnerability are highest in urban areas, followed by urban fringe areas, and lowest in rural areas. In contrast, the urban–rural differences in hazard are not significant. Temporally, the mean values of risk, hazard, exposure, and vulnerability show an increasing trend in both urban and rural areas.

Figure 8.4 shows the average heat risk differences for each LCZ type from 2000 to 2020. There are significant differences in average heat risk between LCZs each year, with the maximum difference reaching 0.49. Built-up types (LCZ 1–10) generally have higher average risk values than land cover types (LCZ A–G). LCZs with more high-rise buildings or higher IS ratios usually face higher heat risk, such as LCZ 1 (compact high-rise), LCZ 2 (compact mid-rise), LCZ 4 (open high-rise), and LCZ 7 (lightweight low-rise) among the built-up types and LCZ E (bare rock or paved) among the land cover types. LCZ 10 (heavy industry) has a large amount of anthropogenic heat emissions, significantly affecting the surface energy balance, leading to consistently high heat risk values. Conversely, LCZ 6 (open low-rise) and LCZ 9 (sparsely built) primarily feature permeable surfaces, higher proportions of cooling spaces, and better air circulation, along with lower population density, resulting in lower risk values compared to other built-up types. Across the urban–rural gradient, the heat risk trend for each LCZ is similar, gradually decreasing from urban to rural areas due to the higher population density and exposure in urban areas, leading to higher risk.

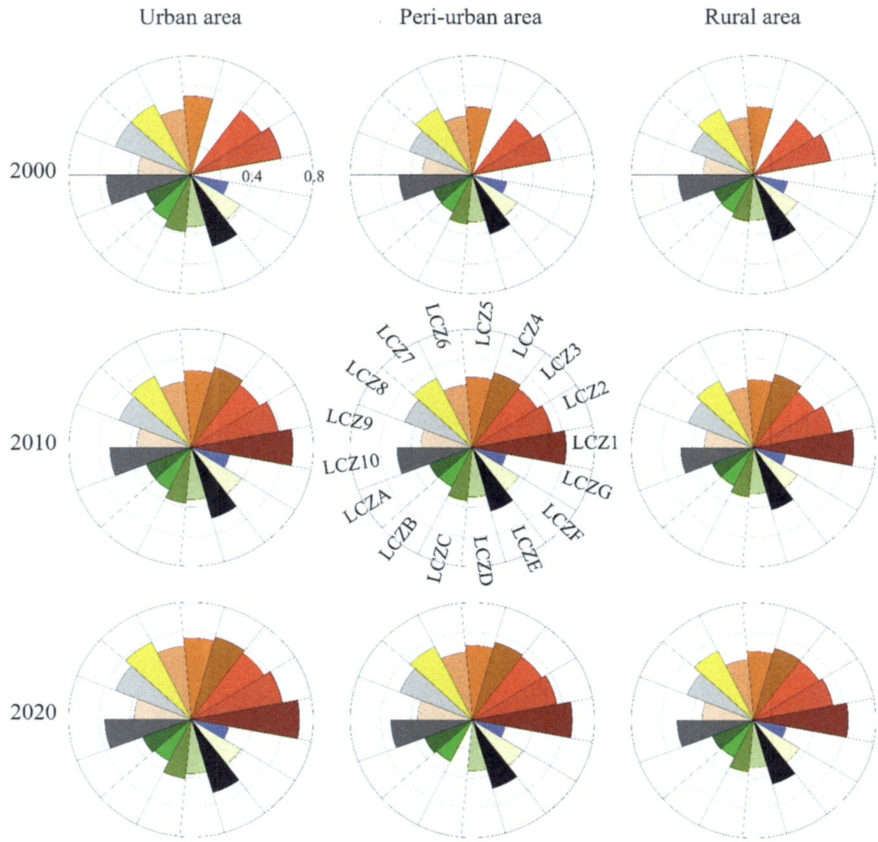

Fig. 8.4 The average risk values of different LCZs under urban–rural gradients in 2000, 2010, and 2020

8.5 Changes of Heat Risk Due to LCZ Conversion

Figure 8.5 shows the relationship between LCZ transitions and changes in heat risk values over different periods. When other types convert to LCZ 1 (compact high-rise), there is a significant increase in heat risk, with a maximum rise of 0.30. Conversions to LCZ 2 (compact mid-rise), LCZ 4 (open high-rise), LCZ 7 (lightweight low-rise), and LCZ 10 (heavy industry) also result in slight increases in risk values. Conversely, when other types convert to LCZ 9 (sparsely built), there is a notable decrease in heat risk, with a maximum reduction of 0.18. Conversions to LCZ 6 (open low-rise) and LCZ 8 (large low-rise) show a slight decrease in risk values.

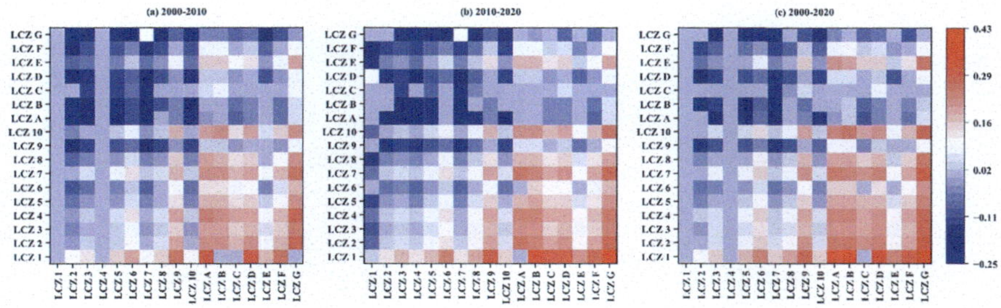

Fig. 8.5 Heat map of changes in heat risk values caused by LCZ conversions in different years (horizontal axis represents the original type, and vertical axis represents the converted type)

8.6 Street-Level Heat Risk Patterns and Mitigation Strategies

The heat risk at a street level can provide more precise spatial targets for mitigating heat risk. This study calculated the mean risk values for each street over different periods and classified them into five categories using the natural breaks method (Fig. 8.6). The results show that the mean risk values of streets form a concentric structure around the urban centers of Xi'an and Xianyang. The number of streets at very high and high risk increased from 36 in 2000 to 60 in 2020, accounting for 56.6% of all streets in the study area. It is urgent for cities to adopt localized measures to alleviate heat risk.

The streets in innermost ring with extremely high risk are closely related to high population exposure, as shown in Fig. 8.3. According to WorldPop age-segmented population data, these streets have a large elderly population. Under similar conditions, elderly individuals, many of whom suffer from cardiovascular and respiratory diseases, are more susceptible to high-temperature hazards. Therefore, it is advisable to identify and register elderly people living alone on these streets. During periods of high temperatures, regular phone calls or visits to check their health status should be made. Additionally, preventive health education on heat risks for the elderly and their caregivers should be strengthened to scientifically prevent heatstroke from the source. These streets, located in city centers, are characterized by compact mid-to high-rise buildings (LCZ 1–2), with scarce land resources. Green space can be increased through wall greening and green roofs to cool the surroundings.

The streets at high-risk level are concentrated in areas with high hazard values, containing numerous large factory buildings and steel structures, identified as LCZ 8 (large low-rise) on the LCZ map. LCZ 8 areas are primarily paved and have low heat capacity, which can be mitigated by replacing roofing materials to reduce heat. Moreover, the commercial and residential areas on these streets typically consist of open mid-to high-rise buildings with some open green spaces. It is essential to enhance the construction of green spaces, increasing the quality, quantity, and accessibility of green areas.

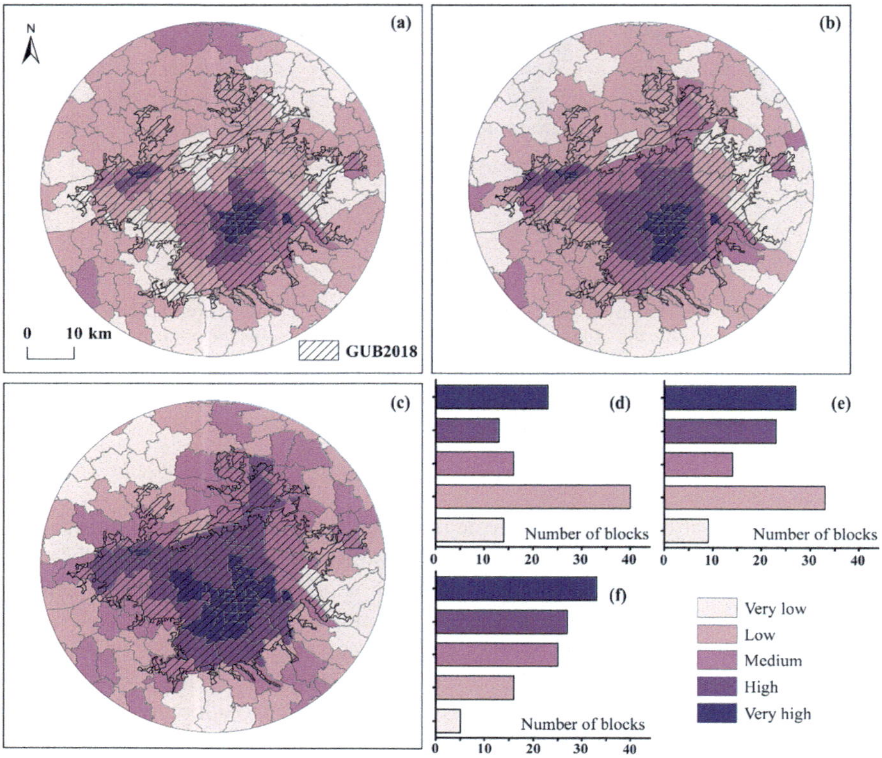

Fig. 8.6 The heat risk at a street level for 2000, 2010, and 2020 (**a**–**c**) and statistical characteristics (**d**–**f**)

Streets located in the boundary area between Xi'an and Xianyang generally have lower heat risk levels compared to the city center. These streets have relatively lower population density, but due to their advantageous geographic location, convenient transportation, and rapid urbanization, local residents can easily access resources from both cities, especially critical resources like medical services. This enables residents to receive timely treatment during heat risk events, reducing the likelihood of heat-related mortality. Therefore, promoting the "Xi'an–Xianyang integration" can help disperse the overly concentrated population in the city center, foster industrial and economic development in the boundary area, and play a crucial role in reducing the urban–rural disparity in heat risk, thereby achieving stable heat risk management across the entire region.

8.7 Conclusions

This chapter proposes a heat storage capacity index based on the LCZ framework and applies it to the IPCC risk assessment framework to explore the spatiotemporal changes in heat risk in the Xi'an metropolitan area for the years 2000, 2010, and

2020. It also examines the relationship between different LCZs and heat risk. The results of this study indicate:

1. The construction areas expanded concentrically around the central areas of Xi'an and Xianyang and gradually merged. The most prevalent built-up type converted from LCZ 7 (lightweight low-rise, 21.0%) to LCZ 5 (open mid-rise, 23.8%) and finally to LCZ 4 (open high-rise, 19.9%). The study area experienced significant vertical growth, with compact buildings increasing from 6.3% to 16.3%, indicating dense urban growth.
2. The spatial distribution of risk areas closely resembled that of vulnerability indicators, showing an outward expansion trend centered around the urban centers of Xi'an and Xianyang. The areas with risk values exceeding the annual average accounted for 44.2%, 52.9%, and 58.2% of the study region in 2000, 2010, and 2020, respectively. Spatially, risk, exposure, and vulnerability indicators exhibited significant urban–rural gradient effects, with mean values highest in urban areas, followed by urban fringe areas, and lowest in rural areas. Temporally, the mean values of risk, hazard, exposure, and vulnerability indicators showed an increasing trend in both urban and rural areas.
3. There were significant differences in average heat risk among different LCZs, with the maximum difference reaching 0.49. Built-up type LCZs generally had higher average risk values than land cover types. LCZs with more high-rise buildings or higher IS ratios typically faced higher heat risks. The changes in heat risk across various LCZs followed a similar urban–rural gradient, decreasing gradually from urban to rural areas. During LCZ conversions, converting other types to LCZ 1 (compact high-rise) significantly increased heat risk, with a maximum rise of 0.30, while converting to LCZ 9 (sparsely built) alleviated heat risk, with a maximum decrease of 0.18. When building types were converted to land cover types, risk values generally decreased, and similarly, risk values generally increased when land cover types were converted to built-up types.

References

Aubrecht, C., & Özceylan, D. (2013). Identification of heat risk patterns in the U.S. National Capital Region by integrating heat stress and related vulnerability. *Environment International, 56*, 65–77. https://doi.org/10.1016/j.envint.2013.03.005

Bechtel, B., Alexander, P., Böhner, J., Ching, J., Conrad, O., Feddema, J., Mills, G., See, L., & Stewart, I. (2015). Mapping local climate zones for a worldwide database of the form and function of cities. *ISPRS International Journal of Geo-Information, 4*, 199–219. https://doi.org/10.3390/ijgi4010199

Elizabeth Loughnan, M., Tapper, N. J., Phan, T., & Mcinnes, J. A. (2014). Can a spatial index of heat-related vulnerability predict emergency service demand in Australian capital cities? *International Journal of Emergency Services, 3*, 6–33. https://doi.org/10.1108/IJES-10-2012-0044

Estrada, F., Botzen, W. J. W., & Tol, R. S. J. (2017). A global economic assessment of city policies to reduce climate change impacts. *Nature Climate Change, 7*, 403–406. https://doi.org/10.1038/nclimate3301

Grigorescu, I., Mocanu, I., Mitrică, B., Dumitraşcu, M., Dumitrică, C., & Dragotă, C.-S. (2021). Socio-economic and environmental vulnerability to heat-related phenomena in Bucharest metropolitan area. *Environmental Research, 192*, 110268. https://doi.org/10.1016/j.envres.2020.110268

Guo, F., Wu, Q., & Schlink, U. (2021). 3D building configuration as the driver of diurnal and nocturnal land surface temperatures: Application in Beijing's old city. *Building and Environment, 206*, 108354. https://doi.org/10.1016/j.buildenv.2021.108354

Hu, L., Wilhelmi, O. V., & Uejio, C. (2019). Assessment of heat exposure in cities: Combining the dynamics of temperature and population. *Science of the Total Environment, 655*, 1–12. https://doi.org/10.1016/j.scitotenv.2018.11.028

IPCC. (2021). *Climate change 2021: The physical science basis*. Contribution of Working Group I to the Sixth Assessment Report of the Intergovernmental Panel on Climate Change. IPCC.

Koopmans, S., Heusinkveld, B. G., & Steeneveld, G. J. (2020). A standardized physical equivalent temperature urban heat map at 1-m spatial resolution to facilitate climate stress tests in The Netherlands. *Building and Environment, 181*, 106984. https://doi.org/10.1016/j.buildenv.2020.106984

Lemoine-Rodríguez, R., Inostroza, L., & Zepp, H. (2022). Intraurban heterogeneity of space-time land surface temperature trends in six climate-diverse cities. *Sci of the Total Environment, 804*, 150037. https://doi.org/10.1016/j.scitotenv.2021.150037

Magnan, A. K., Pörtner, H.-O., Duvat, V. K. E., Garschagen, M., Guinder, V. A., Zommers, Z., Hoegh-Guldberg, O., & Gattuso, J.-P. (2021). Estimating the global risk of anthropogenic climate change. *Nature Climate Change, 11*, 879–885. https://doi.org/10.1038/s41558-021-01156-w

Oke, T. R. (1982). The energetic basis of the urban heat Island. *Quarterly Journal of the Royal Meteorological Society, 108*, 1–24. https://doi.org/10.1002/qj.49710845502

Stewart, I. D., & Oke, T. R. (2012). Local climate zones for urban temperature studies. *Bulletin of the American Meteorological Society, 93*, 1879–1900. https://doi.org/10.1175/BAMS-D-11-00019.1

Ye, X., Wolff, R., Yu, W., Vaneckova, P., Pan, X., & Tong, S. (2012). Ambient temperature and morbidity: A review of epidemiological evidence. *Environmental Health Perspectives, 120*, 19. https://doi.org/10.1289/ehp.1003198

Zhou, L., Yuan, B., Mu, H., Dang, X., & Wang, S. (2021). Coupling relationship between construction land expansion and PM2.5 in China. *Environmental Science and Pollution Research, 28*, 33669–33681. https://doi.org/10.1007/s11356-021-13160-w

Chapter 9
Urban Heat Risk Mitigation Through Green Infrastructure

9.1 Introduction

The combined impacts of climate change and ongoing urbanization have led to increasing surface temperatures in urban areas (Akbari et al., 2016), exacerbating the urban heat island effect (Chakraborty & Lee, 2019). Urban warming not only threatens residents' health but also causes a series of adverse effects on the urban atmosphere, water environments, and infrastructure safety. Thus, continually rising temperature in urban areas is one of the most common barriers to improve residents' well-being and achieve the sustainable development goal 11: Sustainable cities and communities. In addition, urban warming and urbanization have further increased urban greenhouse gas emissions, posing a great challenge to achieve the global target of limiting temperature increase to 1.5 °C, as outlined in the Paris Climate Agreement.

In response to these challenges, a variety of heat mitigation strategies have been proposed and put into effect by the academic and policy sectors. Green infrastructure (GI) has received significant attention due to its eco-friendly, sustainable, and multifunctional characteristics (Gago et al., 2013). GI comprises a network of interconnected natural and semi-natural areas and other environmental components, including familiar features such as urban forests, parks, gardens, and street trees, as well as less common elements like green walls and green roofs. GI operates at multiple scales and serves various functions, providing a range of ecosystem services at different levels, including national, regional, and urban scales. In the urban context, the cooling effect is one of the most valuable services of GI in urban areas, which is one part of climate change adaptation services. Extensive research has been conducted on the cooling effect of urban green infrastructure (UGI) at various scales, highlighting its significant potential to reduce urban temperatures and alleviate heat stress for residents.

L. Zhou et al., *Urban Surface Thermal Environments*, The Urban Book Series,
https://doi.org/10.1007/978-3-031-94499-4_9

Recent review articles have assessed the progress on the cooling effects of UGI in terms of UGI elements, study area and period, research methods, data sources, cooling mechanisms, influencing factors, and implementation strategies. However, although these literatures have adequately discussed the research methodology at the experimental design level, there is still a need for a comprehensive discussion of the specific methods for quantifying cooling effects and analyzing the influencing factors at the operational level.

Moreover, previous review studies mainly focused on UGI with clearly defined spatial extent, such as parks and street trees. The cooling effect of vegetation cover at the neighborhood scale received less attention. Conversely, studies of urban thermal environment (UTE) drivers based on remote sensing have extensively explored the role of vegetation cover at the neighborhood scale, however, usually ignoring the UGI-based perspective. Neighborhood is one of the scales at which urban planning and design activities mainly operate. For example, cities can regulate greenness and building density within neighborhoods through detailed planning. Therefore, summarizing scientific findings on the cooling effect of UGI at this scale is essential for encouraging urban planning authorities to effectively manage thermally vulnerable areas. Furthermore, only a few reviews addressed the cooling effects of UGI at different scales simultaneously and the scheme of UGI's spatial scale was simply discussed. For instance, Bartesaghi et al. (2018) categorized UGI cooling effect studies into microscale, local scale, and mesoscale based on urban climatology. However, the primary factors of UGI cooling effects can vary significantly at different microscales, such as sites, buildings, and streets.

This chapter attempts to synthesize the existing literature to answer the following questions: (1) What spatial scales have previous studies focused on? (2) What specific research methods have been used in previous studies to measure the cooling effect and to investigate the influencing factors? (3) What are the main factors affecting the cooling effect of UGI at different scales, especially at the neighborhood scale? (4) What are the remaining limitations and potential directions in the study of UGI cooling effects? Firstly, this chapter summarizes the research scales considered in different studies, followed by summarizing and comparing the specific methods utilized in previous studies in the three stages of temperature data acquisition, cooling effect measurement, and influencing factor analysis. After that, based on the discussions of the main physical mechanisms on which the cooling effect relies, this chapter summarizes and discusses the main controlling factors of the cooling effect of UGI at different scales. Finally, we discussed the limitations in the current study and future research directions. In the context of Chinese cities continuously transforming toward green development, the study expects to provide certain scientific references for green infrastructure planning and design activities in cities and to point out the current scientific issues that need to be urgently addressed for subsequent studies.

9.2 Research Scales of UGI Cooling Effects

The cooling effect of urban green infrastructure (UGI) primarily pertains to the reduction of canopy air and surface temperatures within and around UGI areas. This cooling effect is achieved through multiple mechanisms, such as shading, evapotranspiration, and the advection of cold air masses, resulting in lower air and surface temperatures compared to adjacent urban built-up areas or the conditions prior to UGI implementation. In some research, UGI's impact on metrics related to human thermal comfort, such as physiologically equivalent temperature or average radiation temperature, is also considered a cooling effect (Taleghani et al., 2016). Due to the differences in the concerned UGI elements research objectives, researchers examined the cooling effect at various spatial scales. Urban climatology categorizes the spatial scales of studies into microscale, local scale, and urban scale (mesoscale). However, this classification is often insufficient to adequately describe the scale differences between studies, especially regarding the cooling effect of UGI. Moreover, previous reviews have sometimes mixed the spatial grain (unit) and extent (study area) as scale, leading to some uncertainties in scale descriptions.

To address these issues, this chapter uses the assumed "impacted spatial scope of cooling effect" to classify the scale of UGI cooling effect into six spatial levels: (1) Site scale: Studies at this level focuses on the cooling effect of UGI within the observed source area of the site, particularly the cooling effect of individual UGI elements like street trees, green walls, and green roofs in their immediate neighborhoods. (2) Building scale: Studies primarily concentrate on how UGI influences building indoor temperature and cooling energy consumption by regulating heat transfer between the indoor and outdoor environments. Many of these studies explore how UGI affects the exterior wall temperature as a proxy to assess its impact on indoor temperature reduction or energy savings. (3) Street scale: Research at this scale examines how UGI affects the average air temperature of streets or the average surface temperatures of street components such as walls and pavements. (4) Green space (patch) scale: This level focuses on the cooling effect of larger individual green spaces (patches), typically parks, and their impact on the surrounding areas. (5) Neighborhood scale: Studies at this scale analyze the relationship between the discretely distributed of UGI and the average temperature within a localized area of the city, such as a neighborhood or community. It particularly considers spatial configuration of UGI. (6) Urban scale: This scale has been less frequently explored but focuses on the impact of UGI on the overall cooling of an entire city. Generally, a part of research at this scale focused on the overall effects of specific UGI strategies, such as increasing the proportion of green space or green roof coverage at the urban scale, through scenario simulation methods. Another section of the studies investigates the comprehensive factors influencing urban heat island intensity using statistical methods, while the vegetation characteristics like overall urban vegetation coverage and the difference between urban and rural NDVI were assumed as one of the main factors.

Consistent with urban climatology, the various scales used in the study of UGI cooling can be categorized as microscales, such as sites, buildings, and streets in the context of urban climatology. Additionally, local scales encompass green spaces and neighborhoods, while mesoscales encompass entire cities. It is essential to acknowledge that while this classification system typically suits most UGI cooling studies, a few studies pose challenges in determining their specific scale. For instance, certain high-resolution simulation studies have explored the impact of UGI deployment strategies on temperature and thermal comfort at the neighborhood level. However, the cooling effect can be affected by local factors like street geometry and the limited cooling distance, leading to variability within the neighborhood. Thus, these studies can alternatively be considered at the site scale, although focusing on the UGI's impact on the whole neighborhood. Similarly, some simulation studies have addressed differences in cooling effects resulting from various UGI implementation measures at the urban scale. Nevertheless, due to the grid resolution set at the neighborhood scale, the variations in cooling effects attributed to UGI measures can also be observed across neighborhoods.

Furthermore, previous research has revealed that UGI can induce cooling effects of varying magnitudes at different spatial scales simultaneously, and the influencing factors may exhibit distinct impacts at these scales, indicating the presence of a "scale effect." For instance, Santamouris (2014) found that green roofs not only enhanced the immediate thermal environment but also significantly reduced the urban heat island effect at the urban scale. Zhang et al. (2017) reported that optimizing the distribution of additional green space could locally reduce temperatures by 1–2 °C and decrease the overall area's temperature by 0.5 °C. Moreover, they identified that concentrated green spaces exerted a more substantial cooling effect at the neighborhood scale compared to the urban scale. Conversely, Yan et al. (2019) observed that enhancing edge density and shape complexity in green spaces increased the temperature of certain green space patches while reducing the average temperature at the neighborhood scale. This phenomenon, known as the "scale effect," may be linked to the localized nature of UGI cooling and its intricate interactions with diverse urban elements. Consequently, it is imperative to conduct a comprehensive examination of UGI's cooling effects and its influencing factors at different scales concurrently to gain a deeper understanding and explore potential trade-offs in harnessing this cooling capacity.

9.3 Research Methods of UGI Cooling Effects

9.3.1 Data Acquisition

To assess the cooling effect of UGI, existing studies primarily employed three distinct methodologies: field observation, remote sensing observation, and model-based predictions, to gather data related to temperature, humidity, wind speed,

Table 9.1 Comparison of methods on data acquisition

Acquisition methods	Applied scales	Temporal resolution	Spatial resolution	Advantage	disadvantage
Field observation	Fixed stations	Minute level	Meter level	Data can truly reflect the cooling effect, easy access to site information	The spatial coverage is small, and it is difficult to fully reflect the cooling effect at macroscopic scales such as blocks and cities
	Temporary stations				
	Mobile observation				
Remote sensing observation	Starborne	Daily–weekly	Meter–kilometer level	The spatial scope of data acquisition is wide and the cost is low	The spatial resolution of commonly used data is relatively low, and only instantaneous temperatures can be obtained, with poor temporal continuity
	Airborne				
	Ground-based				
Model prediction	Mechanic	Minute–hourly	Meter–kilometer level	Experiments can be conducted without spatial or temporal constraints, and the effects of various factors on temperature can be separated	Data accuracy is limited by model assumptions and input parameters, and some studies lack validation of the results
	Statistics				

radiation, and other pertinent parameters (Table 9.1). The field observation approach entails on-site monitoring of environmental responses to UGI-induced cooling, facilitated by meteorological and radiation monitoring instruments stationed at relevant locations. In contrast, the remote sensing observation method involves leveraging satellite-based or airborne sensors to capture surface temperatures, enabling an examination of surface thermal alterations attributable to UGI cooling. The "model-based prediction approach" encompasses the utilization of models and associated scenarios to simulate how the environment would react to UGI cooling effects. Notably, models can be classified into two broad categories: Mechanism-based and statistics-based, with the former experiencing significant advancements after 2000 (Table 9.2).

Comparing the three aforementioned data acquisition methods in terms of their applicable scales, field observation methods are the most localized, typically implemented at microscales encompassing stations and streets. Remote sensing observations, conversely, extend to larger spatial domains, predominantly focusing on local scales, including green spaces and neighborhoods, and at broader urban scales. Model prediction methods, while versatile in scale, are subject to limitations in resolution and cooling mechanism representation based on the selected scale. Spatially, model prediction methods can achieve the finest resolution, reaching up to 0.5 m, as exemplified by the ENVI-met model. Field observation methods follow

Table 9.2 List of mainly used physical models

Applied scales	Models	Model characteristics
City/medium scale	WRF-NOAH land surface model	The most widely used mesoscale model, which fully characterizes the interaction between vegetation and atmospheric soil
	MUKLIMO_3	Highest spatial resolution (100 m) as a mesoscale model
Block scale	VUCM	The first urban canopy model that considers the influence of trees in built-up areas
	PUCM	The first urban canopy model that fully integrates hydrological processes and examines the cooling effect of green roofs
	TUCM	First single-level urban canopy model that fully considers the impact of trees
	BEP-tree	The first multilevel urban canopy model that fully considers the impact of trees
	CWG	First multilevel canopy model that combines building energy model, hydrological model, and complete vegetation effect
Site/building/street scale	ENVI-met	The most widely used microscale model to simulate the effects of almost all types of UGI
	TRNSYS	Widely used building energy model that simulates the effects of green roofs and green walls
	Green-CTTC	Early proposed numerical model to simulate the cooling effect of trees in streets
	SOLWIG	Three-dimensional microscale model to simulate the effects of UGI radiation

in terms of spatial resolution, operational at the site scale, albeit with relatively poor spatial continuity in the resulting data. Remote sensing observation techniques generally produce data with the lowest spatial resolution. Although airborne sensor data can achieve resolutions less than 10 m, remote sensing studies primarily rely on easily accessible satellite-based sensor data. Commonly used satellite thermal band images, like Landsat-7, offer a spatial resolution of only up to 60 m.

Considering the temporal resolution, field observations, and model simulations support near-continuous data collection, facilitating the exploration of daily dynamics and inter-day variations in UGI cooling effects. In contrast, remote sensing observation methods are typically limited to instantaneous measurements at specific times. Some satellite sensors, like MODIS and meteorological satellites with low-spatial resolutions (1 km and up to 250 m, individually), can provide multiple observations within a day. However, many widely used medium-resolution satellite-based sensors only offer revisit periods of several days or even up to a dozen days. It is noteworthy that the recently available ECOSTRESS data have been employed to study the daily dynamics of cooling effects in wooded areas (Zhou & Shepherd, 2010). This dataset holds significant potential for investigating the daily dynamics of UGI cooling effects over large areas.

Moreover, current studies face significant challenges in obtaining high-resolution surface temperatures at scales of meters or even sub-meters and neighborhood-scale air temperatures. In the former case, site observation techniques can only capture temperature data at a limited number of sampling points. Conversely, remote sensing observation methods, which commonly rely on satellite thermal band imagery, are typically restricted to obtaining temperature data at or above local scales, such as green spaces and neighborhoods scales with 10-meter or 100 meter resolution. Model simulation approaches also need to balance computational costs and the potential loss of simulation accuracy stemming from scene and physical process parameterization. A potential solution to this issue lies in the increasing importance of low-altitude unmanned aerial vehicle (UAV) images as UAV technology advances in the future.

In the case of neighborhood-scale air temperatures, there are two feasible approaches. One method involves establishing a statistical relationship between near-surface air temperature and easily observable urban environmental variables or surface temperatures based on sample point data and then applying this relationship globally to estimate temperature distributions. The other approach entails obtaining air temperature distributions through model simulation methods. While the second method is limited by physical mechanisms and urban scenario parameterization schemes incorporated into the simulation, the first approach is constrained by factors such as the quantity and distribution of sample sites and the transient nature of that derived associations.

9.3.2 *Quantification and Measurement*

There are many different cooling effect measures because different studies differ in their goals, scope, and data collection techniques. These measures can be categorized as spatially based, temporally or scenario-based, and statistically based measures depending on the reference objects.

1. Spatially based reference. This type of measure generally designates two types of areas or sites in space as the experimental object and the reference object, respectively, to evaluate the effect of UGI on temperature. For example, the difference between the mean temperature value of a green patch and its neighborhood is calculated as the cold island intensity of a green patch at the patch scale, or the temperature difference between a UGI site and a reference site is calculated at the site scale. The difference in mean temperature values between different land types has also been studied at the neighborhood scale as a measure of cooling effect. In addition, for the cooling effect of UGI at the green patch scale, in addition to temperature differences, other measures such as maximum cooling distance, cooling impact area, and cooling efficiency have been examined from a spatial perspective in previous studies. Whether the extrinsic elements unrelated to UGI in the experiment and reference objects can be properly controlled is the assumption of studies using spatially referenced measures. However, this

premise has not been considered in the majority of prior investigations. For example, at the site scale, studies usually assume that the external factors affecting the observed and reference sites are the same due to spatial proximity, while at the neighborhood scale, direct comparisons of temperatures across land types mean treating the effects of external factors as random noise and thus they can be eliminated by calculating the difference in means. Although some of these assumptions are reasonable, future research needs to limit the impact of outside factors more tightly.

2. Time-based or scenario-based reference. This measure can represent either the temporal changes in temperature before and after the construction and removal of UGI in reality or the temperature difference between different scenarios and the reference scenario in the scenario simulation study. Since studies based solely on temporal comparison data are vulnerable to interference from external influences that are difficult to account for, such data was typically only been used for neighborhood-scale studies. In contrast, scenario simulation studies with strictly controlled experimental conditions and lower costs have become more frequently used to measure cooling effects in recent years. This type of study generally specifies a reference scenario as the baseline for comparison and compares it with the temperature distribution of several designed scenarios to obtain the temperature difference as a cooling measure. Although the accuracy of this method is affected by the model assumptions and the accuracy of the input data, it is still valuable due to its reduced manipulation cost. In addition, although time-based studies are susceptible to interference from external factors, they still need to be valued as the only way to measure the cooling effect of UGI measures.
3. Statistical (regression) relationship-based. Although statistical relationship-based studies also focus on the cooling effect of UGI, they do not directly specify UGI and reference objects but instead investigate the quantitative response of local average temperature or urban heat island intensity to the proportion or other characteristics of UGI through statistical methods. For example, Kong et al. (2014) discovered that the surface temperature can drop by 0.83 °C for every 10% increase in the proportion of vegetation cover. In addition, several researchers have used nonlinear models to analyze the cooling effect of green space ratio from 0 to 1.

Each of these three types of methods has its own applicable scale. For example, spatially based methods are more suitable for the site scale because it is easier to find suitable reference sites at this scale, while temporally based methods require a certain scale of the spatial extent of the observed objects or they are vulnerable to environmental factors and are therefore more often used at the green patch and neighborhood scales. Statistical measurements are usually performed at the neighborhood scale, since UGIs are spatially discrete green elements that lack distinct object boundaries at this scale, applying the other two methods is more challenging. In addition, current studies based on statistical relationships mostly use surface temperature data, and its spatial continuity makes it easy to find enough analysis units as samples to support the statistical analysis within the city.

Furthermore, in addition to the classification based on the reference object, cooling effect measures can be categorized into various types based on specific physical indicators. These indicators encompass canopy air temperature, surface temperature, average radiation temperature, and thermal equivalent temperature, of which the latter two have direct relevance to human thermal comfort. For instance, Farhadi et al. (2019) observed that an increase in street canopy cover significantly reduced surface temperature and enhanced thermal comfort but had no notable impact on canopy air temperature. It is worth noting that findings obtained using different measures may occasionally diverge.

In addition to directly evaluating the cooling effect through environmental responses like temperature and thermal comfort indices, some studies indirectly assess this effect by examining UGI performance in various cooling mechanisms. For instance, Chen et al. (2019) investigated the transpiration cooling effect of different tree species by monitoring trunk sap flow. Similarly, in research on green roof cooling effects, some researchers primarily focus on parameters such as roof albedo, leaf temperature, and substrate temperature to gauge the capacity of this type of UGI in reducing net radiation and mitigating heat gain in buildings (Malys et al., 2014).

9.3.3 Analysis of Influencing Factors

The process of influencing factor analysis is systematically designed to quantitatively evaluate the relationship between cooling effect measures and their influencing factors, with the aim of identifying the dominant elements among them. In some parametric simulation studies, the effects of each influencing factor can be directly obtained by varying individual model parameters. However, for observational studies, the effects of influencing factors are typically assessed using statistical methods. Although traditional statistical techniques like correlation analysis and multiple regression analysis are still commonly employed, machine learning methods are gradually gaining traction (Gage & Cooper, 2017). Machine learning approaches offer an advantage over multiple regression analysis in which they are better suited to capturing the nonlinear responses of dependent variables to explanatory variables and interactions among these variables. Nonetheless, machine learning methods may not be as effective as multiple regression methods in terms of explaining and attributing response relationships. It is important to note that machine learning methods generally require a larger sample size, making them more suitable for research utilizing remote sensing data, which can yield a substantial number of samples.

Additionally, in previous studies, to identify the key factors influencing the cooling effect of UGI, researchers have primarily quantified the relative importance of factors using various methods (Table 9.3). While methods such as hierarchical decomposition, random forest, machine learning, and geographic detector approaches are gaining traction, the standardized coefficient method based on linear

Table 9.3 Methods of evaluating the relative importance of influencing factors

Methods	Advantage	Disadvantage
Standardization factor (Ziter et al., 2019)	Simple calculation and clear meaning	Lack of ability to deal with multicollinearity and common variance
Geodetector (Yu et al., 2018)	The effect of explanatory variables on temperature and the interaction between variables can be studied without considering specific relationships	Lack of standard methods for dealing with continuous variables
Hierarchical decomposition (Chen et al., 2014)	Independent importance of each explanatory variable can be obtained	Excessive computational cost required to deal with large number of explanatory variables
Projection-based variable importance (Jaganmohan et al., 2016)	Large number of explanatory variables can be handled efficiently by projection method	Inability to effectively strip common variance
Random forest (Gage & Cooper, 2017)	Nonlinear effects can be considered	Inability to exclude the interference of multicollinearity among variables and the need for more samples
Variance decomposition (Li & Zhou, 2019)	Importance of cluster variables can be handled	Generally, only a few groups of variables can be handled, and the implementation method relies on regression equations and cannot consider nonlinear effects and interactions

regression models remains the primary technique for analyzing relative importance. Beyond identifying dominant factors, researchers have also begun to investigate nonlinear characteristics, including threshold effects and moderating effects within relationships. A threshold effect occurs when there is a significant change in the response of the dependent variable before and after a certain threshold of the independent variable. For instance, Ziter et al. (2019) found that the maximum decrease in temperature resulting from an increase in vegetation cover was achieved when the vegetation cover exceeded 40%. Conversely, moderating effects involve the influence of a moderating variable on the strength and direction of the relationship between another explanatory variable and the target variable. While researchers are interested in exploring the moderating effects of categorical variables on the cooling effect of UGI, such as climate zone type and season, relatively few studies have specifically investigated the moderating effects of continuous variables.

9.4 Impacting Factors of UGI Cooling Effects

9.4.1 Mechanism of UGI Cooling Effects

The principal mechanisms underlying the physical processes used in current research to elucidate the cooling effect of UGI include shading, latent heat cooling, and cold air mass advection. Shading entails the obstruction of solar radiation by the vegetation canopy, creating shadows on various urban surfaces. This reduces the net radiation received by these surfaces, subsequently lowering their surface and air temperatures or directly decreasing the shortwave radiation exposure to humans. The primary determinants of shading effects of tree canopies are tree height, leaf area, and canopy radius. However, for green walls, the leaf area index (LAI) plays a significant role in influencing shading effects. Evapotranspiration cooling refers to the process by which leaves of vegetation and moist soil surfaces evaporate, converting some of the net radiation they receive into latent heat in the form of water vapor. This process reduces the sensible heat flux and mitigates the warming of near-surface air. The key factors determining UGI evapotranspiration encompass the aerodynamic conductivity of the canopy, stomatal conductivity of the leaf surface, soil water potential per unit of leaf area, net radiation received by the canopy, and air vapor pressure deficit. Cold air mass advection, which involves the movement of cold air masses within the vegetation-covered area into the surrounding urban environment through wind systems, serves as the foundation for the cooling effect of UGI on the external built environment. This mechanism plays a more prominent role in green space and neighborhood-scale studies compared to the previous two microscopic mechanisms. Some studies also consider relatively high albedo in the case of green roofs and the thermal isolation provided by green walls and green roofs as key mechanisms in achieving the cooling effect.

9.4.2 Comprehensive Influencing Factors of UGI Cooling Effects

Studies have identified various influences at different scales, revealing that as the scale increases, spatial structure factors become more significant compared to vegetation morphology and internal species composition. This is attributed to variations in observational scales, data collection constraints, and the scale-dependent nature of UGI cooling as an ecological process.

1. Site scale: Factors influencing sub-canopy temperature at this scale include vegetation species type, canopy morphology, leaf area index, transpiration rate, background temperature, weather conditions, and sky visibility. However, there is some disparity in the role of transpiration rate; some studies suggest its significant effect on sub-canopy air temperature.

2. Trees scale: Trees in direct contact with the external environment, like street trees, exhibit suppressed daytime cooling by evapotranspiration due to strong air mixing, leading to potentially greater cooling at night. Green roofs are affected by weather, climatic background, and soil moisture, whereas surface temperature is influenced by canopy leaf area index and leaf inclination angle distribution. In the case of green walls, their external cooling effect is influenced by background climate and proximity to the wall.
3. Building scale: The cooling and cooling energy-saving impact of green roofs on the building's interior environment primarily relates to the type of green roof (intensive or extensive), leaf area index, substrate layer thickness, soil moisture content, number of building stories, urban density, and the climatic zone of the study area. For green walls, the cooling effect on building interiors depends on factors like leaf area index, vegetation species type, green wall orientation, substrate thickness and water content, cavity thickness, and isolation materials.
4. Street scale: Increasing the proportion of street trees or reducing tree spacing increases air temperature, surface temperature, and thermal comfort. Characteristics such as tree canopy width, tree height, and leaf area index also influence these effects. The geometric configuration of the street, including its aspect ratio and orientation, as external environmental factors, can significantly impact tree-based cooling.
5. Green space scale: The cooling effect of green space patches on temperature is mainly related to patch area and season, while its impact on surface temperature is influenced by interior and external factors, including patch area, shape complexity, proportion of patches occupied by various vegetation types, proportion of different types of patches, shape complexity in the surrounding area, and climatic and meteorological conditions.
6. Neighborhood scale: At this level, spatial structure of UGI becomes more critical than its internal composition. The cooling effect extends beyond green spaces and their immediate surroundings to the entire neighborhood unit. Spatial structure includes cover ratio and spatial configuration, with most studies emphasizing the importance of the former, though some results vary in hot and dry climates.
7. Urban scale: For SUHII, key factors include differences in urban and suburban green space ratios and green volume. Controlling for these differences, landscape configuration characteristics of urban green spaces, such as edge density, mean patch shape index, and patch density, significantly affect heat island intensity. Factors such as season, time of day, and climate zone also play a significant role in this relationship.

In summary, it was discovered that the factors influencing the cooling effect of UGI can be classified into three categories: External environmental factors, UGI spatial configuration, and UGI species composition and vegetation morphology (Fig. 9.1). Of these, species composition and vegetation morphology and spatial configuration within UGI are the dominant factors controlling UGI, while external environmental factors such as meteorological and climatic factors and built-up area

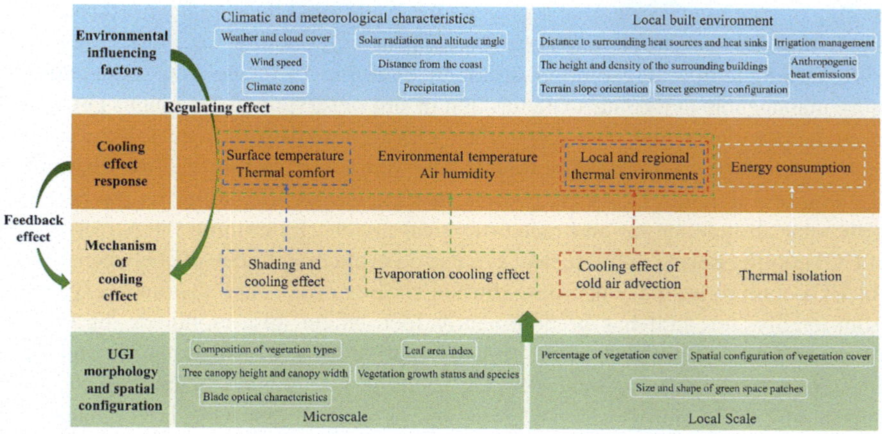

Fig. 9.1 Illusion on the mechanisms and impacting factors of UGI cooling effects

characteristics can influence the realization of the cooling effect mechanism by external environmental factors such as meteorological and climatic factors and built-up area characteristics which indirectly influence the cooling effect of UGI and its response to the effect of internal factors. In addition, the environmental response to the cooling effect of UGI can also feedback to regulate the magnitude of the cooling effect of UGI through the cooling effect mechanism, especially the evapotranspiration cooling effect. These results emphasize the need for future research to more fully account for the effects of external influences and to better interpret them in light of cooling processes and physiological perspectives of vegetation.

9.5 Limitation and Future Perspectives

9.5.1 Spatiotemporal Factors

This study faces constraints related to the spatial and temporal resolution of temperature data, which limits the investigation at various scales. These limitations also vary depending on the type of temperature data used. For example, at the spatial scale, it remains challenging to accurately determine surface temperature distribution through observations, particularly for urban facades at the microscale of sites, buildings, and streets. This limitation restricts the exploration of how different UGI measures influence surface temperature distribution and energy balance at the surface scale within cities. Similarly, acquiring temperature distribution data at the neighborhood scale is difficult, impeding the exploration of how temperature is influenced by the spatial composition and configuration of UGI within neighborhoods. Furthermore, current research has largely overlooked the variations in UGI cooling at continental or global scales, which could shed light on how macroscopic urban morphology and climate zone type impact UGI cooling.

Regarding temporal resolution, although obtaining fine-scale temperature data is more manageable at the site scale through in situ observations, it remains challenging to acquire intra- and inter-day surface temperature distributions at local scales, such as green spaces and neighborhoods. This limitation hinders the investigation of the temporal dynamics of UGI cooling at these scales, including its interactions with solar altitude angles and meteorological factors. Examining the performance of UGI cooling during extreme heat events and evaluating trade-offs between different UGI configurations concerning diurnal cooling effects are also underexplored. Additionally, research has predominantly focused on intra-month or intra-week average UGI cooling effects, with limited attention to long-term dynamics and their driving factors. Studying longer time periods is vital for understanding the true impact of UGI establishment and removal, as well as assessing how changes in internal composition, external climate, and human management activities influence the growth and cooling effects of UGI vegetation. Furthermore, there is a lack of research on the effectiveness of UGI cooling strategies under future climate change and urban growth due to the absence of suitable urban climate modeling tools and operational challenges associated with existing tools. Investigating these aspects is crucial for cities to strategically plan the distribution and configuration of UGI.

9.5.2 Mechanism of Cooling Effects

Despite the extensive research into cooling mechanisms, there remain notable research gaps and lacks. It is not yet clear how different cooling mechanisms operate at varying scales and how they collectively contribute to the overall cooling effect of UGI, especially at larger scales like neighborhoods and cities, where conducting experiments can be challenging. The relative importance of cooling mechanisms by different UGI vegetation types at the site and building scales remains unclear. Additionally, it is important to investigate the combined impacts of meteorological factors, soil water content, UGI canopy morphology, and leaf characteristics, as well as whether interaction effects exist among them. At the street scale, researchers should explore how street morphology and anthropogenic heat from traffic influence the energy balance within UGI, including shading and evapotranspiration. At the green patches, neighborhood, and urban scales, there is a need to understand how the cold air mass advection mechanism enables UGI to exert cooling effects and how the location and urban morphology of UGI influence this mechanism. These scientific questions require further exploration. Moreover, certain macroscale UGI cooling characteristics should be verified through more detailed mechanism experiments. These include the edge effect within green patches at the patch scale, the multiscale effects resulting from varying the size of analysis units at the neighborhood scale, the interplay among analysis units at the neighborhood level, and the moderating effect of building morphology and UGI spatial structure on the cooling effect of UGI.

Beyond their impact on urban management, the results of mechanism studies are crucial for accurately simulating UGI cooling effects. However, current simulation studies often oversimplify the consideration of various physical processes and cooling mechanisms. This is evident in the rudimentary representation of UGI and its associated elements, such as disregarding the leaf and soil matrix characteristics of UGI. Additionally, many studies only account for cooling effects under well-irrigated conditions, and there is a need to better convey the effects of irrigation measures in a more precise and efficient manner. At the neighborhood and urban scales, accurately representing the advection mechanism of cold air masses remains challenging, as it relies on parameterizing complex urban surfaces and modeling the horizontal and vertical transport processes of cold air masses. Further consideration is required to examine the influence of building maintenance facilities, like green roofs and green walls, on anthropogenic heat emissions from buildings and their effects on the outdoor thermal environment.

9.5.3 Comprehensive Impacting Factors

As mentioned previously, factors influencing the cooling effect of UGI can be categorized into three main groups: External environmental conditions (comprising climatic, meteorological factors, and urban geometry), spatial configuration, and internal aspects such as species composition and vegetation morphology. However, research that comprehensively investigates all three categories concurrently remains relatively scarce, particularly at larger scales, such as local and urban levels, where spatial configuration of UGI tends to be the primary focus. For instance, while the influence of meteorological factors has been established at the site scale, it has received less attention in studies conducted at the local and urban scales. Similarly, internal factors, including the types of vegetation species, have been somewhat overlooked in these studies. These limitations stem from challenges in acquiring inter-day temperature data at larger scales and the difficulty in obtaining information on the distribution of vegetation species over extensive areas. Addressing these limitations hinges on advances in observational tools and data sources.

Moreover, although the effects of urban geometry are better incorporated at microscales like sites and streets, they are seldom observed in studies conducted at the neighborhood scale. This deficiency can be attributed to the current lack of efficient tools for characterizing neighborhood morphology. Some recent studies have made initial assessments of how local climate zone types influence the cooling effect of UGI using remote sensing data. Future research could extend these investigations by considering this neighborhood morphology classification system.

9.5.4 Perspectives of Ecosystem Services

Previous studies have primarily focused on assessing the reduction or cooling impact of UGI on physical environmental parameters, such as air temperature, surface temperature, surface sensible heat flux, and mean radiant temperature. These measurements are indicative of improved urban thermal conditions or enhanced thermal comfort. However, these studies often fail to consider the heterogeneity, distribution, and movement behavior of urban residents or outdoor pedestrians in the built environment as the actual beneficiaries of UGI's cooling services. Consequently, the cooling effect of UGI is often regarded more as an ecological function than as a delivered ecosystem service.

Current research has explored the cooling services of UGI at different scales based on demographic and social attributes; however, this area of study remains in its early stages. For example, Mcdonald et al. (2018) conducted a cost–benefit analysis to calculate the total benefits (cooling intensity multiplied by the population) to the residents in the vicinity of urban tree patches in the context of PM_{10} pollution and heat mitigation by urban trees. Baró et al. (2019) employed the i-Tree Eco tool to investigate the site-scale cooling services provided by street trees across different social groups.

Moreover, viewing cooling as an ecosystem service has important implications for equity. For instance, Drescher (2019) criticized the proposal by Ziter et al. (2019) to increase vegetation cover in areas with more than 40% vegetation cover to achieve maximum cooling efficiency. The absence of UGI makes the thermal environment harsher in areas with low-income residents who have limited capacity to adapt to extreme heat. However, current UGI cooling studies often consider equity of access to cooling services. A significant topic for discussion is whether cities should primarily focus on creating large green spaces to achieve substantial cooling effects on the surroundings or opt for a distributed network of smaller green spaces to ensure more equitable access to cooling services for urban residents. Striking a balance between equity and the effectiveness of cooling services is a key aspect that future research should address.

From the perspective of ecosystem services, UGI serves multiple functions, including temperature regulation, rainwater storage, and air purification. The cooling function of UGI offers co-benefits, such as its interaction with rainwater storage. UGI's cooling function reduces the surface evaporation of stored rainwater in the soil, while the saved rainwater contributes to additional water supply for UGI's transpiration. However, there are also trade-offs associated with the cooling function and other UGI functions. For example, while trees contribute to air purification by filtering particles, the accumulation of particles on leaves may lead to the blockage of leaf stomata, affecting water transpiration and the normal metabolic processes of vegetation, which, in turn, impacts the cooling effect. Furthermore, UGI can have disservices, such as certain tree species releasing higher concentrations of biogenic organic volatiles that are harmful to humans. A high density of street trees in urban areas may reduce wind speeds, allowing air pollutants to accumulate and impeding the effective dissipation of heat from buildings at night.

Additionally, the configuration of UGI plays a crucial role in achieving its cooling effects. While some studies indicate that small networks of closely spaced green spaces can achieve better regional cooling, the fragmentation of UGI may have adverse effects on other ecosystem services, such as biodiversity and carbon sequestration. Therefore, in future simulations and planning practices concerning the cooling effects of UGI, it is essential for researchers to consider the multifunctionality of UGI and its potential negative effects in an integrated manner. This approach will help maximize the overall ecosystem services provided by UGI.

9.5.5 *Prospects for Future Research*

Cities represent complex and open giant systems. Various urban elements are coupled with each other, and sometimes incoherence in the coupling induces the emergence of various "urban diseases," while cities are also facing multiple external shocks and uncertainties such as epidemics, macroeconomic downturns, and local conflicts. Thus, there is an urgent need to enhance urban resilience. As climate change increasingly dominates the future development of cities and becomes the primary threat to urban health, climate change resilience has become a key component of urban resilience. UGI is an important means to improve climate resilience, reduce urban health risks, and achieve a green and healthy urban future. Therefore, the role of UGI needs to be highlighted in urban planning and governance. Thus, it is necessary to systematically study several UGI ecosystem services to provide effective scientific support for urban planning and urban resilience. Among them, the cooling effect is a key UGI ecological service, which deserves further research in the following aspects.

1. The application of new data sources, including drones and sensor networks, for collecting information on temperature, vegetation, and built environment characteristics is vital to address current research limitations. For example, unmanned aerial vehicle (UAV) technology is gaining traction in this field. Several studies have employed UAVs to monitor the surface temperature of small-scale UGI. UAVs offer high-spatial-resolution surface temperature monitoring, the ability to measure temperature distribution on the three-dimensional surfaces of buildings, and intra-day multitemporal observations. UAVs also allow for the inversion and identification of vegetation characteristics, such as volume and species, providing valuable information for cooling effect studies. Moreover, UAV technology can create 3D models of the built environment, offering accurate building geometry configurations and surface material features for microscale UGI cooling effect simulations, thereby enhancing accuracy. The implementation of sensor networks, in alignment with China's smart city initiatives, can help gather tem-

perature data across larger areas, reducing uncertainties and offering significant potential for development.

2. Research on combined cooling strategies is essential. While the previous focus was mainly on individual UGI types, it is crucial to recognize that UGI facilities such as trees, grasses, green walls, and green roofs can be deployed in combination to enhance cooling benefits. UGI can also be integrated with other cooling strategies, including water features (blue spaces) and albedo-based methods like cool roofs and cooling paving. Future research should investigate potential synergies or antagonisms between these strategies due to cooling mechanisms, determine the controlling factors, and optimize spatial arrangements to maximize synergistic effects or minimize antagonistic effects, ultimately improving the effectiveness of hybrid cooling strategies.
3. The cooling effect of urban-scale UGI systems is an area that requires more attention. While previous research primarily focused on the relationship between UGI proportion and cooling effects, UGI can serve as a network of interconnected green elements. This network not only improves its own temperature and the surrounding built environment but also acts as a barrier and buffer between adjacent heat island areas, preventing the formation of extensive heat islands and extreme hotspots. The UGI network achieves this function through source points and connecting corridors. It is important to analyze and optimize the cooling effect of urban-scale UGI distribution patterns from a network perspective. Measures like urban edge green belts, wedge-shaped green belts, and large-scale regional rooftop greening also mitigate the urban heat island effect through ventilation corridors and cold air mass advection. In the context of green development in Chinese cities, expanding these effects by developing green belts and optimizing ventilation should be considered. It is essential to assess the loss of cooling services due to urban expansion into vegetation cover areas and prioritize vegetation protection.

9.6 Conclusions

UGI plays a crucial role in ecological city construction and sustainable urban development, offering significant potential for mitigating urban warming and its adverse impacts. However, research on the cooling effects of UGI has been limited, and this chapter aims to provide a comprehensive review of this subject, with a special focus on the neighborhood scale. Here are the key findings:

1. **Research scales**: Previous studies have examined the cooling effects of UGI across various scales, including site, building, street, green space, neighborhood, and urban scales. The site, green space, and neighborhood scales have received more attention, while urban-scale studies have been relatively rare.
2. **Data acquisition methods**: Data for UGI cooling studies are primarily collected through three methods: field observation, remote sensing, and model simulation.

Each method has its strengths and limitations in terms of data continuity and temporal resolution.

3. **Measurement methods**: Cooling effects are measured based on spatial reference, time or scenario reference, and statistical relationships. Researchers use various techniques, such as correlation analysis, multiple regression, and increasingly, machine learning methods, to assess the influence of different factors.
4. **Cooling mechanisms**: The cooling effect of UGI is primarily achieved through three mechanisms: Shading, evapotranspiration, and cold air mass advection. External environmental factors and internal factors, such as vegetation form, species composition, and spatial layout, interact to influence the cooling effect. The importance of spatial structure increases as the scale of analysis expands.
5. **Research limitations**: Current research has limitations related to spatial and temporal scales, cooling mechanisms, factor combinations, and the perspective of ecosystem services. New technologies, such as UAV remote sensing and sensor networks, should be used to gather data on urban temperature and vegetation characteristics, explore the combined effects of multiple cooling strategies, and investigate urban-scale UGI systems. Addressing these limitations will lead to more efficient, comprehensive, and equitable UGI cooling functions for a greener, more inclusive, resilient, and sustainable urban future.

This chapter highlights the need for continued research in the field of UGI cooling effects and the potential for improving urban environments through green infrastructure.

References

Akbari, H., Cartalis, C., Kolokotsa, D., Muscio, A., Pisello, A. L., Rossi, F., Santamouris, M., Synnefa, A., Wong, N. H., & Zinzi, M. (2016). Local climate change and urban heat Island mitigation techniques—The state of the art. *Journal of Civil Engineering and Management, 22*, 1–16. https://doi.org/10.3846/13923730.2015.1111934

Baró, F., Calderón-argelich, A., Langemeyer, J., & Connolly, J. J. T. (2019). Under one canopy? Assessing the distributional environmental justice implications of street tree benefits in Barcelona. *Environmental Science & Policy, 102*, 54–64. https://doi.org/10.1016/j.envsci.2019.08.016

Bartesaghi, C., Osmond, P., & Peters, A. (2018). Evaluating the cooling effects of green infrastructure: A systematic review of methods , indicators and data sources. *Solar Energy, 166*, 486–508. https://doi.org/10.1016/j.solener.2018.03.008

Chakraborty, T., & Lee, X. (2019). A simplified urban-extent algorithm to characterize surface urban heat islands on a global scale and examine vegetation control on their spatiotemporal variability. *International Journal of Applied Earth Observation and Geoinformation, 74*, 269–280. https://doi.org/10.1016/J.JAG.2018.09.015

Chen, A., Yao, X. A., Sun, R., & Chen, L. (2014). Effect of urban green patterns on surface urban cool islands and its seasonal variations. *Urban Forestry & Urban Greening, 13*, 646–654. https://doi.org/10.1016/j.ufug.2014.07.006

Chen, X., Zhao, P., Hu, Y., Ouyang, L., Zhu, L., & Ni, G. (2019). Urban Forestry & Urban Greening Canopy transpiration and its cooling effect of three urban tree species in a subtropical city-

Guangzhou, China. *Urban Forestry & Urban Greening, 43*, 126368. https://doi.org/10.1016/j.ufug.2019.126368
Drescher, M. (2019). Urban heating and canopy cover need to be considered as matters of environmental justice. *PNAS, 116*, 26153–26154. https://doi.org/10.1073/pnas.1917213116
Farhadi, H., Faizi, M., & Sanaieian, H. (2019). Mitigating the urban heat Island in a residential area in Tehran: Investigating the role of vegetation, materials, and orientation of buildings. *Sustainable Cities and Society, 46*, 101448. https://doi.org/10.1016/j.scs.2019.101448
Gage, E. A., & Cooper, D. J. (2017). Relationships between landscape pattern metrics, vertical structure and surface urban Heat Island formation in a Colorado suburb. *Urban Ecosystem, 20*, 1229–1238. https://doi.org/10.1007/s11252-017-0675-0
Gago, E. J., Roldan, J., & Ordóñez, J. (2013). The city and urban heat islands : A review of strategies to mitigate adverse effects. *Renewable and Sustainable Energy Reviews, 25*, 749–758. https://doi.org/10.1016/j.rser.2013.05.057
Jaganmohan, M., Knapp, S., Buchmann, C. M., & Schwarz, N. (2016). The bigger, the better? The influence of urban green space design on cooling effects for residential areas. *Journal of Environmental Quality, 145*, 134–145. https://doi.org/10.2134/jeq2015.01.0062
Kong, F., Yin, H., James, P., Hutyra, L. R., & He, H. S. (2014). Effects of spatial pattern of greenspace on urban cooling in a large metropolitan area of eastern China. *Landscape and Urban Planning, 128*, 35–47. https://doi.org/10.1016/j.landurbplan.2014.04.018
Li, X., & Zhou, W. (2019). Optimizing urban greenspace spatial pattern to mitigate urban heat Island effects: Extending understanding from local to the city scale. *Urban Forestry & Urban Greening, 41*, 255–263. https://doi.org/10.1016/j.ufug.2019.04.008
Malys, L., Musy, M., & Inard, C. (2014). A hydrothermal model to assess the impact of green walls on urban microclimate and building energy consumption. *Building and Environment, 73*, 187–197. https://doi.org/10.1016/j.buildenv.2013.12.012
Mcdonald, R., Kroeger, T., Mcdonald, R. I., Boucher, T., Zhang, P., & Wang, L. (2018). Where the people are: Current trends and future potential targeted investments in urban trees for PM10 and temperature mitigation in 27 U. S. Cities Landscape and Urban Planning Where the people are: Current trends and future potential targeted invest. *Landscape and Urban Planning, 177*, 227–240. https://doi.org/10.1016/j.landurbplan.2018.05.014
Santamouris, M. (2014). Cooling the cities—A review of reflective and green roof mitigation technologies to fight heat Island and improve comfort in urban environments. *Solar Energy, 103*, 682–703. https://doi.org/10.1016/j.solener.2012.07.003
Taleghani, M., Sailor, D., & Ban-weiss, G. A. (2016). Micrometeorological simulations to predict the impacts of heat mitigation strategies on pedestrian thermal comfort in a Los Angeles neighborhood. *Environmental Research Letters, 11*, 24003. https://doi.org/10.1088/1748-9326/11/2/024003
Yan, J., Zhou, W., & Jenerette, G. D. (2019). Testing an energy exchange and microclimate cooling hypothesis for the effect of vegetation configuration on urban heat. *Agricultural and Forest Meteorology, 279*, 107666. https://doi.org/10.1016/j.agrformet.2019.107666
Yu, Z., Xu, S., Zhang, Y., Jørgensen, G., & Vejre, H. (2018). Strong contributions of local background climate to the cooling effect of urban green vegetation. *Scientific Reports, 8*, 6798. https://doi.org/10.1038/s41598-018-25296-w
Zhang, Y., Murray, A. T., & Turner, B. L. (2017). Optimizing green space locations to reduce daytime and nighttime urban heat Island effects in Phoenix, Arizona. *Landscape and Urban Planning, 165*, 162–171. https://doi.org/10.1016/j.landurbplan.2017.04.009
Zhou, Y., & Shepherd, Æ. J. M. (2010). Atlanta's urban heat Island under extreme heat conditions and potential mitigation strategies. *Natural Hazards, 52*, 639–668. https://doi.org/10.1007/s11069-009-9406-z
Ziter, C. D., Pedersen, E. J., Kucharik, C. J., & Turner, M. G. (2019). Scale-dependent interactions between tree canopy cover and impervious surfaces reduce daytime urban heat during summer. *PNAS, 116*, 7575–7580. https://doi.org/10.1073/pnas.1817561116

Chapter 10
Conclusions and Outlook

10.1 Overall Summary

This book is dedicated to an in-depth exploration of the UTE and its contributing factors across China's Guanzhong Plain, with a focus on a multiscale approach. Its core objectives are to conduct a comprehensive investigation into the long-term dynamics of LST, study the potential drivers, assess the adverse impacts of UHI on human health, and propose effective UHI mitigation strategies. This book aims to provide a profound understanding of the UTE, with a particular emphasis on Xi'an, an ancient inland city.

To investigate the UTE comprehensively and systematically in the Guanzhong Plain, this book organizes its research into four primary categories, encompassing a total of ten chapters. These categories include *characteristics* of UTE (Chap. 3), *drivers* of UTE (Chapters 4, 5, 6, and 7), *heat risk* of urban residents (Chap. 8), and *mitigation strategies* of UHI effects (Chap. 9).

In Chap. 1, this book initiates its exploration by introducing two significant factors that exacerbate the UTE: Climate change and urbanization. This chapter offers a thorough review of the drivers and impacts of these factors, along with an examination of the future challenges associated with the adaptation to climate change and the development of sustainable cities. Furthermore, the chapter delves into the interconnected relationship between climate change and urbanization, with a particular focus on their impacts on the current state of the UTE.

Chapter 2 provides a comprehensive overview of the progress made in the study of the UTE and systematically synthesizes the relevant knowledge in the field. The chapter begins by introducing key terminologies, such as UTE and UHI. It proceeds to dissect the historical evolution of UTE into distinct periods, including the early phase, the accumulation period, the modern foundation, and the redevelopment phase, offering a meticulous review of each era. The subsequent sections of the

L. Zhou et al., *Urban Surface Thermal Environments*, The Urban Book Series,
https://doi.org/10.1007/978-3-031-94499-4_10

chapter focus on summarizing the datasets and methodologies employed to characterize UHI. Notably, temperature datasets are predominantly sourced from in situ monitoring, remote sensing imagery, and numerical simulations. To address inherent limitations in these temperature datasets, two prevalent strategies are frequently employed, namely gap-filling and spatiotemporal downscaling. Furthermore, this chapter provides an insightful account of the ongoing progress in investigating the characteristics and driving factors of UHI. It also delves into the far-reaching implications of UHI on both human systems and natural systems.

Chapter 3 initiates with an in-depth introduction to the study area, encompassing the Guanzhong Plain, Xi'an metropolitan area, and the core district of Xi'an. This comprehensive overview encompasses various aspects, including local climate, urbanization trends, and socioeconomic characteristics. Subsequently, the chapter delves into an extensive analysis of the spatiotemporal dynamics of land cover and LST, both daytime and nighttime, across the Guanzhong Plain. This analysis spans a period from 2000 to 2018, with intervals of 5 years. The findings unveil a notable trend in the substantial connection and concentration of high-temperature zones within the Xi'an metropolitan area. The analysis further identifies the specific land cover types that contribute positively or negatively to temperature variations. Notably, cropland, water bodies, urban land, and rural settlements are identified as positive contributors to elevated temperatures, while forestland and grassland exhibit cooling effects and are classified as negative contributors. These results offer significant insights into potential strategies for mitigating the regional thermal environment's deterioration. Specifically, the findings suggest that controlling urban sprawl can be an effective measure in limiting the exacerbation of high temperatures at a regional scale. Additionally, the active cultivation and protection of forestlands can enhance their cooling effect, offering a complementary approach to address the thermal challenges.

Chapter 4 delves into the spatial hierarchy of the UTE and elucidates the intricate relationship between landscape structure, quantified through landscape metrics and LST across different spatial levels of the UTE. The analysis leverages satellite imagery, local contour tree algorithms, and the XGBoost model. This comprehensive study identifies 21 primary UHI areas and their respective hierarchies within the UTE. Notably, the primary urban district of Xi'an exhibits the most complex structure, characterized by six distinct temperature levels. Additionally, the chapter underscores the significant interaction effects between landscape metrics within specific value ranges and LST, further enriching our understanding of the UTE.

Chapter 5 undertakes a quantitative assessment of the individual and combined influences of urban morphological parameters and landscape indices of UGS on LST. Utilizing the boosted regression tree method, the chapter investigates the relative contributions and marginal effects of these driving factors on LST. Moreover, it discerns the warming and cooling effects of buildings and UGS. The findings emphasize that the building coverage ratio predominantly contributes to urban warming, while an increase in building areas tends to suppress the cooling effects of UGS.

Chapter 6 focuses on exploring the spatial distribution characteristics of LST and relative humidity in urban areas across different seasons, revealing the regulatory role of dry and wet conditions on LST and their seasonal variations. The results indicate that LST shows significant differences across different seasons and varying dry and wet environments, influenced by the combined effects of vegetation cover, building types, and humidity conditions. Therefore, multiple factors need to be considered comprehensively to regulate the urban thermal environment.

Chapter 7 delves into the examination of 3D landscape structure's influence on LST during the summer, contrasting heat wave and non-heat wave periods. The results reveal that the impact of landscape structure remains consistent between these two periods, underscoring the significance of building height and impervious surface characteristics for landscape-optimized thermal mitigation during heat wave events. Furthermore, the case study demonstrates that many landscape optimization-based thermal mitigation measures effective during non-heat waves can be similarly useful during heat waves.

Chapter 8 proposes a heat storage capacity index to explore the spatiotemporal changes in heat risk. The results indicate the spatial distribution of risk areas closely resembled that of vulnerability indicators, showing an outward expansion trend centered around the urban centers of Xi'an and Xianyang. Spatially, risk, exposure, and vulnerability indicators exhibited significant urban–rural gradient effects, with mean values highest in urban areas, followed by urban fringe areas, and lowest in rural areas. Temporally, the mean values of risk, hazard, exposure, and vulnerability indicators showed an increasing trend in both urban and rural areas.

Chapter 9 provides a comprehensive review of the progress made in understanding the cooling effects of urban green infrastructure. This review spans research methods, physical mechanisms, and driving factors systematically. It addresses the current gaps and challenges associated with spatiotemporal scales, physical mechanisms, comprehensive driving factors, and the ecosystem service perspective.

10.2 Contributions

While considerable progress has been achieved in the field of urban thermal environments, few books provide a comprehensive investigation spanning the *characteristics* of the UTE, *drivers* of the UTE, *heat risk* of urban residents, and *mitigation strategies* of UHI effects. This book stands out for its multidisciplinary approach, considering multiscale, multifactorial, and three-dimensional perspectives. It offers several innovative contributions:

First, this book differentiates itself by emphasizing UTE characteristics, a departure from previous studies that predominantly focused on the two-dimensional perspective of horizontal land cover changes. It introduces the LCZ scheme, which not only considers the two-dimensional aspects but also the three-dimensional characteristics, including surface property heights. This book conducts a comprehensive, long-term investigation of the consequences of LCZ conversion on the UTE. It uncovers intra-city variations in UHI and elucidates inter-annual dynamics. Notably, regarding the twin cities of Xi'an and Xianyang in the study area, this book reveals that the integration of these two cities is ongoing, encompassing not only built-up expansion but also the spread of high-temperature areas. These findings provide valuable insights for informed urban planning and future urban renewal strategies.

Second, this book conducts a thorough examination of the drivers behind UHI. It introduces an analytical framework that explores the hierarchical nature of UHI, delving into the implications of driving factors on UHI and the interplay among these factors. This framework underscores the importance of integrating the hierarchical structures of UHI into studies of the UTE. Furthermore, this book uncovers the relative contributions and marginal effects of both 2D and 3D urban morphological parameters, offering quantitative insights into mitigating UHI effects at the block and tract levels. This book also quantifies the nonlinear influence of 3D landscape structures on LST during heat wave and non-heat wave periods, shedding light on mitigation strategies for individual blocks.

Third, this book introduces a novel approach to identify potentially vulnerable individuals who may be exposed to high levels of heat. A case study illustrates that the LCZ scheme holds promise for studying thermal comfort and enhancing the overall UTE. This is especially relevant in the context of adapting to increasingly frequent extreme heat events in the future. At the same time, this book focuses on the in-depth exploration of non-coastal cities and city clusters in China, providing case studies for cognate urban regions globally.

Lastly, through a comprehensive review of the literature on the cooling effects of urban green infrastructure (UGI), this book explores the potential of new data acquisition methods, the synergy of multiple cooling measures, and the cooling effects of UGI at the city scale. In particular, this book dissects the scale change law of thermal environment changes from four dimensions: City clusters, metropolitan areas, cities, and blocks. This is of significant importance and value for better understanding scale evolutions from macro to micro.

10.3 Limitations

While this book has made significant progress in exploring the UTE from a multiscale perspective, the case studies it presents have some limitations.

First, datasets and data preprocessing could introduce uncertainties. The LST data used in this book were captured at specific times, usually around 11:19 a.m. for Landsat imagery. This timing is due to the long revisit period (16 days) of the Landsat 8 satellite, cloud contamination, and weather conditions, which limit the availability of suitable images for representing seasonal variations in LST. This issue can result in uncertainties when modeling the relationship between LST and its driving factors.

Second, while this book comprehensively explores the relationship between driving factors and LST from both 2D and 3D perspectives, the physical mechanisms behind these relationships in complex urban systems remain unclear. In particular, there may be interactions or connections among multidimensional urban properties, leading to competition and synergy in underlying thermal processes. These interactions often vary across different analysis units, making it challenging to apply findings practically for mitigating the UHI effect. Therefore, it is insufficient to solely characterize these relationships; additional efforts should be directed toward investigating the physical mechanisms of thermal processes.

As a final note, this book pays relatively less attention to meteorological datasets. Considering the direct link between meteorological parameters and human health and overall well-being, the complex impacts associated with weather conditions should be quantified more appropriately within the context of human health. It is suggested that future scientists verify with each other by coupling multi-source remote sensing data and actual site monitoring data in their research in this area, which will lead to more accurate and reliable conclusions. Furthermore, the human mobility data will be the most direct data support for future thermal environment research and the key to mitigating the thermal environment. We must actively embrace and face the opportunities and adjustments brought about by mobile individual big data and AI science.

10.4 Outlook

As research on the quantitative evaluation of the urban thermal environment continues to advance, several aspects related to data and techniques can be further explored and enhanced:

1. Enhanced ground observation network: To address issues like uneven distribution and the lack of representativeness of ground meteorological stations, a promising direction is to expand and densify the network of observation stations. This can lead to a more refined evaluation of the urban thermal environment by providing comprehensive and detailed data.
2. Spatial resolution improvement: Existing thermal images from satellites such as Landsat, ASTER, and MODIS typically have a spatial resolution ranging from 60 m to 1000 m. Future research can focus on developing spatial downscaling methods that combine data from multiple satellites. This approach can significantly enhance the spatial resolution of thermal infrared imagery, enabling it to capture richer surface thermal information for small-scale outdoor thermal environment studies.
3. Fine-scale thermal monitoring: The use of infrared cameras mounted on unmanned aerial vehicles (UAVs) or direct ground-based infrared cameras can provide fine-scale thermal property observations at the block or even single-building scale. This high-spatial-resolution data can be instrumental in fine-scale forecasting of the thermal environment, including city-wide and smaller-scale assessments.

In addressing urban heat environment challenges, several effective measures can be implemented:

1. Ecosystem greening and restoration: Increasing the presence of green spaces, parks, gardens, and green belts can help mitigate urban heat. Protecting and restoring natural ecosystems around cities, such as wetlands and forests, can regulate temperature and enhance air quality.
2. Building and urban planning strategies: Techniques like green roofs, vertical greening, eco-friendly buildings, and white roofs can be used to reduce heat absorption and radiation from buildings. Rational layout planning, expanded street widths, ventilation channels, and green corridors can also improve the urban heat environment.
3. Water resource management: Utilizing water resource techniques, such as constructing artificial lakes or reservoirs, street sprinkling, and rainwater collection, can reduce urban heat. Scientific management of urban water resources, including water recycling and protection measures, can also lessen the impact of the urban heat environment.
4. Green transportation and energy efficiency: Encouraging green transportation options like electric buses, subways, and shared bicycles can reduce heat generated by traffic emissions. Promoting clean energy and energy-saving technologies can decrease energy consumption and carbon emissions. These strategies reduce energy consumption, limit heat generation, and foster low-carbon transport systems, thereby cooling the city.
5. Public participation: Strengthening public awareness and engagement through education and awareness campaigns is crucial. Encouraging individuals and communities to take collective actions such as energy conservation and tree planting can contribute significantly to mitigating urban heat environment issues.

Even as the world transitions to green energy systems, climate change will continue to increase anomalous heat events globally. As the world is now mostly urban and continues to urbanize, the strategic greening of urban landscapes remains a priority toward advancing multiple United Nations sustainable development goals (SDGs). For example, extreme heat can exacerbate poverty as the most vulnerable are most impacted by extreme heat's impact on the type of work and number of hours worked. Strategic greening efforts can therefore advance SDG 1, no poverty. Similarly, SDG 3, good health and well-being, is advanced through enhancing urban green spaces by reducing cardio-pulmonary disease vis-a-vis improved air quality and minimizing the ill effects of heat stress through increased shade from tree cover. To the extent green spaces reach underserved neighborhoods, SDG 10, reduced inequalities, can also be advanced. SDG 11, sustainable cities and communities, is, of course, at the heart of improving and increasing urban green spaces. Similarly, urban green spaces help fight climate change by absorbing more greenhouse gas emissions and by decreasing libido associated with urban heat islands, thereby serving SDG 13, climate action. Likewise, by providing wildlife habitat, life on land, SDG 15, is catalyzed through efforts to increase urban green spaces. Lastly, partnerships for the goals, SDG 17, is critical as the majority of the world is now urban and is becoming more urban, making green spaces a priority investment for global partnerships.

To conclude, this in-depth study from China's Guanzhong Plain offers an empirical and methodological base from which future studies may borrow in examining options for monitoring and implementing urban greening solutions to mitigate the human impact of climate change-related extreme heat events. Although some of the characteristics of the Guanzhong Plain may be unique, many of the topics examined in this book remain applicable to urban landscapes globally. With improved methodological tools and creative problem solving, other major urban areas globally may learn important lessons from this book toward improving the coupled well-being of humans and nature.

Index

L. Zhou et al., *Urban Surface Thermal Environments*, The Urban Book Series,
https://doi.org/10.1007/978-3-031-94499-4

The manufacturer's authorised representative in the EU is Springer Nature Customer Service Centre GmbH, Europaplatz 3, 69115 Heidelberg, Germany. If you have any concerns regarding our products, please contact ProductSafety@springernature.com

Printed and bound by CPI Group (UK) Ltd, Croydon, CR0 4YY
05/10/2026
02228127-0002